Advances in Groundwater Flow and Solute Transport

Advances in Groundwater Flow and Solute Transport

Pushing the Hidden Boundary

Special Issue Editors

Hongbin Zhan
Quanrong Wang
Zhang Wen

MDPI • Basel • Beijing • Wuhan • Barcelona • Belgrade

MDPI

Special Issue Editors

Hongbin Zhan
Texas A&M University
USA

Quanrong Wang
China University of Geosciences
China

Zhang Wen
China University of Geosciences
China

Editorial Office
MDPI
St. Alban-Anlage 66
4052 Basel, Switzerland

This is a reprint of articles from the Special Issue published online in the open access journal *Water* (ISSN 2073-4441) from 2017 to 2019 (available at: https://www.mdpi.com/journal/water/special_issues/hidden_boundary)

For citation purposes, cite each article independently as indicated on the article page online and as indicated below:

LastName, A.A.; LastName, B.B.; LastName, C.C. Article Title. *Journal Name* **Year**, *Article Number, Page Range.*

ISBN 978-3-03921-074-9 (Pbk)
ISBN 978-3-03921-075-6 (PDF)

Contents

About the Special Issue Editors

Hongbin Zhan, Professor and Holder of Endowed Dudley J. Hughes Chair in Geology and Geophysics, Texas A&M University. Dr. Hongbin Zhan is a Professor of Geology and Geophysics, a Professor of Water Management and Hydrological Science, a Professor of Energy Institute, all at Texas A&M University (TAMU). He is currently the Holder of Endowed Dudley J. Hughes Chair in Geology and Geophysics at TAMU, and he was the Holder of Endowed Ray. C Fish Professorship in Geology at TAMU from 2010-2018. He is the Distinguished Chang-Jiang Scholar from Ministry of Education of China (2009). He is the recipient of many honors, including Distinguished Achievement Award in Faculty Teaching (2009) and Distinguished Achievement Award in Faculty Research (2016) from College of Geosciences at TAMU, Distinguished Oversea Young Scientist Award from National Natural Science Foundation of China (2004), Fellow of Geological Society of America (2006), to name a few. He has published more than 200 papers, with a Web of Science citation of 2383 and H-index of 29. His teaching and research interests are primarily in fundamental processes of groundwater hydrology, flow and transport in geological formations, and their applications in water resources management and geological, environmental, and petroleum engineering. He is recently interested in unconventional subsurface flow and transport processes, with the studied media changing from permeable porous and fractured ones to much less permeable ones such as clay and shale, and the studied pore sizes also changing from millimeters to micro-meters or even nano-meters.

Quanrong Wang, Professor, China University of Geosciences (Wuhan). Dr. Quanrong Wang is a Professor of Hydrogeology in China University of Geosciences (CUG), Wuhan, and he is the Holder of the CUG Scholar. His teaching courses include Numerical Modeling of Groundwater Flow, Field Hydrogeology, and Reactive Transport in Subsurface. His research interests are mainly focused on the mechanics and the numerical modeling of the surface water and groundwater interactions, reactive transport in subsurface, and heat transport in the aquifer. He and his co-authors developed some new models of the reactive radial transport around the injection/withdraw wells, and some models of pipe flow in the porous media.

Zhang Wen, Professor, China University of Geosciences (Wuhan). Dr. Zhang Wen is a Professor of Water Resources and Hydrogeology, at China University of Geosciences, Wuhan (CUG). He has published more than 30 papers. He is the recipient of many honors, including Nomination award of outstanding Ph.D. dissertation in China (from Ministry of Education of China), Natural Science Award from Ministry of Education of China, and Distinguished Achievement Award in Young Faculty Teaching (2017) at CUG. His teaching and research interests mainly focus on groundwater dynamics, solute transport process in the groundwater, and their applications in groundwater resources evaluation and remediation. He is recently interested in non-Darcian flow and non-Fickian transport in low permeability media, well hydraulics with considering multiple controlling factors, technology of the groundwater circulation well (GCW) for environmental restoration.

![water logo] *water*

Editorial

Editorial of Special Issue "Advances in Groundwater Flow and Solute Transport: Pushing the Hidden Boundary"

Hongbin Zhan [1,2,*], **Quanrong Wang** [2,3] **and Zhang Wen** [2]

[1] Department of Geology and Geophysics, Texas A&M University, College Station, TX 77843-3115, USA
[2] School of Environmental Studies, China University of Geosciences, Wuhan 430074, China; wangqr@cug.edu.cn (Q.W.); wenz@cug.edu.cn (Z.W.)
[3] Laboratory of Basin Hydrology and Wetland Eco-restoration, China University of Geosciences, Wuhan 430074, China
* Correspondence: zhan@geos.tamu.edu

Received: 22 February 2019; Accepted: 1 March 2019; Published: 5 March 2019

Abstract: The theme of this special issue is to explore the new territories beyond conventional subsurface flow and transport theories. We have selected 12 articles in this special issue and these articles cover a wide range of problems including (1) Non-Fickian chemical transport in various environments; (2) Non-Darcian flow; (3) Flow and transport in low-permeability media; (4) Vadose zone process; (5) Regional scale groundwater flow and groundwater-surface interaction; (6) Innovative numerical methods. The major contributions of these papers are summarized in this editorial.

1. Non-Fickian Chemical Transport in Various Environments

For this subject, Pannone [1] has adopted a stochastic approach to deal with an evolving-scale heterogeneous formation using power-law semi-variograms. Pannone [1] has analytically shown that dispersion in such a hierarchy system can be ergodic and Fickian or non-ergodic and super-diffusive, based on the scaling exponent value and the magnitude of Peclet number, which was defined in this study as the ratio of the product of the ensemble mean velocity at the initial plume size to the local dispersion. Specifically, a large Peclet number will make the transport process closer to asymptotically ergodic-Fickian conditions. In contrast, a higher scaling exponent will make the transport process closer to a non-ergodic super-diffusive regime. The finding of this article is quite different from what has been reported in previous studies on the same topic.

Also concerning non-Fickian transport, Chen et al. [2] has employed well designed laboratory experiments on flow and transport in synthetic single smooth and rough fractures using a conservative tracer of Brilliant Blue FCF (bis {4-(N-ethyl-N-3-sulfophenylmethyl) aminophenyl}-2-sulfophenyl methylium disodium salt) dye. This study provided visible evidence that the classical advection–dispersion equation failed to capture the long-tailing of breakthrough curves (BTCs), and the continuous time random walk (CTRW) model was better at explaining the long-tailing of BTCs. Furthermore, the coefficient β in the CTRW model was found to be most relevant for characterizing the heterogeneity of the rough single fractures.

Concerning the deep geological repositories for nuclear wastes, Suzuki et al. [3] has pointed out that mass transport of radioactive contaminants displayed anomalous behaviors and often produced power-law tails in BTCs due to the spatial heterogeneities in hosting fractured rocks. This paper proposed a new equation involving an integral term of convolution and fractional derivatives to account for mass exchange between a fracture and porous rock matrix.

2. Non-Darcian Flow

In terms of non-Darcian flow, Wang et al. [4] has conducted a series of experimental flow tests for bimsoils with various slenderness ratios. This study stated that the sample height had a strong influence on the flow characteristics of bimsoils, and the degree of non-Darcy flow decreased with the increase of sample height.

Also focusing on the experimental work of non-Darcian flow, like Wang et al. [4], Chen et al. [2] has conducted flow tests in synthetic single smooth and rough fractures. This study showed that non-Darcian Forchheimer flow was evident in both smooth and rough fractures, and it showed that the non-Darcian coefficient β_c in the Forchheimer equation was most relevant for characterizing the heterogeneity of the rough single fractures.

3. Flow and Transport in Low-Permeability Media

Flow and transport in low-permeability media has become a focal point of research in recent years [5,6]. In this regard, Lu et al. [7] has conducted a field-scale research in Ningchegu site located east of Tianjin of China to investigate flow and transport in continental silty clay, mud-silt clay, and marine silty clay deposits. After analyzing the hydraulic conductivity data collected from 52 boreholes, Lu et al. [7] reported that a Levy stable distribution was a better choice to describe the hydraulic conductivity distribution statistics than the log-normal, normal, Weibull, or gamma distributions.

4. Vadose Zone Process

Vadose zone process (VZP) has been studied more than five decades and still has many controversies. This is partially due to the highly nonlinear flow and transport processes in this zone and partially due to various driving forces coexisting in this zone. Vadose zone serves as an important intermediate buffer zone between the Earth atmosphere and groundwater. Besides hydrological processes, biological and geochemical processes also play active roles in VZP. Three papers of this special issue are focused on VZP. Cheng et al. [8] has reported a field study of influence of irrigation on desert farmland soil moisture dynamics. It included a new type of lysimeter installed below the 150 cm soil profile to continuously measure the so-called deep soil recharge (DSR). This study showed that farmland consisting of an upper 50 cm plough soil and a lower 100 cm filled clay soil can save more water, which is useful in agricultural and water resource management in arid regions.

Also focusing on VZP, Liu and Zhan [9] have proposed a new method of calculating the steady-state evaporation for an arbitrary matrix potential at bare ground surface. This solution expands our present knowledge of evaporation at bare ground surface to more general field conditions, and can be very useful for quick assessment of the amount of evaporation at bare ground.

Concerning the variability of soil depths which are controlled by many natural and environmental factors, Yu et al. [10] has proposed a simple model to describe the relationship between soil depth and infiltration flux taking into account of a non-Gaussian distribution of rock biogeochemical weathering rates. This model demonstrates the importance of fundamental principles of physics to quantify the coupled effects of five major soil-forming factors of Dokuchaev.

5. Regional Scale Groundwater Flow and Groundwater—Surface Interaction

Similar to Suzuki et al. [3] for deep geological repository of high-level radioactive nuclear waste, Cao et al. [11] has focused on the hydrogeological conditions of the Beishan area in China, a repository site of high-level radioactive nuclear waste in China. This study illustrated the special features of regional-scale groundwater flow in the Beishan area, which will be considered in designing the repository facility in this area. In particular, the model is capable of evaluating the influence of the extreme climate and regional faults on the groundwater flow pattern, factors that matter for long-term safe operation of the repository site.

Hydrologic exchange flux (HEF) is a crucial component of hydrological cycle and its strength closely affects the biogeochemical and ecological processes in the hyporheic zone. Focusing on a scale of 1000 m river reach, Zhou et al. [12] has used self-recording thermistors for measuring the vertical thermal profiles and a time series of hydraulic gradients derived from river stage and inland water levels monitoring to estimate HEFs. This method is capable of providing a high-resolution spatial and temporal variation of HEF rates over a large river reach, information that is crucial for understanding the hyporheic zone process.

6. Innovative Numerical Methods

Although numerical methods are routinely used for studying subsurface flow and transport at present, new advancements in this area have never been haltered. Two papers in this special issue represented two examples of continuous endeavor for pushing the boundary of advanced numerical simulation techniques. In one of the two papers, Masciopinto et al. [13] has proposed an innovative approach to model flow and salt transport in fractured coastal aquifers affected by seawater intrusion in Bari, Italy. The model was based on a stochastic method to transfer all real medium heterogeneities into the numerical model. This model provided a reliable estimation of local advancements of the freshwater/saltwater wedge in coastal aquifers, and the numerical model results were corroborated by the non-invasive geophysical measurements including the electrical resistivity tomography (ERT) method at the site.

Focusing on tackling the groundwater flow uncertainties issues, Dong et al. [14] has proposed a new method using the Interval Parameter Perturbation (IPP) principle. The IPP method avoids the dilemma faced by many other statistical and stochastic methods in which the statistical characteristics (such as mean, variance, covariance, etc.) of random variables of concerns must be known a priori, which is usually not feasible in real-world applications. The new IPP method used in Dong et al. [14] did not require the complete statistical characteristics of the random variables. Instead, it only needed the bounded uncertain intervals of variables of concern. The benefit of this method is its capability of analyzing the uncertainties of groundwater flow when it is difficult to obtain the complete statistical characteristics of the hydrogeological systems.

7. Summary

It is our hope that this special issue can stimulate long-lasting interests among the hydrological community to explore the new frontiers of subsurface hydrology in areas that are often either untouched or overlooked before. The advancement of hydrology relies on constant push of knowledge boundary, and this special issue represents one step forward in this direction. We thank all the authors, reviewers, and editorial staff members for producing this special issue.

Conflicts of Interest: The authors declare no conflict of interest

References

1. Pannone, M. An analytical model of Fickian and non-Fickian dispersion in evolving-scale log-conductivity distributions. *Water* **2017**, *9*, 751. [CrossRef]
2. Chen, Z.; Zhan, H.B.; Zhao, G.Q.; Huang, Y.; Tan, Y.F. Effect of roughness on conservative solute transport through synthetic rough single fractures. *Water* **2017**, *9*, 656. [CrossRef]
3. Suzuki, A.; Fomin, S.; Chugunov, V.; Hashida, T. Mathematical modeling of non-Fickian diffusional mass exchange of radioactive contaminants in geological disposal formations. *Water* **2018**, *10*, 123. [CrossRef]
4. Wang, Y.; Li, C.H.; Wei, X.M.; Hou, Z.Q. Laboratory investigation of the effect of slenderness effect on the non-Darcy groundwater flow characteristics in bimsoils. *Water* **2017**, *9*, 676. [CrossRef]
5. Zaheer, M.; Wen, Z.; Zhan, H.B.; Chen, X.L.; Jin, M.G. One-dimensional solute transport at small scale in low-permeability homogeneous and saturated soil columns. *Geofluids* **2017**, 6390607. [CrossRef]

6. Cai, J.C.; Sun, S.Y.; Zhang, Z.E.; Pan, Z.J. Editorial to the special issue: Modeling and characterization of low permeability (tight) and nanoporous reservoirs. *Transp. Porous Media* **2019**, *126*, 523–525. [CrossRef]

7. Lu, C.P.; Qin, W.; Zhao, G.; Zhang, Y.; Wang, W.P. Better-fitted probability of hydraulic conductivity for a silty clay site and its effects on solute transport. *Water* **2017**, *9*, 466. [CrossRef]

8. Cheng, Y.B.; Li, Y.L.; Zhan, H.B.; Liang, H.R.; Yang, W.B.; Zhao, Y.M.; Li, T.J. New comparative experiments of different soil types for farmland water conservation in arid regions. *Water* **2018**, *10*, 298. [CrossRef]

9. Liu, X.; Zhan, H.B. Calculation of steady-state evaporation for an arbitrary matrix potential at bare ground surface. *Water* **2017**, *9*, 729. [CrossRef]

10. Yu, F.; Faybishenko, B.; Hunt, A.; Ghanbarian, B. A simple model of the variability of soil depths. *Water* **2017**, *9*, 460. [CrossRef]

11. Cao, X.Y.; Hu, L.T.; Wang, J.S.; Wang, J.R. Regional groundwater flow assessment in a prospective high-level radioactive waste repository of China. *Water* **2017**, *9*, 551. [CrossRef]

12. Zhou, T.; Huang, M.Y.; Bao, J.; Hou, Z.S.; Arntzen, E.; Mackley, R.; Crump, A.; Goldman, A.E.; Song, X.H.; Xu, Y.; et al. A new approach to quantify shallow water hydrologic exchanges in a large regulated river reach. *Water* **2017**, *9*, 703. [CrossRef]

13. Masciopinto, C.; Liso, I.S.; Caputo, M.C.; de Carlo, L. An integrated approach based on numerical modelling and geophysical survey to map groundwater salinity in fractured coastal aquifers. *Water* **2017**, *9*, 875. [CrossRef]

14. Dong, G.M.; Tian, J.; Zhan, H.B.; Liu, R.Y. Groundwater flow determination using an interval parameter perturbation method. *Water* **2017**, *9*, 728. [CrossRef]

water

MDPI

Article

A Simple Model of the Variability of Soil Depths

Fang Yu [1], **Boris Faybishenko** [2], **Allen Hunt** [3,*] **and Behzad Ghanbarian** [4]

[1]   Department of Earth and Environmental Sciences, Wright State University, 3640 Colonel Glenn Highway, Dayton, OH 45435, USA; yu.39@wright.edu

[2]   Earth Sciences Division, E. O. Lawrence Berkeley Laboratory, 1 Cyclotron Road, Berkeley, CA 94720, USA; bafaybishenko@lbl.gov

[3]   Department of Physics and Department of Earth & Environmental Sciences, Wright State University, 3640 Colonel Glenn Highway, Dayton, OH 45435, USA

[4]   Department of Petroleum and Geosystems Engineering, University of Texas at Austin, Austin, TX 78712, USA; b.ghanbarian@gmail.com

*   Correspondence: allen.hunt@wright.edu

Received: 27 March 2017; Accepted: 19 June 2017; Published: 26 June 2017

Abstract: Soil depth tends to vary from a few centimeters to several meters, depending on many natural and environmental factors. We hypothesize that the cumulative effect of these factors on soil depth, which is chiefly dependent on the process of biogeochemical weathering, is particularly affected by soil porewater (i.e., solute) transport and infiltration from the land surface. Taking into account evidence for a non-Gaussian distribution of rock weathering rates, we propose a simple mathematical model to describe the relationship between soil depth and infiltration flux. The model was tested using several areas in mostly semi-arid climate zones. The application of this model demonstrates the use of fundamental principles of physics to quantify the coupled effects of the five principal soil-forming factors of Dokuchaev.

Keywords: soil formation; percolation; infiltration; erosion

---

1. Introduction

The concepts of soil formation have been extensively examined, starting from the beginning of the 19th century (Justus von Leibig: see http://www.madehow.com/knowledge/Justus_von_Liebig.html), and thereafter modified and refined by many world-renowned soil scientists, e.g., Charles Darwin [1] in England, Vasily Dokuchaev [2] in Russia, and Grove Karl Gilbert [3], George Nelson Coffey [4], and Eugene W. Hilgard [5] in the United States. The conceptual approaches to a pedogenic theory proposed by many scientists are fundamentally different, and have been revisited many times [6–8]. Although these theories are conceptually different, they all generally converge over the idea of Dokuchaev's five natural soil-forming factors: biota impact, climate impact, initial material, terrain, and time, as quoted in both Glinka [9] and Jenny [10]. Dokuchaev, as quoted in Glinka [9], emphasized the necessity to determine the solution of the soil-forming factor equation, stating that:

> In the first place we have to deal here with a great complexity of conditions affecting soil; secondly, these conditions have no absolute value, and, therefore, it is very difficult to express them by means of figures; finally, we possess very few data with regard to some factors, and none whatever with regard to others. Nevertheless, we may hope that all these difficulties will be overcome with time, *and the soil science will truly become a pure science.*

The consideration of these theories and factors provide together a more comprehensive view of soil formation than either can do alone [11] for different types of landscapes, including those dominated by exposed bedrock, or fertile soils.

For later reference, soils are most commonly divided into three horizons or layers: O, A, and B, although the E, P, and C horizons are also fairly commonly discussed. In short, the O layer, present in forests, but not grasslands, is dominated by organic material, e.g., decaying plant or animal matter. The A and B layers were defined originally by Dokuchaev; the A horizon being the topsoil or humus, which is typically brown or black due to its high organic content, while the B layer is called the subsoil, and is typically more brightly colored due to the presence of clay minerals and iron oxides. However, even the full traditional classification scheme does not capture the modern understanding of soil evolution completely.

The evaluation of the lower boundaries of soil is important for many scientific and practical applications, such as agricultural and hydrological studies [12]. The depth of the soil mantling the Earth's surface tends to vary from a few centimeters to several meters or even tens of meters [13], depending on many natural and environmental factors. For example, Hillel [14] described soil as the "top meter or so of the Earth's surface, acting as a complex biophysical organism". It is well recognized that an understanding and a prediction of how and where water infiltrates from the land surface and moves through the vadose zone within a landscape controls the process of biogeochemical weathering, and soil hydrology can be used to explain soil morphology and an ecosystem's dynamical functions [12]. It has been proven that soil is being transformed globally from natural to human-affected material, the lower boundary of soil is much deeper than the solum historically confined to the O to B horizons, and most soils are a kind of paleosol, being products of many soil-forming processes that have ranged widely over the lifespans of most soils. In other words, a soil's polygenesis is dependent on fluxes of matter and energy, which are thermodynamically transforming soil systems [15,16]. Nevertheless, when one compares predictions with data for soil depth, it is necessary at least to hypothesize the relevance of theory to a particular boundary, which we have consistently chosen to be the bottom of the soil Bw horizon, an oxidation depth.

We pose the question: are these soil fluxes dependent on infiltration from the land surface, and does the spatial variability of infiltration underlie soil depth variability?

Water flow, soil erosion (and deposition), and soil formation all affect soil depth. Soil erosion is chiefly accomplished through advective processes such as overland flow and rainsplash [17], though soil creep and a number of other processes contribute as well. Biogeochemical weathering, a basis of soil formation, requires water to carry reaction reagents to the weathering front, and reaction products away [18], and thus relates to deep infiltration. Erosion rates vary over about 4 orders of magnitude, from a fraction of a meter per million years in the interior of Australia, to a maximum of over 1000 m per million years in the Himalayas [19]. Precipitation rates vary from 2 mm/year in the Atacama Desert to 10 m/year in the New Zealand Alps (and in many other regions). The soil production rate is linearly proportional to precipitation [18,20], and has been reported as an exponentially diminishing function of soil depth [21], though the "humped" function was reported in several recent studies; see Heimsath et al. [22]. Soil erosion and soil production must be correlated, which is a stipulation guaranteed by the conditions of an apparent steady-state landscape evolution, but is also possible if the fundamental physical processes controlling the soil formation processes are related, even when steady-state conditions do not apply.

Historically, soil formation s has been represented in terms of a single formula, in which the principle factors of formation $s = f(cl, o, r, p, t)$, are represented independently from each other, e.g., cl, for climate, o for organisms, r for topography (relief), p for parent material, and t for time [9,10]. A guiding convention has been that soil is predominantly formed due to biogeochemical weathering, as a combination of physical, chemical, thermal, and biological processes together causing the disintegration of rocks, an evolutionary process that does not stop with the initial formation of soil. These processes, themselves, are limited by the atmosphere–rhizosphere, subsurface interaction, and in particular, infiltration from the land surface and solute transport in the unsaturated (vadose) zone. For example, organisms and precipitation supply the CO_2 necessary to drive silicate-weathering processes [23]. Soil processes are also affected by erosion, deposition, (e.g., aeolian) plant root uptake, microbial processes, infiltration, and evapotranspiration.

In the following, we present the general model. Then, we test it: first to see whether the values of the typical input variables generate values in accord with typical soil depths around the globe, then to see whether it reproduces variability in soil depth in accord with that observed over variable climatic input and what is known about parent material particle size variability with respect to topography (slope). Finally, we consider the implications of our model treatment and its wide range of applicability for landscape evolution concepts and discussions of agricultural sustainability.

2. General Model

The model for soil formation derives from percolation theory for solute transport in porous media. Percolation theory can be applied to enumerate the dominant flow paths when the medium is highly disordered. Chemical weathering in situ is shown to be transport-limited [24,25], but the lower boundary of chemical weathering is the bottom of the B horizon. The soil depth, neglecting erosion, is taken to be the distance of solute transport [18]. Percolation concepts that relate solute transport time and distance thus relate soil age and depth through the process of transport-limited chemical weathering [24].

It has been shown [18] that, when erosion (and deposition) can be neglected,

$$x = x_0 \left(\frac{t}{t_0} \right)^{1/D_b}.$$

(1)

Equation (1) describes the evolution of soil depth x as a function of time t. This expression has been derived based on the results of field and laboratory investigations of solute transport [25,26] associated with the chemical weathering of soil. It has also been assumed that this equation can be used to describe the bottom depth of the Bw horizon. Here, $D_b = 1.87$ is the fractal dimensionality of the percolation backbone for vertical flow, with 1.87 valid for full saturation and three-dimensional connectivity. In percolation theory, the backbone is obtained from the dominant, optimally connected flow paths by trimming off portions that connect only at one spot, called dead-ends. The mass fractal dimension of the backbone provides the scaling exponent relating time and distance. Predominantly downward flow occurs also under wetting conditions, but in this case the correct exponent is only slightly different, with $D_b = 1.861$, an insignificant difference from 1.87. In this expression, x_0 is a characteristic particle size of the soil parent material, and $(x_0/t_0) \equiv I/\varphi$, with

$$I = P - AET + run\text{-}on - run\text{-}off,$$

where I is the net infiltration rate; φ is the soil porosity (used to change the Darcy velocity to the pore-scale velocity); P is precipitation; and AET is the average evapotranspiration. The soil production function is then given by

$$dx/dt = (1/D_b)(x_0/t_0)(x_0/x)^{D_b - 1}.$$

Erosion is assumed to be taken into account by subtracting a constant, E, from the right-hand side of the differential equation. When soil erosion and soil production processes are equal in magnitude, $dx/dt = 0$, and the soil depth, x, is given by the equation

$$x = x_0 \left(\frac{I}{1.87\varphi E} \right)^{1.15}.$$

(2)

The power 1.15 of Equation (2) is $1/(D_b - 1)$.

Equations (1) and (2) implicitly represent a combination of the effects of climate, topography, and evapotranspiration. Equation (2) does not contain a time variable, since this equation represents the solution of Equation (1), consistent with an asymptotical convergence of the soil formation depth to a steady state value.

3. Predicting a Typical Soil Depth

If a typical soil depth is a depth you would expect to measure, this equates the term, in a narrow sense at least, to a mean soil depth. However, the actual values of soil depths vary from zero to tens of meters. What is a typical soil depth? Batjes [26], who considered the 4353 soils in the World Inventory of Soil Emission (WISE), used it to build up UNESCO's (United Nations Educational, Scientific, and Cultural Organization) database of 106 soil types presented on its soil map, assuming a characteristic soil depth of 1 m. Montgomery [27] gives a mean soil depth of 1.09 m, with a mean soil depth of 2.74 m for native vegetation, and 2.01 m for soil production areas. Hillel suggested that 1 m is a typical soil depth. What value would be suggested by Equation (2)?

Take x_0 = 30 μm, the size of a typical silt particle. Silt is the middle particle size (geometric mean) class in all soil classification schemes, and 30 μm is the middle (arithmetic mean) of the silt range. The same value, 30 μm, is also the geometric mean of the individual arithmetic means of the three principal soil particle classes, clay, silt, and sand. This particular length scale relates most closely to parent material, whether the soil is weathering from a bedrock with a specific mineral size, or whether it is forming on, e.g., an alluvial deposition. To calculate a mean infiltration rate, we must consider not only the precipitation, but also the water lost to evaporation and transpiration as well as what runs off. These variables relate to climate, the hydraulic conductivity of the substrate, and to the role of plants in the water cycle. Schlesinger and Jasechko [28] estimate that, globally, transpiration constitutes 61% of *AET*, and returns approximately 39% of *P* to the atmosphere. Thus, *AET* represents a mean fraction (0.39/0.61) = 64% of *P*. Lvovich's [27] estimation that *AET* = 65% of *P* is almost identical, and he also gives a global mean precipitation of 834 mm. Lvovich [29] estimates that a global mean of 24% of *P* travels to streams by overland flow, leaving only 11% of *P* for deep infiltration. The mean terrestrial *P* is reported as between 850 mm and 1100 mm [30], with a mean of 975 mm. Sixty-four percent of 975 mm is 624 mm, leaving 351 mm for *P* − *AET*. However, 11% of 975 mm is only 102 mm. On any local site, however, the difference between the *run-on* and the *run-off* can be either positive or negative. Thus, these estimates suggest that the amount of water reaching the base of the soil should be a column of water somewhere between 102 mm and 351 mm. Alternatively, we can consider the mean global *AET* over cold, temperate, and tropical, forested and non-forested, regions. Using the six different values given by Peel et al. [31] for these biomes generates an *AET* value of 654 mm, which is fairly close to the value inferred from Schlesinger and Jasechko [28], and implying *P* − *AET* = 321 mm. The actual infiltration rate is obtained from *I* through division by the porosity. We assume a typical porosity of 0.4, leading to values of I/φ between 255 mm/year and 878 mm/year (using a combination of Schlesinger and Willmott's numbers), or between 225 mm/year and 735 mm/year (using the numbers of Lvovich). These values average to 566 mm/year, or 480 mm/year, depending on the particular estimates applied, and are reasonable. A typical erosion rate of about *E* = 30 m/Myear ≈ [(1 m/Myear) (1000 m/Myear)]$^{0.5}$, is obtained from the geometric mean of the range of erosion rates discussed in Bierman and Nichols [19]. Using x_0 = 0.00003 m, I/φ = 806 mm/year, and *E* = 30 m/Myear, and the first range of *I* values given, the result for *x* is 0.48 m < *x* < 1.81 m, while for the second range of *I* values given, 0.42 m < *x* < 1.53 m. Both the arithmetic and geometric means of both ranges cluster around 1 m.

4. What Can We Say about the Variability of Soil Depths?

Let us consider first the ratio *I/E*, which, raised to the power 1.15, has the potential to produce the greatest variation (range) in soil depths. In fact, this ratio should be quite insensitive to *P*, since both *I* and *E* tend to increase with increasing precipitation. For example, Dunne et al. [32] found a linear relationship between *I* and *P*, in general accord with the previously cited tendency for *AET* to be roughly half of *P*. Reiners et al. [33] reported a linear relationship between *P* and erosion, *E*. Their study utilized a rainfall gradient at similar temperatures across the Cascade Mountains in Washington State, United States. Along this transect one should thus expect roughly constant soil depths, and the ratio *I/E* should, in the absence of steep topography, remain relatively invariant.

What about trends with temperature? Data from Sanford and Selnick [34] revealed a tendency for the fraction of precipitation lost to AET to increase with increasing temperature, particularly in conjunction with aridity. Thus, the conclusions of Heimsath et al. [35] regarding the results of their Australian measurements, "[t]he suite of results from different field sites indicates that erosion rates generally increase with increasing precipitation and decreasing temperature," indicate that the processes of soil formation may be dependent on evapotranspiration. Consequently, the water potentially available for either infiltration or overland flow, $(P - AET)$, serves as a predictor of E and soil formation, rather than simply P. We, therefore, hypothesize that both the numerator and denominator in Equation (1) would contain a proportionality to the quantity $(P - AET)$, meaning that weather conditions (within a specific climatic zone) would have far less influence on soil depth than commonly assumed. However, I and E can be expected to have a complementary dependence on the partitioning of water to overland flow, which brings in the effect of topography. The relationship between the potential evaporation, evapotranspiration, precipitation, and runoff was considered in great detail by Budyko [36], and many soil scientists and hydrologists followed Budyko's approach, e.g., Gentine et al. [37].

Concerning topography, regions with steeper topography will tend to have higher overland flow, and thus higher erosion rates, resulting in lower infiltration and soil formation rates. As an example, Burbank et al. [38] found that erosion rates and precipitation in the Himalayan mountains were not correlated (in contrast to Reiners et al. [33]), and attributed their *anomalous* result to the strong tendency for the precipitation to decline where the slope was increasing. Notably, however, the declining precipitation with increasing slope should lead to a diminution in soil production compared with erosion, and a higher probability of bedrock exposure, as is indeed the case in this region. Divergent topography, with concomitant divergence in surface water flux and therefore soil transport, will produce thinner soils than convergent topography, as noted in Heimsath et al. [39], a tendency intensified by steeper topography generally. Our reasoning, though it may be accentuated in reality by lateral soil transport [39], does not depend on such transport, and is merely a consequence of the greater infiltration values in topography that is convergent and not so steep that soil covering is missing entirely.

How does soil depth depend on the particle size of original sediments? This question is more nuanced than the previous question. Soil depths should nominally be proportional to particle sizes. However, erodibility has a strong dependence on particle size, first increasing with increasing size from clay to silt, and then decreasing with increasing size at larger sizes. The seeming anomaly at small particle sizes is due to the cohesive forces between clay grains, which are typically charged. Thus, as long as I is principally precipitation-limited, a decrease in particle size below silt size tends to reduce soil erosion, on account of the increasing cohesive forces between the grains. Therefore, x_0, for finer soils at least, should be positively correlated with the erosion rate, E, and the two stated influences will tend to cancel. However, at larger particle sizes, increasing particle sizes should tend to decrease E, accentuating the tendency for soils to be deeper; although, if the argument is turned around, a greater importance of erosion will tend to remove finer components, leading to a coarser soil. Sandy soils should thus have the deepest weathering horizons, although for larger particle sizes, the term soil is not characteristically employed. If I is principally hydraulic conductivity-limited, however, then greater precipitation rates, P, will not tend to increase I, but will tend to increase water run-off and erosion, E, leading to much thinner soil depth, regardless of particle size. Thus, unfractured crystalline rock with very low hydraulic conductivity values will tend to be exposed, unless it is buried through, e.g., fluvial deposition. Similar conclusions hold for increased slope angles, which will increase E and, more likely, decrease I, both of which should lead to thinner soils.

Finally, it is worth noting that, especially for very low soil formation rates and erosion rates in, e.g., continental interiors such as Australia [40], soil formation rates do tend to be larger than soil erosion rates, consistent with the predicted power-law decay of the soil production function (rather than the oft-assumed exponential form of Heimsath et al. [41]). The slow decay toward a steady-state soil production value leads to an increased tendency of soils in arid regions not to attain steady-state conditions [40], and for their depths to be smaller than that predicted from steady-state landscape evolution assumptions.

5. Comparison With Data: Mainly Climate

Below, we use data of White et al. [42], He et al. [43], Egli et al. [44] the Heimsath group [35,39,45–47], to confirm the relative consistency of soil depths across climatic gradients, but not across a variation in topography. The San Gabriel Mountain data of Southern California [45] demonstrate the variation of soil depths along a gradient in topographic relief, and thus erosion rates, but not of climate. We have found particle size data for only five of the data sets below, and even in some of these cases we had to generate a median particle size from graphic representations of what are considered to be typical distributions of particle diameters for a given texture [48].

In southeastern Australia, where many of the Heimsath group's field sites are located, precipitation tends to increase inland up to the escarpment, and then decrease with increasing altitude. The decrease in *P* with increasing altitude is mirrored by a diminution in *AET* from between 600 mm/year and 700 mm/year, to between 500 mm/year and 600 mm/year [49]. The Frog's Hollow and Brown Mountain sites at about 1000 m elevation have more limited vegetation cover and cooler temperatures compared with Nunnock River and Snug, both factors that tend to reduce evapotranspiration.

In the San Gabriel Mountains, "the landscape varies from gentle, soil mantled and creep dominated in the west to steep, rocky and landslide dominated in the east", accompanied by an increase in erosion rates from about 35 m/Myear to over 200 m/Myear, with the boundary to landslide dominated at about 200 m/Myear. The actual soil depth extremes were taken from Figure 3 of Heimsath et al. [45], and restricted to non-landslide-dominated slopes. On landslide-dominated slopes, the soil depth was "patchy", a scenario not addressed here. Thus, the variation in soil depth from west to east along the San Gabriel Mountains is a result of a variation in the erosion rate due to changes in mountain slope, rather than a variation in, e.g., climatic variables. Net infiltration rates are calculated as $I = P - AET - Run\text{-}off$, given the fact that run-off tends to be higher than run-on, and can play a role in water loss on site (according to Lvovich [29], 24% of precipitation flows into the ocean through surface run-off globally, while only 11% goes into deep infiltration). A summary of predicted and observed soil depths over two orders of magnitude of erosion rates is given in Table 1. Our predicted mean soil depth across 12 sites on four continents is 1.14 m, while the observed mean soil depth across those sites is 0.81 m.

In Figure 1, we compare predicted and observed soil depths, forcing the linear fit to go through the origin (for the San Gabriel Mountains, mean values are used here). We have an overall 43% overestimation for the mean soil depths across 12 sites, with less than 15% discrepancy at 3 sites, and 5 out of 12 underestimations (~22% on average) along with 7 overestimations (~90% on average). A large fraction of the overestimation comes from Merced River (84%), which has a slow average erosion rate and might not have reached a steady-state condition, and from east of the San Gabriel Mountains (200% to 500%) with a very shallow observed soil depth of 3 cm, which contributes a large discrepancy to the percentage, if not the actual discrepancy. There are a number of other potential reasons for overestimation. Our choice of an arithmetic mean for the observed soil depths tends to minimize the influence of shallower soil depths in the reporting of regional values for soil depth, but younger, shallower soils still tend to reduce a mean depth compared with a predicted steady-state value. Other sources of errors could come from reducing an entire particle size distribution to a median particle size, the accuracy of *P*, *AET*, and *Run-off* rates, and the porosity of the soils, particularly since we do not have a means to address local variability for most sites. Note that removing Merced River (maximum predicted value) will result in reducing R^2 from 0.934 to 0.679, while decreasing the numerical pre-factor from 1.66 to 1.05, making the relationship nearly one-to-one.

Table 1. Predicted soil depths from reasonable infiltration and given erosion rates.

Station	Region [a]	E (m/Myr)	I/φ [b] (m/yr)	P (m/yr)	Predicted [c] x (m)	Observed Mean x (m)	Reference Number [d]
Brown Mountain	AU	14	0.22	0.69	0.95	0.62	[35,49]
Frog's Hollow1	AU	10	0.2	0.72	1.29	1.5	[35,46,49]
Frog's Hollow2	AU	27	0.2	0.72	0.41	0.43	[35,46,49]
Snug	AU	35	0.26	0.87	0.42	0.61	[35,49]
Nunnock River	AU	35	0.45	0.91	0.78	0.62	[35,49]
Coos Bay	OR	119	1.58	1.68	0.80	0.56	[47,50]
Gongga Mountain	CH	2500	0.78	1.95	0.14	0.203	[43,51–55]
European Alps	EU	127	0.55	1.55	0.22	0.33	[44,56,57]
Tennessee Valley	N.CA	35	0.7	0.92	0.43	0.4	[39,48,50,58]
Merced River	C.CA	13.75	0.1	0.31	7.12	3.87	[42,50]
San Gabriel Mountain-west	S.CA	35	0.55	0.81	0.65 to 1.3	0.55	[45,50,59]
San Gabriel Mountain-east	S.CA	200	0.55	0.81	0.09 to 0.18	0.03	[45,50,59]

Notes: [a] Region: AU = Australia, OR = Oregon, N. CA = North California, C. CA = Central California, S. CA = South California, CH = China, EU = Europe. [b] $I = P - AET - Run\text{-}off$. P, AET and *Run-off* rates used to calculate the infiltration rate in the United States were obtained from the Cal-adapt website [50], and in Australia, P values from Heimsath group [35], AET and *Run-off* values from the Bureau of Meteorology [49], P for Gongaga Mountain is from He et al. [43], AET was given by Gao et al. [51], given the fact that 80% of the total run-off from Gongga Mountain comes from glacial melting [54], 20% of run-off rates from Lin and Wang [55] was estimated as run-off lost from precipitation. For the European Alps, P is from Egli et al. [44], AET from the evapotranspiration map of Europe on the IMPACT2C web-atlas [56], run-off values from Wehren et al. [57]. Here we take $\varphi = 0.4$ as typical porosity. [c] For all data sets except Gongga Mountain and the 4 Californian sites, a typical particle size of 30 μm was chosen to calculate predicted depth, since no particle size data were reported. For Tennessee Valley, 10 μm was taken as x_0, since most of the soils at Tennessee Valley fit into the clay loam category [58] with median particle size 10 μm [48]. He et al. [43] report a typical particle size of 400 μm for Gongga Mountain; White et al. [42] give 530 μm for Merced River; for San Gabriel Mountains, soils are mainly loams on the hillslopes [59], which has median particle size ranging from 20 to 40 μm. Here we take both values to obtain a range of soil depths in the San Gabriel Mountains.

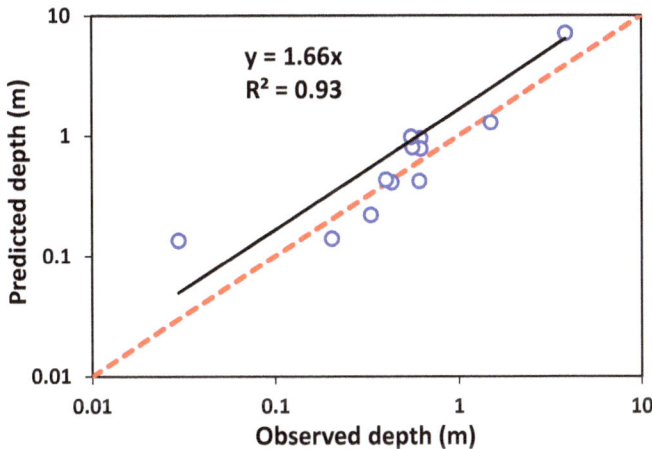

$$y = 1.66x$$
$$R^2 = 0.93$$

Figure 1. Comparison of the predicted soil depth via Equation (2) versus the observed depths for 12 sites (open circuits) from all around the world. The dashed red line represents the 1:1 line. See Table 1 for further details.

6. Comparison with Data: Slope Angle

Let us consider specifically the slope angle dependence of soil depth as exhibited by the data from the San Gabriel Mountains [45] and a result from Norton and Smith from 1930 as reported in Jenny [10]. We apply Equation (2) with the known value of $P - AET - run\text{-}off$ from Table 1, and a 30 μm median particle diameter. In order to address the slope dependence of soil depth, Equation (2) requires a slope-dependent erosion rate. Although it is not within our capability to predict such a function, Mongomery and Brandon [60] reported in their Figure 1 an empirical function for the slope angle dependence of erosion rates in the Olympic Mountains in Washington. Incorporating this empirical input makes it possible to use Equation (2) to predict the slope angle dependence of soil depth. The comparison is shown in Figure 2. In order to make this equation predictive, the input of the erosion rate function (Montgomery and Brandon [60]) is critical. This function tends to produce a rapid reduction in soil depths to nearly zero as slopes of about 30 degrees are exceeded (since zero values cannot be plotted on a logarithmic graph, such values were converted to 0.001 for both axes). In spite of the considerable scatter in field values, it appears that our prediction captures the essential trends accurately. Note that the use of either 20 μm or 30 μm for the fundamental particle size will result in an overestimation of the soil depth at zero slope, but an underestimation at larger slopes, since the latter values are deeper than the zero slope depths, a result attributed by the authors [45] to the effects of soil deposition.

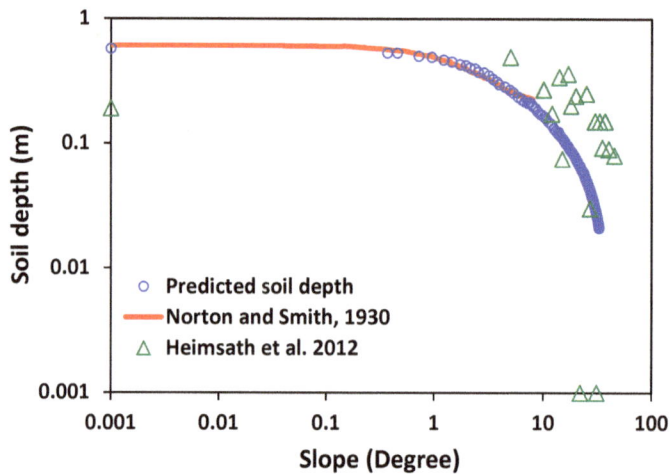

Figure 2. Predicted and observed soil depth as a function of slope. The Norton and Smith data were digitized from Jenny [10], but extend only to a slope of 8 degrees. The Heimsath et al. [45] data were reported in a Table and extend over the range of 6 degrees to 32 degrees. The erosion function input was digitized from Figure 1 of Montgomery and Brandon [60]. *I* was given in Table 1 here, and a typical particle size of 30 μm, close to the 20 μm value considered to be most likely to characterize the San Gabriel mountain slopes, was applied. In order to reduce the scatter in the reported data, we give the mean soil depth at any particular slope value, although this will attach additional weight to the locations with deeper soils.

7. Implications for Geomorphological Studies of Natural and Agricultural Landscapes

The existing landscape evolution models require a large number of inputs, such as soil production and soil transport as functions of depth, parent material, topography, climate, and organisms, while delivering several quantities of interest. The outputs include: (1) regional denudation rates, which are important for understanding (neo) tectonics; (2) spatial variability of soil erosion; (3) spatial variability of soil depth; and (4) spatial variability of landforms. When adapted to landscapes with human interference, such as agriculture, the second and third of these products may take on additional significance regarding sustainability. The first and fourth results typically involve larger time scales than the human time scales resolved by agricultural practices. Many processes are relevant to the evolution of landscapes, and these can have different impacts at different time scales and in different locations. We point out that the choice of soil production model has an impact on the results of such landscape evolution models, and that our soil production function may help to resolve some difficult problems in landscape evolution modeling.

When an exponential model for soil production is assumed, which delivers a maximum soil production rate modulated by an exponential decay, one finds for a steady-state soil thickness the negative of the logarithm of the ratio of the erosion rate to the maximum soil production rate, $-\ln(E/R_m)$ [61]. For a wide range of typical values for E and maximum soil production rate, R_m (as reported by Heimsath and co-workers), it turns out that this formula yields soil depths ca. 1 m. But the physical reason for a relatively consistent soil depth that lies in the correlation between E and R through the factor $P - AET$ is missing in Roering [61] (both run-off and net infiltration, I, increase with $P - AET$). In Roering's treatment, the relative consistency in the output arises from the logarithmic phenomenology, which is very slowly varying in comparison with our power law. Consider, however, what happens if I, which is the upper limit of soil production rates in our treatment, is substituted for R_m in the Roering soil thickness relationship. Since maximum soil production rates [62] are ca. 3000 m/Myear, but infiltration rates can be approximately three orders of magnitude larger, even with Roering's logarithmic dependence, substituting I for R_m would increase the predicted soil depth considerably. Nevertheless, replacement of R_m by I in Roering's result generates the same argument in the logarithm as appears in our power law. This is a significant correspondence. Our result expressed in Equation (2) generates, in principle, a much more sensitive function of parameters such as infiltration and erosion rates to the soil depth. However, the tendency of each to increase with increasing $P - AET$ makes this ratio rather insensitive to changes in climate. However, see what happens if the erosion rate changes by over an order of magnitude due to topography, such as in the San Gabriel Mountains. Our result predicts a better than order of magnitude change in soil depth, as observed, as well as the approximate functional form of the soil depth as a function of slope angle (Figure 2), whereas that of Roering predicts a variation less than a factor 2 (compare the actual soil depth distinction between the western and eastern provinces of a factor approximately 20).

Although the most important topic may thus relate to absolute values of soil production and erosion, issues in the local variability of soil production and erosion relate to the shapes of the topography as well, and here understanding is also lacking. More generally, the use of common landscape models [63–65] does not allow for the prediction of the wide range of observed shapes of landscapes. From Roering [61], "linear transport models [use of the diffusion equation for soil downslope transport] produce constant curvature, not planar slopes, necessitating integration of various downslope transport mechanisms. Put simply, the [introduced] flux-slope nonlinearity enables nearly steep and planar (low convexity) sideslopes to erode at rates commensurate with highly convex hilltops." As Roering points out, such problems have been addressed by incorporating soil depth-dependent transport as well as soil production into landscape evolution models [66,67], which allows an increase in soil transport rates downslope even in the case of planar slopes. However, with our result of a soil formation function which is highly dependent on infiltration, the low infiltration rate on hilltops and slopes (as compared to hollows) will tend to produce a smaller soil production rate, which could, without this strong dependence on infiltration, otherwise be interpreted as resulting from a larger erosion rate. Perhaps a portion of the difficulties encountered by landscape evolution models is that they do not incorporate sufficient local variability in soil production rates due to the convergence/divergence of surface flow.

Our results also have implications for agricultural systems. In particular, we can predict what a new steady state soil depth will be if the soil erosion rate is increased by an order of magnitude or more, as results from traditional agricultural practices. While it is possible to write an accurate result for the soil production in terms of the instantaneous depth and the erosion rate, which allows a more rigorous prediction of the time frame over which the soil adapts to the new erosion conditions, several factors suggest that it may be better to calculate this time scale simply by taking the quotient of a typical original soil depth (say 1 m) and dividing by the erosion rate. One complicating factor is that the soil production function may change when the soil is very shallow, at least if the bedrock does not have a hydraulic conductivity comparable to that of the soil. In such a case, the relevant infiltration rate may be much smaller in a very shallow soil, and the soil production function comparably reduced.

Such a situation could lead to a "humped" soil production function [22], an instability resulting in total soil loss, and, essentially, a two state system, where it may be difficult for natural systems to evolve between the two, i.e., once the soil is lost, it does not return. Results for our predicted steady-state soil depths are given in Table 2 below. In the case of calculations for the time required to strip a landscape of soil, we used an arbitrary starting depth of 1 m, in approximate accord with our general predictions, Roering's [61] equation, and, as it turns out, with steady-state depths calculated from Equation (2) in accord with the input erosion rates given by Montgomery [27].

Table 2. Predicted steady-state soil depths for conditions reported by Montgomery [27].

Status or Condition	$<E>$ [a] (mm/yr)	Predicted Depth [a] (m)	Time [a] (yr)	E [b] (mm/yr)	Predicted Depth [b] (m)	Time [b] (yr)
Traditional agriculture	3.934	0.0038	254	1.537	0.011	651
Conservation agriculture	0.124	0.21	8064	0.082	0.33	12,200
Natural vegetation	0.053	0.54		0.013	2.74	
Soil production [c]	0.036	0.85		0.017	2.01	
Geological erosion	0.173	0.14		0.029	1.09	
Mean soil depth [d]		0.51			1.95	

Notes: [a] $<E>$ denotes mean erosion rates. Predicted depth is calculated using $<E>$, $I = 0.5$ m/year, and $x_0 = 0.00003$ m. T is the corresponding time (approximate) to reach steady state starting form a soil depth of $x = 1$ m, $T = x/<E>$; [b] E denotes median erosion rates. Predicted depth is calculated using E, $I = 0.5$ m/year, and $x_0 = 0.00003$ m. T is the corresponding time (approximate) to reach steady state starting form a soil depth of $x = 1$ m, $T = x/E$; [c] The rate is not an erosion rate, but the soil production rate, and the depth is the soil depth that would generate such a production rate; [d] Mean soil depth is calculated from averaging individual values in the column above it.

Using a mean erosion (or soil production) rate, one finds a mean soil depth of 0.51 m, while using a median erosion rate, the result is 1.95 m, consistent with our understanding that 1 m is a typical soil depth. An important result is that, for traditional agriculture, using a typical soil depth of 1 m generates a time until virtually complete soil loss of between 250 and 650 years, a result that is consistent with Montgomery's [68] assertion that limits on agriculture placed by soil loss were critical in setting the period of domination of a number of classical civilizations at about 500 years.

Consider some of the individual values of soil depths in Table 2. With native vegetation, the range of depths extends from 54 cm to 2.74 m. The soil production values suggest a range of soil depths from 85 cm to 2.01 m. However, the geologic erosion rates indicate steady-state soil depths from 14 cm to 1.09 m. (Note that soil degradation, i.e., soil depletion, is taking place due to a combination of factors, such as deforestation (30%), agricultural activities (28%), overgrazing (35%), overexploitation (7%), and industrialization (1%) [69]. Among factors of physical degradation are water erosion (55%), wind erosion (29%), chemical degradation (12%), and compaction/crusting (4%) [69]. Some of the more interesting results, however, may be the implied steady-state soil depths for normal agriculture, which lie between 4 mm and 11 mm, whereas the corresponding depths for conservation agriculture range from 21 cm to 33 cm. Thus, it is clear that, while normal agriculture is not sustainable, conservation agriculture is also quite marginal, as most crops need more than 33 cm of soil to thrive.

8. Summary

Based on the concept that the soil forming processes combined are dependent on the infiltration rate, we developed a simple model for the prediction of the soil depth. The model is verified by comparison of predicted and actual soil depths for a range of climates with emphasis on semi-arid zones, as well as along a slope gradient. Discrepancies between the data and model calculations remain, but they are comparatively small, taking into account the variety of soil forming factors: topography, climate, organisms, and parent material. When expressed as a function of time, it is shown to describe the temporal development of soil production. However, according to Dokuchaev and later soil scientists, it is equally important to be able to express a range of other soil forming factors, such as carbon and nitrogen content and other chemicals.

In our discussion and comparison with data, some details are still missing, such as (in most cases) site-specific porosity, or run-off values. The relevant porosity may be the effective value, which excludes pores that do not connect, or connect to flow paths at only one point, and likely excludes also internal water adsorption into clay minerals. We also have not addressed issues of climate change, which could require significant alterations of parameter values over time. It is to be hoped that a more detailed investigation of such parameter values and their potential temporal variation will improve the accuracy of our predictions. Additionally, we have not addressed specific flow paths, which may depend on a wide variety of factors, such as the slope angle, or whether the soil is graded or layered. All of these factors may introduce as yet unaccounted for variability. In spite of these omissions, the theoretical approach appears to account properly for soil formation factors over a wide range of climates and slope angles.

Author Contributions: Yu (writing, soil data location and digitization, climate and run-off data search, analysis), Hunt (conception), Faybishenko (writing, relationships to fundamental soil formation models and history of soil science), Ghanbarian (writing and analysis).

Conflicts of Interest: The authors declare no conflict of interest.

References

1. Darwin, C. *The Formation of Vegetable Mould through the Action of Worms, with Observations on Their Habits*; John Murray: London, UK, 1881.
2. Dokuchaev, V.V. Russian Chernozem. In *Selected Works*; Monson, S., Ed.; Israel Program for Scientific Translations Ltd.: Moscow, Russia, 1948; Volume 1, pp. 14–419.
3. Gilbert, G.K. The convexity of hilltops. *J. Geol.* **1909**, *17*, 344–350. [CrossRef]
4. Coffey, G.N. *A study of the Soils of the United States, Bulletin 85*; USDA Bureau of Soils: Washington, DC, USA, 1912; p. 114.
5. Hilgard, E.W. *Soils*; The Macmillan Company: New York, NY, USA, 1914.
6. Cline, M.G. Historical highlights in soil genesis and classification. *Soil Sci. Soc. Am. J.* **1977**, *41*, 250–254. [CrossRef]
7. Richter, D.D.; Markewitz, D. *Understanding Soil Change: Soil Sustainability Over Millennia, Centuries and Decades*; Cambridge University Press: Cambridge, UK, 2001.
8. Richter, D.D.; Yaalon, D.H. "The Changing Model of Soil" Revisited. *Soil Sci. Soc. Am. J.* **2011**, *76*, 766–778. [CrossRef]
9. Glinka, K.D. *Dokuchaiev's Ideas in the Development of Pedology and Cognate Sciences*, 1st ed.; Series: Akademiia Nauk SSR. Russian pedological investigations; The Academy: Leningrad, USSR, 1927.
10. Jenny, H. *Factors of Soil Formation: A System of Quantitative Pedology*; Dover: New York, NY, USA, 1941.
11. Johnson, D.L.; Schaetzl, R.J. Differing views of soil and pedogenesis by two masters: Darwin and Dokuchaev. *Geoderma* **2014**, *237*, 176–189. [CrossRef]
12. Schoeneberger, P.J.; Wysocki, D.A. Hydrology of soils and deep regolith: A nexus between soil geography, ecosystems and land management. *Geoderma* **2005**, *126*, 117–128. [CrossRef]
13. Glinka, K.D. *Pochvovedenie, Moscow, 1931 (Translated Treatise on Soil Science by Israel Program for Scientific Translations, Jerusalem)*; National Science Foundation: Washington, DC, USA, 1931.
14. Hillel, D. Soil: Crucible of life. *J. Nat. Resour. Life Sci. Educ.* **2005**, *34*, 60–61.
15. Chizhikov, P.N. The lower boundary of soil. *Sov. Soil Sci.* **1968**, *11*, 1489–1493.
16. Richter, D.D.; Bacon, A.R.; Brecheisen, Z.; Mobley, M.L. Soil in the Anthropocene. *Earth Environ. Sci.* **2015**, *25*. [CrossRef]
17. Anderson, R.S.; Anderson, S.P. *Geomorphology: The Mechanics and Chemistry of Landscapes*; Cambridge Press: New York, NY, USA, 2010.
18. Hunt, A.G.; Ghanbarian, B. Percolation theory for solute transport in porous media: Geochemistry, geomorphology, and carbon cycling. *Water Resour. Res.* **2016**, *52*, 7444–7459. [CrossRef]
19. Bierman, P.R.; Nichols, K.K. Rock to sediment—slope to sea with 10Be–rates of landscape change. *Annu. Rev. Earth Planet. Sci.* **2004**, *32*, 215–255. [CrossRef]
20. Amundson, R.; Heimsath, A.; Owen, L.; Yoo, K.; Dietrich, W.E. Hillslope soils and vegetation. *Geomorphology* **2015**, *234*, 122–132. [CrossRef]

21. Carson, M.A.; Kirkby, M.J. *Hillslope Form and Process*; Cambridge University Press: Cambridge, UK, 1972; p. 475.

22. Heimsath, A.M.; Fink, D.; Hancock, G.R. The 'humped' soil production function: eroding Arnhem Land, Australia. *Earth Surf. Process. Landf.* **2009**, *34*, 1674–1684. [CrossRef]

23. Hilley, G.E.; Porder, S. A framework for predicting global silicate weathering and CO2 drawdown rates over geologic time-scales. *Proc. Natl. Acad. Sci. USA* **2008**, *105*, 16855–16859. [CrossRef] [PubMed]

24. Hunt, A.G.; Ghanbarian-Alavijeh, B.; Skinner, T.E.; Ewing, R.P. Scaling of geochemical reaction rates via advective solute transport. *Chaos* **2015**. [CrossRef] [PubMed]

25. Yu, F.; Hunt, A.G. Damköhler number input to transport-limited chemical weathering and soil production calculations. *ACS Earth Space Chem.* **2017**, *1*, 30–38. [CrossRef]

26. Batjes, N.H. Total carbon and nitrogen in the soils of the world. *Eur. J. Soil Sci.* **1996**, *47*, 151–163. [CrossRef]

27. Montgomery, D. Soil erosion and agricultural sustainability. *Proc. Natl. Acad. Sci. USA* **2007**, *104*, 13268–13272. [CrossRef] [PubMed]

28. Schlesinger, W.H.; Jasechko, S. Transpiration in the global water cycle. *Agric. For. Meteorol.* **2014**, *189*, 115–117. [CrossRef]

29. Lvovitch, M.I. The global water balance: U.S. National Committee for the International Hydrological Decade. *Natl. Acad. Sci. Bull.* **1973**, 28–42. [CrossRef]

30. Willmott, C.J.; Robeson, S.M.; Feddema, J.J. Estimating continental and terrestrial precipitation averages from rain-gauge networks. *Int. J. Climatol.* **1994**, *14*, 403–414. [CrossRef]

31. Peel, M.C.; McMahon, T.A.; Finlayson, B.L. Vegetation impact on mean annual evapotranspiration at a global catchment scale. *Water Resour. Res.* **2010**. [CrossRef]

32. Dunne, T.; Zhang, W.; Aubrey, B.F. Effects of rainfall, vegetation, and microtopography on infiltration and run-off. *Water Resour. Res.* **1991**, *27*, 2271–2285. [CrossRef]

33. Reiners, P.W.; Ehlers, T.A.; Mitchell, S.G.; Montgomery, D.R. Coupled spatial variations in precipitation and long-term erosion rates across the Washington Cascades. *Nature* **2003**, *426*, 645–647. [CrossRef] [PubMed]

34. Sanford, W.E.; Selnick, D.L. Estimation of evapotranspiration across the conterminous United States using a regression with climate and land-cover data. *J. Am. Water Res. Assoc.* **2013**, *49*, 217–230. [CrossRef]

35. Heimsath, A.M.; Chappell, J.; Fifield, K. *Eroding Australia: Rates and Processes from Bega Valley to Arnhem Land*; Bishop, P., Pillans, B., Eds.; Australian Landscapes, Geological Society: London, UK, 2010; Volume 346, pp. 225–241. [CrossRef]

36. Budyko, M.I. *Climate and Life*; Academic Press: Orlando, FL, USA, 1974.

37. Gentine, P.; D'Odorico, P.; Lintner, B.R.; Sivandran, G.; Salvucci, G. Interdependence of climate, soil, and vegetation as constrained by the Budyko curve. *Geophys. Res. Lett.* **2012**, *39*. [CrossRef]

38. Burbank, D.W.; Blythe, A.E.; Putkonen, J.; Pratt-Sitaula, B.; Gabet, E.; Oskin, M.; Barros, A.; Ojha, T.P. Decoupling of erosion and precipitation in the Himalayas. *Nature* **2003**, *426*, 652–655. [CrossRef] [PubMed]

39. Heimsath, A.M.; Dietrich, W.E.; Nishiizumi, K.; Finkel, R.C. Cosmogenic nuclides, topography, and the spatial variation of soil depth. *Geomorphology* **1999**, *27*, 151–172. [CrossRef]

40. Yu, F.; Hunt, A.G. An examination of the steady-state assumption in soil development models with application to landscape evolution. *Earth Surf. Process. Landf.* **2017**, under review.

41. Heimsath, A.M.; Dietrich, W.E.; Nishiizumi, K.; Finkel, R.C. The soil production function and landscape equilibrium. *Nature* **1997**, *388*, 358–361. [CrossRef]

42. White, A.G.; Blum, A.E.; Schulz, M.S.; Bullen, T.D.; Harden, J.W.; Peterson, M.L. Chemical weathering rates of a soil chronosequence on granitic alluvium: I. Quantification of mineralogical and surface area changes and calculation of primary silicate reaction rates. *Geochimi. Cosmochim. Acta* **1996**, *60*, 2533–2550. [CrossRef]

43. He, L.; Tang, Y. Soil development along primary succession sequences on moraines of Hailuogou Glacier, Gongga Mountain, Sichuan, China. *Catena* **2008**, *72*, 259–269. [CrossRef]

44. Egli, M.; Dahms, D.; Norton, K. Soil formation rates on silicate parent material in alpine environments: Different approaches–different results? *Geoderma* **2014**, *213*, 320–333. [CrossRef]

45. Heimsath, A.M.; DiBiase, R.A.; Whipple, K.X. Soil production limits and the transition to bedrock-dominated landscapes. *Nat. Geosci.* **2012**, *5*, 210–214. [CrossRef]

46. Skaggs, T.H.; Arya, L.M.; Shouse, P.J.; Mohanty, B. Estimating particle-size distribution from limited soil texture data. *Soil Sci. Soc. Am. J.* **2001**, *65*, 1038–1044. [CrossRef]

47. Australian Bureau of Meteorology, Commonwealth of Australia, Bureau of Meteorology. Available online: http://www.bom.gov.au/water/landscape/ (accessed on 1 May 2017).

48. Heimsath, A.M.; Chappell, J.; Dietrich, W.E.; Nishiizumi, K.; Finkel, R.C. Late Quaternary erosion in southeastern Australia: A field example using cosmogenic nuclides. *Quat. Int.* **2001**, *83–85*, 169–185. [CrossRef]

49. Heimsath, A.M.; Dietrich, W.E.; Nishiizumi, K.; Finkel, R.C. Stochastic processes of soil production and transport: Erosion rates, topographic variation and cosmogenic nuclides in the Oregon coast range. *Earth Surf. Process. Landf.* **2001**, *26*, 531–552. [CrossRef]

50. Cal-Adapt Website. Available online: http://cal-adapt.org/data/tabular/ (accessed on 21 May 2017).

51. Gao, G.; Chen, D.; Xu, C.; Simelton, E. Trend of estimated actual evapotranspiration over China during 1960–2002. *J. Geophys. Res. Atoms.* **2007**, *112*. [CrossRef]

52. Ouimet, W.B.; Whipple, K.; Granger, D.E. Beyond threshold hillslopes: Channel adjustment to base-level fall in tectonically active mountain ranges. *Geology* **2009**, *37*, 579–582. [CrossRef]

53. Zhang, X.B.; Quine, T.A.; Walling, D.E.; Wen, A.B. A study of soil erosion on a steep cultivated slope in the Mt. Gongga region near Luding, Sichuan, China, using the 137Cs technique. *Acta Geol. Hisp.* **2000**, *35*, 229–238.

54. Cao, Z.T. The characteristics of glacier hydrology in the area of Gongga Mountain. *J. Glaciol. Geocryol.* **1995**, *17*, 73–83.

55. Lin, Y.; Wang, G.X. Scale effect on runoff in alpine mountain catchments on China's Gongga Mountain. *Hydrol. Earth Syst. Sci.* **2010**, *7*, 2157–2186. [CrossRef]

56. IMPACT2C Web-Altas, Evapotranspiration of Europe. Available online: https://www.atlas.impact2c.eu/en/climate/evapotranspiration/ (accessed on 15 December 2016).

57. Wehren, B.; Weingartner, R.; Schädler, B.; Viviroli, D. General characteristics of Alpine Waters. *Alp. Waters* **2010**, *6*, 17–58.

58. Yoo, K.R.; Amundson, A.; Heimsath, M.; Dietrich, W.E. Erosion of upland hillslope soil organic carbon: Coupling field measurements with a sediment transport model. *Glob. Biogeochem. Cycles* **2005**, *19*, GB3003. [CrossRef]

59. Rulli, M.C.; Rosso, R. Modeling catchment erosion after wildfires in the San Ganbriel Mountains in the sourthern California. *Geophys. Res. Lett.* **2005**, *32*. [CrossRef]

60. Montgomery, D.R.; Brandon, M.T. Topographic controls on erosion rates in tectonically active mountain ranges. *Earth Planet. Sci. Lett.* **2002**, *201*, 481–489. [CrossRef]

61. Roering, J. How well can hillslope models "explain" topography? Simulating soil transport and production with high resolution topographic data. *Geol. Soc. Am. Bull.* **2008**, *120*, 1248–1262. [CrossRef]

62. Larsen, I.J.; Almond, P.C.; Eger, A.; Stone, J.O.; Montgomery, D.R.; Malcolm, B. Rapid soil production and weathering in the Southern Alps, New Zealand. *Science* **2014**, *343*, 637–640. [CrossRef] [PubMed]

63. Kirkby, M.J. Modelling cliff development in South Wales: Savigear re-reviewed. *Z. Geomorphol.* **1984**, *28*, 405–426.

64. Anderson, R.S.; Humphrey, N.F. Interaction of Weathering and Transport Processes in the Evolution of Arid Landscapes. In *Quantitative Dynamic Stratigraphy: Englewood Cliffs*; Cross, T.A., Ed.; Prentice Hall: Upper Saddle River, NJ, USA, 1989; pp. 349–361.

65. Howard, A.D. Badland morphology and evolution: Interpretation using a simulation model. *Earth Surf. Process. Landf.* **1997**, *22*, 211–227. [CrossRef]

66. Furbish, D.J.; Fagherazzi, S. Stability of creeping soil and implications for hillslope evolution. *Water Resour. Res.* **2001**, *37*, 2607–2618. [CrossRef]

67. Heimsath, A.M.; Furbish, D.J.; Dietrich, W.E. The illusion of diffusion: Field evidence for depth-dependent sediment transport. *Geology* **2005**, *33*, 949–952. [CrossRef]

68. Montgomery, D. *Dirt, The Erosion of Civilization*; University of California Press: Berkeley, CA, USA, 2007; p. 295. ISBN 9780520248700.

69. Cunningham, W.; Saigo, B. *Environmental Science: A Global Concern*; WCB/McGraw-Hill: Boston, MA, USA, 1998.

water

MDPI

Article

Better-Fitted Probability of Hydraulic Conductivity for a Silty Clay Site and Its Effects on Solute Transport

Chengpeng Lu [1,2,*] Wei Qin [2], Gang Zhao [3], Ying Zhang [4] and Wenpeng Wang [2]

[1] State Key Laboratory of Simulation and Regulation of Water Cycle in River Basin, China Institute of Water Resources and Hydropower Research, Beijing 100038, China
[2] State Key Laboratory of Hydrology-Water Resources and Hydraulic Engineering, Hohai University, Nanjing 210098, China; qwhhu@hhu.edu.cn (W.Q.); wangwp1983@gmail.com (W.W.)
[3] Hydraulic Research Institute of Jiangsu Province, Nanjing 210017, China;18936006590@163.com
[4] Water Resources Planning Bureau of Jiangsu Province, Nanjing 210029, China; yingzhang.hhu@gmail.com
* Correspondence: luchengpeng@hhu.edu.cn; Tel.: +86-258-378-7683

Received: 21 April 2017; Accepted: 23 June 2017; Published: 27 June 2017

Abstract: The heterogeneous hydraulic conductivity of a subsurface medium is vital to the groundwater flow and solute transport. Probability is efficient for characterizing and quantifying the field characterization of hydraulic conductivity. Compared with sandy mediums, silty clay is paid less attention to due to its low hydraulic conductivity. For long-term solute transport and seawater intrusion, the low-permeable medium is considered as a remarkably permeable medium. This study reports on a comprehensive investigation on the hydraulic conductivity field of the Ningchegu site, located east of Tianjin City of China. Four layers recognized by 52 boreholes, plain fill, continental silty clay, mud–silt clay and marine silty clay, were deposited from the top to the bottom. The hydraulic conductivities measured via permeameter tests ranged from 2×10^{-6} m/d to 1.6×10^{-1} m/d, which corresponded to the lithology of silty clay. The magnitude and the range of the hydraulic conductivity increased with the depth. Five probability distribution models were tested with the experimental probability, indicating that a Levy stable distribution was more matched than the log-normal, normal, Weibull or gamma distributions. A simple analytical model and a Monte Carlo technique were used to inspect the effect of the silty clay hydraulic conductivity field on the statistical behavior of the solute transport. The Levy stable distribution likely generates higher peak concentrations and lower peak times compared with the widely-used log-normal distribution. This consequently guides us in describing the transport of contaminations in subsurface mediums.

Keywords: Levy stable distribution; permeameter test; hydraulic conductivity; silty clay; solute transport

1. Introduction

The emphasis on hydraulic conductivity (K) is contributed by Darcy's law, and the field-measured values of K are of well-known heterogeneity with over 13 orders of magnitude [1]. Groundwater contamination has become one of the most important environmental issues all over the world. It is necessary to predict groundwater flow and solute transport in order to protect groundwater quality. However, the heterogeneity of porous media and the incomplete knowledge of data information lead to difficulties in the estimation of hydraulic properties and geophysical variables, and thus create difficulty in estimating or predicting the subsurface flow and transport.

The representation of hydraulic conductivity distributions is an important issue in predicting the groundwater flow field [2] and subsurface contaminant transport [3]. Several equations of the contaminant transport make assumptions regarding the properties of hydraulic conductivity, including the probability density function (PDF) [4]. The hydraulic conductivity of aquifer materials has most often been found to be log-normally distributed [5–11]. Sanchez-Vila et al. [12] demonstrated numerically that randomly varying transmissivities could exhibit a scale effect because of deviations from the log-normal distribution. Hyun [13] reported that the log permeability data at the Apache Leap Research Site (ALRS) near Superior, Arizona, were indeed represented more accurately by a Levy stable distribution than by a Gaussian distribution, and inferred that the permeability scale effect was probably true at many other sites.

The lithologies of aquifer mediums from MADE (Macro-Dispersion Experiment test site) [14], Borden [9], and Cape Code [7], regarded as the three most popular hydrogeological test sites, are generally sand, gravel, and sandy material mixed with lenses of silt. Vereecken and Doring [15] investigated the spatial variability of basic aquifer parameters and hydraulic conductivities using data from the clay and silt from Krauthausen experiment.

Most of the studies on the effect of heterogeneous hydraulic conductivity on solute transport were focused on the heterogeneous aquifer. However, PDFs of low K and the effect on flow and solute transport were not reported through literature. This could have resulted from testing difficulties in terms of low-K mediums with weak migration and solute abilities [11,16].

Stochastic methods were developed and used to deal with these difficulties [17].The stochastic simulation technique was applied to quantitatively study the impact of hydraulic conductivity heterogeneity on groundwater solute transport [18]. The study by Kohlbecker et al. [2] indicated that Levy stable distributions with increments in the log conductivity gave rise to Levy stable distributions with increments in the logarithm of the velocity (ln u) using Monte Carlo and MODFLOW (U.S. Geological Survey modular finite-difference flow model) techniques. Wang and Huang [3] analyzed the impact of hydraulic conductivity on solute transport processes in a highly heterogeneous aquifer. Few analyses were reported on the characteristic parameters of solute transport using stochastic test methods.

The twofold objectives of this study were to investigate the probability distribution of a low K, and the effects of PDFs on solute transport according to the collected data. The field study for the sampling and permeameter tests performed at the Ningchegu (NCG) site are introduced first. Then the statistics of a low-K field are shown, especially the statistical distributions of the low-K data tested by five widely used distributions. Finally, the effects of PDFs on solute transport were analyzed using a simple analytical model. The effects of Levy stable distributions on solute transport were compared with log-normal distributions.

2. Materials and Methods

2.1. Sampling and Lithology

The locations of boreholes are within the eastern Tianjin city, around 5 km west of the Bo Sea. This area was considered as a candidate of the NCG reservoir, resulting in the collection from dozens of boreholes during August to November of 2011 to depict the lithology at different depths. In total, 52 boreholes (Figure 1A) were evenly installed around an area of 6.2 km^2 at an areal density of 8 boreholes/km^2. The average distance of each pair of two boreholes was 345 m. Each borehole was drilled up to a depth of 10 m, and different samples (3~5 cores per borehole) were collected according to the diversity of the lithology. The cylinder sampler used was 20 cm in length and 110 mm in inner diameter. The undisturbed samples were then moved to a laboratory at Hohai University to test their hydraulic conductivities. In terms of the lithology of sediments in the study area dominated by clay and silty clay, a recommended falling head method [1] applied to the fine material was used to analyze the hydraulic conductivity of the core using a permeameter (TST-55; Φ: 61.8 mm $\times$ 40 mm; made in China). The infiltration process for low-K sediments was quite slow, and a successful permeameter test averagely took more than 6 hours in our study. Therefore, eight permeameters were used to test different clay samples, simultaneously. In total, we finished 212 falling head permeameter tests to investigate the hydraulic conductivity of the reservoir sediments.

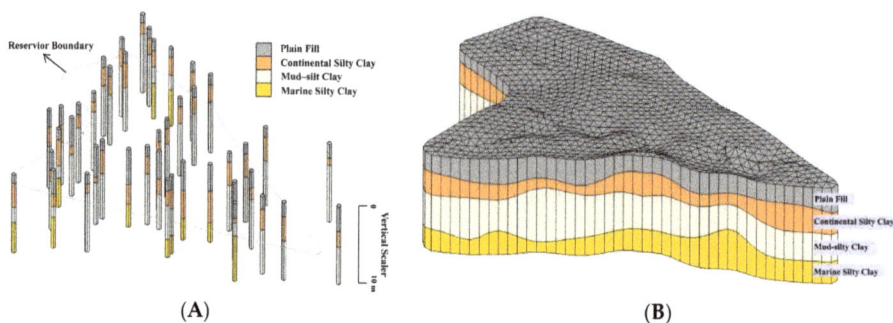

Figure 1. Spatial distribution of boreholes (**A**), and the 3D geological structure (**B**).

A thick, loose Quaternary sediment layer, which is a continental-dominated paralic deposition, covers the eastern area of Tianjin city. The lithology is mainly composed of clay, silty clay, and sand lens, formed from alluvium of the Hai River and Ji Canal [19]. Based on the information collected from borehole drilling, a 3-D lithology model was developed (Figure 1B). Four layers were identified by drilling, including the shallowest layer of plain fill with a thickness of between 1 and 1.5 m. The second shallowest layer of continental silty clay had a thickness of 0 to 3.4 m, indicating that some areas lacked the material of continental silty clay. The third mud–silt clay layer was buried below the second layer, with of thickness of 0 to 7.0 m, and the deepest layer of marine silty clay had a thickness of 0 to 7.3 m. Based on the borehole information, the study area was completely covered with the plain fill, but the other three layers somewhat showed stratum absence. The fourth layer of marine silty clay was only revealed at down to, but not limited to, 10 m of the boreholes, and the lack of the fourth layer locally came from the limited drilling depth of 10 m. The first and second layers were noticeably thinner than the two deeper layers shown in Figure 1B.

2.2. Method of Statistics

The statistical values of hydraulic conductivities from each layer were summarized, including the minimum (min), maximum (max), mean (mean), and standard deviation (SD). Furthermore, finding the probability distribution of hydraulic conductivities for each layer was one key objective of this study, and these were tentatively fitted by the two mostly widely used probability distributions (Gaussian and log-normal distributions), and Levy stable, gamma and Weibull distributions.

The Gaussian distribution is one of the most commonly used statistical models. A Gaussian distribution is associated with a bell shape, with very large/small values appearing in a low probability. However, strongly heterogeneous fields are not uncommon in nature, and very large/small properties may appear frequently in such geologic formations. A log-normal distribution is a continuous probability distribution of a random variable whose logarithm is normally distributed, which is widely used in geophysical variables. The logarithmic transform is a common means to seek a more appropriate distribution for a random variable [20]. μ and σ are the mean and standard deviation of the Gaussian or log-normal distributions, respectively.

To describe the statistical characteristics of heterogeneous fields, a Levy stable distribution has recently been proposed to analyze geological data [2,21,22]. Except for very few cases such as the Gaussian (normal) and Cauchy distributions, the closed-form expressions for density and distribution functions of the Levy stable distribution are not available [23].The characteristic function most often employed in numerical calculations is the following [24]:

$$\varphi_0(t) = \begin{cases} \exp\left(-\gamma^\alpha |t|^\alpha \left[1 + i\beta(\tan\frac{\pi\alpha}{2})(signt)(|\gamma t|^{1-\alpha} - 1)\right] + i\delta t\right), & \alpha \neq 1 \\ \exp\left(-\gamma|t|\left[1 + i\beta\frac{\pi}{2}(signt)(\ln|t| + \ln\gamma)\right] + i\delta t\right), & \alpha = 1 \end{cases} \tag{1}$$

where φ_0 is the characteristic function; α is a Levy index with a range of 0–2; β is a skewness parameter ranging between −1 and 1; γ is a scale parameter in the interval (0, +∞), corresponding to the standard deviation of the Gaussian distribution; δ is a shift parameter; i is the imaginary part of a complex number; and *sign(t)* is a logical function that extracts the sign of a real number t. A symmetric Levy stable distribution with a zero mean is determined when $\beta = 0$. A standard Levy stable distribution has $\gamma = 1$ and $\delta = 0$.

In addition, the gamma and Weibull distributions, which are also popular asymmetric and heavy-tailed distributions, were used to test the probability distribution of the silty clay hydraulic conductivity. The PDFs of the gamma and Weibull distributions are as follows. The PDF of the gamma distribution is

$$f(x) = \frac{1}{b^a \Gamma(a)} x^{a-1} e^{-(x/b)} \tag{2}$$

where $\Gamma(\cdot)$ is the gamma function.

The Weibull PDF is positive only for positive values of x, and is zero otherwise. For strictly positive values of the shape parameter b and scale parameter a, the density is

$$f(x) = \frac{b}{a} \left(\frac{x}{a}\right)^{b-1} e^{-(x/a)^b} \tag{3}$$

Both the Weibull distribution and the gamma distribution can give exponential distributions with particular choices of one of their two parameters. The PDF curves of the Weibull and gamma distributions are more complex compared to those of a standard exponential distribution, mainly because both of the two distributions have two independent parameters.

The primary means to test whether the data follow a specific distribution is by using nonparametric test procedures [25]. Taking the normal distribution as an example, the null hypothesis (H0) for all tests of normality is that the data are normally distributed. Rejection of H0 says that this is doubtful. Failure to reject H0, however, does not prove that the data do follow a normal distribution, especially for small sample sizes. It simply says that normality cannot be rejected with the evidence at hand. The Kolmogorov–Smirnov (K–S) test is widely used to serve as a goodness of fit test [26]. Moreover, the Anderson–Darling (A–D) test is a statistical test for whether a given sample of data is drawn from a given probability distribution [27]. Therefore, these two nonparametric tests were applied to test the similarity with the given probability distributions.

2.3. Random Modeling of Solute Transport to K Distribution

The migration of pollutants satisfies the convection–diffusion equation in a saturated medium. The 1-D steady flow dynamic dispersion model is a relatively simple and widely used method in the convection–diffusion problem. Particularly, in a 1-D infinite column filled with a porous medium, under the condition of an instantaneously injected tracer, the analytical solution of the solute concentration in the medium is described as the following [28]:

$$C(x,t) = \frac{m/w}{2n_e \sqrt{\pi D_L t}} e^{-\frac{(x-ut)^2}{4D_L t}} \tag{4}$$

where x is distance from the injection point in meters, t is time (d), $C(x,t)$ is the tracer concentration (g/L), m is the tracer mass (kg), w is the cross-sectional area (m^2), u is the flow velocity (m/d), n_e is the effective porosity, and D_L is the longitudinal dispersion coefficient (m^2/d).

The time is labeled as the peak time (t_p) when the solute reaches the maximum concentration at a distance from the injection point, and the corresponding concentration is called the peak concentration (C_p). The migration distance x is set when the first derivative equals zero ($dC/dt = 0$); the peak time (t_p) is written as the following:

$$t_p = \frac{-D_L + \sqrt{D_L^2 + x^2 u^2}}{u^2} \tag{5}$$

The peak concentration (C_p) is acquired via the peak time (t_p):

$$C_p = \frac{m/w}{2n_e\sqrt{\pi D_L t_p}} e^{-\frac{(x-ut_p)^2}{4D_L t_p}} \tag{6}$$

The peak time and the peak concentration are two important characteristics used to describe the process of solute transport in a saturated–unsaturated medium. Based on this method, the maximum negative effects and emergency time of the solute (pollutants) can be recognized, to some extent. It is noted that these simple analytic equations were concluded in the conditions of a 1-D infinite domain and instantaneously injected pollutants, which is actually a homogeneous rather than heterogeneous domain. However, in water-environment-related issues, the governor or institution needs to make decisions quickly and accurately, and most of the decisions come from technical support using these simple solutions. Due to the uncertainty of probability distributions of K, the offset of induced outcomes will be addressed herein.

While the PDF cannot always be explicitly expressed, such as for the Levy stable distribution, the Monte Carlo method [29] is an alternative method, compared with inverse function methods, to simulate the random characteristics of pollutants' transport. According to the statistical characteristics of K, combined with the data from the NCG site, the Monte Carlo technique was used to produce random hydraulic fields, to further inspect the effect of random distributions of K on t_p and C_p. In terms of unclosed forms of the Levy stable distribution, the CMS (Chambers-Mallows-Stuck) method involved in a MATLAB toolbox developed by Liang and Chen [23] was applied herein to generate the Levy random number.

This study will focus on analyzing the applicability of the Levy stable distribution for describing heterogeneity characteristics of a low K in this test site. In addition, it is common to apply the log-normal distribution for numerical simulations [30]. In order to further compare the solute transport of two distributions in saturated mediums, a large number (10,000) of the hydraulic conductivity values following the corresponding distribution were produced using the Monte Carlo method, and combined with the actual regional hydrostratigraphic situation to eliminate negative and maximum values (a value of K larger than 0.6 m/d was not considered as a low permeability medium).

3. Results and Discussion

3.1. Hydraulic Conductivity of the Silty Clay Medium

The K values were analyzed according to their identical lithologies, and the general statistics of K values from each layer are listed in Table 1. Figure 1B also illustrates that the different types of geologic sedimentary processes gave rise to different types of stratigraphy facies. The sediments of each layer, from the shallow to the deep respectively, were plain fill, continental silty clay, mud–silt clay and marine silty clay. Given the very different physical/chemical origins of different facies, it likely did not make sense to conduct statistical analyses based on combined data from multiple facies.

The sample sizes of the deepest and the shallowest layers were 112 and 18 respectively, representing the largest and smallest sizes in the four layers. The minimum K values of the four layers had insignificant differences within the same order of magnitude. However, the maximum value increased from 6.8×10^{-4} m/d for the first shallow plain fill layer, to 1.6×10^{-1} m/d for the fourth marine silty clay layer. The range of K values for each layer apparently increased from the shallow to the deep. The average hydraulic conductivity had the same pattern, with an increasing range from 6.7×10^{-5} m/d to 3.9×10^{-3} m/d. The standard deviation also increased from 1.68×10^{-4} m/d for the first layer, to 1.92×10^{-2} m/d for the fourth layer.

Table 1. Tested hydraulic conductivity and the lithology of the NCG site.

Geologic Material		First Layer	Second Layer	Third Layer	Fourth Layer
K (m/d)	Min	2×10^{-6}	4×10^{-6}	9×10^{-6}	6×10^{-6}
	Max	6.8×10^{-4}	1.6×10^{-3}	1.3×10^{-2}	1.6×10^{-1}
	Mean	6.7×10^{-5}	9.5×10^{-5}	5.6×10^{-4}	3.9×10^{-3}
	SD	1.68×10^{-4}	2.79×10^{-4}	2.12×10^{-3}	1.92×10^{-2}
Sample size		18	46	36	112
Lithology		Plain fill	Continental silty clay	Mud–silt clay	Marine silty clay

It is noted that the sample size of the first layer was only 18, and the lithology of the first layer was not silty clay, but plain fill. Therefore, the statistical analysis of the first layer was excluded in the following text. The frequency histograms for the hydraulic conductivity of each layer were plotted (Figure 2). The normal distribution was easily rejected by the low hydraulic conductivity values, such as NCG datasets. Therefore, the horizontal axes are in the natural logarithm scale. The frequency bars were nearly symmetrically distributed, as inspected from Figure 2, indicating these K datasets were graphically confirmed to be logarithmic normal distributions.

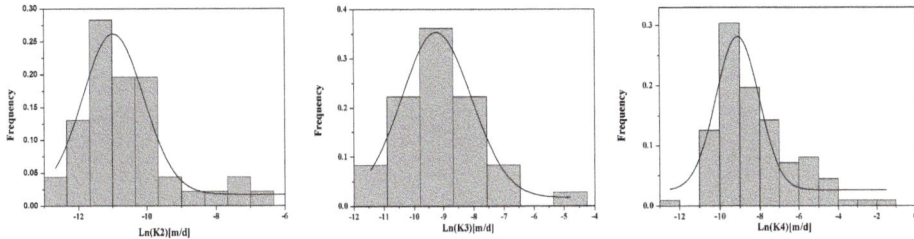

Figure 2. Frequency histograms of K for three layers, and the normal curves.

All the hydraulic conductivities obtained from the four layers' sediments versus the elevations were plotted (Figure 3). An apparently decreasing trend and a horizontal divide (elevation of 0 m a.s.l) is shown by this scatter plot. The shallow sediments for layers 1 and 2 had a relatively small range of K compared to the deep sediments for layers 3 and 4. This overall decreasing trend of K was controlled by the stratigraphic lithology. The marine sediments were general coarser than the continental sediments and plain fill, shown from the larger values and the greater range of K.

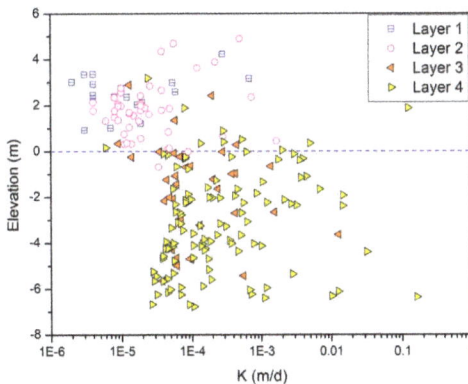

Figure 3. Hydraulic conductivities of mediums, decreasing via elevation.

3.2. Probability Density Function of the Low Hydraulic Conductivity Field

The fitted parameters for these applied PDFs and the results of probability tests are listed in Table 2. The K data from the three layers were not confirmed to follow normal nor gamma distributions. Although Figure 2 shows graphical confirmation of a logarithmic normal distribution of hydraulic conductivities from the three layers, the fourth layer did not follow the log-normal distribution via the K–S nor A–D tests. Only the third layer followed the Weibull distribution via the K–S test. Moreover, all the low hydraulic conductivity values passed the probability of the Levy stable distribution.

Only the normal distribution could be considered to be symmetrically distributed in this study. All the results of the K–S and A–D tests revealed that the K values of low-permeable mediums had asymmetrical probability distributions, and also verified that the Levy stable distribution had more generality compared to the others. This is because the Levy stable distribution can serve as a family of PDFs, and four independent parameters of the Levy model provided better adaptability.

Table 2. Fitting PDFs for the K samples from different layers.

Type of PDF		Second Layer	Third Layer	Fourth Layer
Normal	K–S	No	No	No
	A–D	No	No	No
	Fitted Parameters	$\mu = 9.55 \times 10^{-5}$ $\sigma = 2.70 \times 10^{-4c}$	$\mu = 5.63 \times 10^{-4}$ $\sigma = 2.1 \times 10^{-3}$	$\mu = 0.0039$ $\sigma = 0.0192$
Log-normal	K–S	Yes	Yes	No
	A–D	Yes	Yes	No
	Fitted Parameters	$\mu = -10.59$ $\sigma = 1.33$	$\mu = -9.08$ $\sigma = 1.44$	$\mu = -8.25$ $\sigma = 1.91$
Levy	K–S	Yes	Yes	Yes
	A–D	Yes	Yes	Yes
	Fitted Parameters	$\alpha = 0.46$ $\beta = 1$ $\gamma = 4.48 \times 10^{-6}$ $\delta = 7.47 \times 10^{-6}$	$\alpha = 0.52$ $\beta = 0.39$ $\gamma = 2.10 \times 10^{-5}$ $\delta = 5.30 \times 10^{-5}$	$\alpha = 0.52$ $\beta = 1$ $\gamma = 5.93 \times 10^{-5}$ $\delta = 2.10 \times 10^{-6}$
Gamma	K–S	No	No	No
	A–D	No	No	No
	Fitted Parameters	$a = 0.48$ $b = 0.0002$	$a = 0.41$ $b = 0.0014$	$a = 0.26$ $b = 0.015$
Weibull	K–S	No	Yes	No
	A–D	No	No	No
	Fitted Parameters	$a = 0.0001$ $b = 0.6$	$a = 0.0002$ $b = 0.56$	$a = 0.0007$ $b = 0.43$

The Levy stable distribution is more flexible than the Gaussian distribution for fitting geophysical data obtained from strongly heterogeneous fields [22,31]. Herein, the K values from four layers were successfully modeled by the Levy stable distribution, and the four fitted Levy parameters are compared in Table 2. From layer 2 to layer 4, the Levy index α and scale parameter γ increased from 0.46 to 0.52, and 4.48×10^{-6} to 5.93×10^{-5}, respectively. As α decreases, the frequency of sudden large jumps in the random field increases [32]. The parameter γ is known as the scale parameter. It is equal to half the variance when $\alpha = 2$, and plays a similar role for $\alpha < 2$ (i.e., it is a measure of the width of the distribution). As γ increases, the magnitudes of the sudden large jumps increase [33]. The skewness parameter β values for layers 2 and 4 were the same, equal to 1 (the maximum for this parameter), indicating extremely right-skewed distributions for these datasets. β for layer 3 equaled 0.39, and the degree of skewness was weaker than for the other two layers. δ, the shift parameter, represents the centering of the distribution, and it is equal to the mean of the distribution only when $\alpha > 1$ and $\beta = 0$. The differences of δ in Table 2 show the different centering of the four layers.

Based again on previous research [31,34], larger-scale property variations appear to be distributed normally, while smaller-scale variations follow the Levy stable distribution, and display an increased probability for sudden extreme events such as high or low K values. The measurement scale of NCG was controlled by the permeameter used (Φ: 61.8 mm × 40 mm). Hence, these K datasets could be categorized as small-scale. More extremes for K could be obtained through this permeameter. Another direct reason is that the K field of NCG was highly heterogeneous, as shown in Table 1 and Figure 3.

3.3. The Effects of K PDFs on Solute Transport

The studied case was an assumed 1-D cylinder filled with a homogenous isotropic porous medium, of infinite length. The conservative solute weighing 10 kg was injected to the cylinder. The cylinder cross-section was square at 10 cm × 10 cm. The effective porosity (n_e) was 0.3, and the longitudinal dispersion coefficient (D_L) was 0.05 m^2/d. Although the porosity and dispersion coefficients were indeed of high variations for subsurface mediums, even for the silt and clay materials, we only focused the influence on the solute transport from the variation of K rather than the porosity and dispersion coefficients. The hydraulic gradient was set to a constant of 1.0, and the velocity (u) become a single-valued function of K; then the explicit expressions of t_p-K and C_p-K could be easily obtained from Equations (5) and (6). The monotonic variations, both of increasing C_p and decreasing t_p, with the increasing K value were observed from Figure 4. High t_p and low C_p values were consistent with the small K, and low t_p and high C_p values corresponded to a strong permeability of high K.

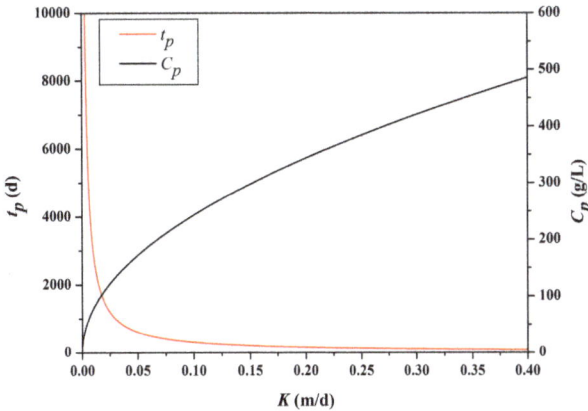

Figure 4. Variation curves of t_p and C_p versus the corresponding K values.

The statistical values of K for the silty clay medium from the third layer at the NCG site were applied to simulate the stochastic K datasets. As mentioned above, it can be concluded that the third layer of the medium followed log-normal and Levy stable distributions. The Monte Carlo method adopting the parameters from Table 2 was performed for the simulation of K. Afterwards, the peak time (t_p) and peak concentration (C_p) could be simulated by Equations (5) and (6).

3.3.1. Effects of K PDFs on Peak Time

As shown in Figure 5, the range of peak times simulated by the Levy stable distribution (50.0 d, 1.0×10^5 d) was a little greater than that simulated by the log-normal distribution (788.3 d, 9.9×10^4 d). The minimum values especially between these two distributions were significantly different. This indicated that the Levy stable distribution could generate greater random values of K and then result in a smaller peak time. Additionally, the mean and median values of simulated peak time values from the Levy stable distribution (8.4×10^4 d and 9.6×10^4 d) were apparently greater than those of the log-normal distribution (7.7×10^4 d and 8.8×10^4 d). The interquartile ranges (IQRs) for the Levy-stable and log-normal distributions were 1.1×10^4 d and 3.7×10^4 d, respectively. From Figure 5, the simulation results of peak times from the Levy stable distribution are gathered in the area of greater value, and random numbers of K generated by the Levy stable distribution concentrate in the area of lower value. These results agreed with the high-peak and heavy-tail characteristics of the Levy stable distribution.

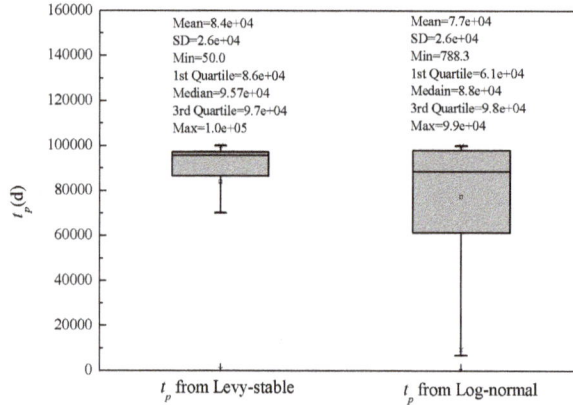

Figure 5. Boxplot of t_p using K values followed Levy stable and log-normal distributions.

According to the differences of peak times and cumulative probabilities from the two distributions, the results shown in Figure 6 can be divided into three different sections. Section I includes the area of peak times less than 3.3×10^4 d. Equation (4) explicitly indicates the negative correlation between peak time and flow velocity. In this study, velocity (u) was proportional to K through the assumed 1-D model, and the peak time (t_p) and K had a negative correlation. Section I reflects the simulation result when K was relatively large. As shown in Figure 5, the results calculated by the Levy stable distribution indicated that the probability of small peak time events (red line) was clearly higher than that calculated by the log-normal distribution (black line).

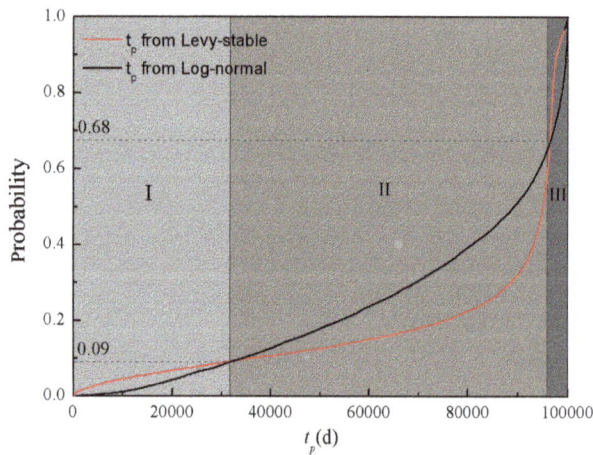

Figure 6. Cumulative frequency distributions of t_p from Levy stable and log-normal distribution K values.

The peak time of Section II was between 3.3×10^4 d and 9.6×10^4 d, and Section II is shown as the middle part of Figure 5. According to the correlation between the peak time and K mentioned above, this section corresponds to the range for which K was moderate (greater than that in Section I and less than that in Section III). The simulation results indicated that the log-normal distribution achieved a higher probability than the Levy stable distribution. The probability of random events with moderate peak times simulated by the Levy stable distribution was less than that for the log-normal distribution. The peak time of Section III was the highest, representing the smallest K. The simulation results indicated that the probability of high peak time events simulated by the Levy stable distribution was larger than for the log-normal distribution.

In conclusion, under the Levy stable distribution, the distribution of t_p calculated by the simple analytical model had more non-uniform characteristics. The probabilities of a greater peak time occurring (Section I) and smaller values occurring (Section III) were both larger than for under the widely used log-normal distribution. However, the comparison of moderate peak time events (Section II) revealed that the probability calculated by the log-normal distribution was clearly greater than that by the Levy stable distribution. Our results indicate that simulated t_p values from log-normally-distributed K are likely ranged within the medium level, and the Levy stable distribution can produce higher or lower extremes much more easily.

3.3.2. Effects of K PDFs on Peak Concentration

As shown in Figure 7, the range of peak concentrations (C_p) simulated by the Levy stable distribution (8.1 g/L, 594.7 g/L) was significantly greater than that simulated by the log-normal distribution (8.1 g/L, 155.9 g/L). The maximum values between these two distributions were significantly different. The average value of the simulated peak concentration from the Levy stable distribution (17.5 g/L) was slightly greater than that from the log-normal distribution (13.5 g/L). However, the median of the Levy stable distribution (9.8 g/L) was slightly less than the counterpart of the log-normal distribution (11.0 g/L). Although the standard deviation of C_p from the Levy stable distribution (43.9 g/L) was much greater than that from the log-normal distribution (7.4 g/L), the Levy stable distribution could generate a concentrated C_p (IQR = 1.9 g/L) compared with the log-normal distribution (IQR = 5.9 g/L), as seen from the shorter box in Figure 7.

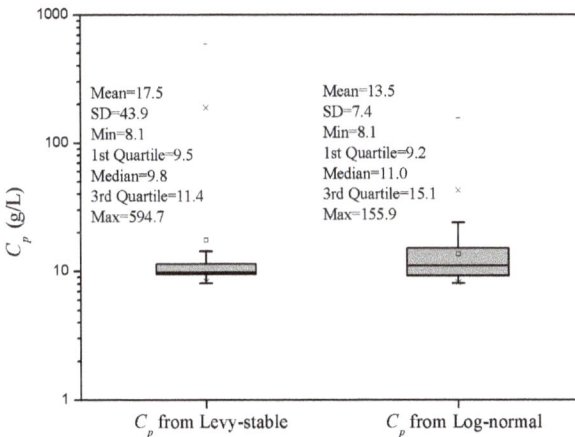

Figure 7. Boxplot of C_p using K values following Levy stable and log-normal distributions.

Similarly to the cumulative frequency of t_p shown in Figure 6, three different sections of C_p are discriminated in Figure 8 (Sections I and II are zoomed in Figure 8B). According to Equation (3), the peak concentration (C_p) had a positive correlation with the flow velocity (u). Since the flow velocity (u) had a positive correlation with the hydraulic conductivity (K), a positive correlation between C_p and K could be inferred. The cumulative probability of random events in Sections I and II contributed approximately 90% of all random samples. In Sections I and II, the difference of cumulative frequency distributions of C_p under two different distributions was small, as shown by Figure 8A. Approximately 90% of simulated peak concentrations were located in the range of 8–25 g/L for both of the two distributions. In Section III, the peak concentration from the log-normal distribution achieved the highest value of 155.9 g/L, and the greater C_p (>155.9 g/L) only appeared under the assumption of Levy-stable-distributed K. The distribution of C_p under the Levy stable distribution was much flatter than that under the log-normal distribution. The Levy stable distribution had better performance to simulate extreme conditions and to reveal the heavy-tailed characteristic of hydraulic conductivity, especially.

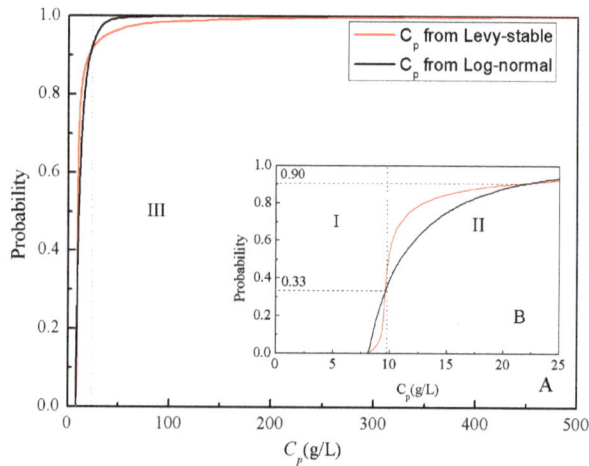

Figure 8. Cumulative frequency distribution of C_p from Levy stable and log-normal distribution K values (8A shows the full extent and 8B shows the part of C_p less than 25 g/L).

4. Conclusions

The hydraulic properties of an aquifer vary irregularly in 3-D space, as a general rule. In this study, a comprehensive field investigation of the NCG site was performed. The lithology of NCG is dominated by silty clay and clay. In total, 212 falling head permeameter tests were performed to investigate the characteristics of hydraulic conductivity for the silty clay site. The statistical patterns of hydraulic conductivity were analyzed, and the probability distributions of K were tested by five models.

One of our main findings was that the hydraulic conductivity of a low hydraulic conductivity medium, such as silty clay, likely comes from the Levy stable distribution. The log-normal distribution could also cover most of the low hydraulic conductivity, and the Weibull distribution could describe part of the samples of clay, but neither the normal distribution nor the gamma distribution could fit any of the experimental probability curves of the NCG site. Therefore, the Levy stable distribution is recommended for depicting the statistics of a low-K field.

The effects of the probability distribution of the hydraulic conductivity on the solute transport were analyzed using a simple analytical model. The peak concentration and its corresponding time were selected to represent the transport process. The Levy stable distribution was apt to generate the extremes of the solute transport (higher peak concentration and lower peak time), compared with the widely used log-normal distribution. These results have a great guiding significance for describing the migration characteristics of contaminations in underground mediums.

Acknowledgments: The researchers would like to extend their thanks to the National Natural Science Foundation of China grants (41201029 and 51279208). This study was also supported by the Open Research Fund of the State Key Laboratory of Simulation and Regulation of Water Cycle in River Basin (China Institute of Water Resources and Hydropower Research), Grant NO. IWHR-SKL-201502, the Fundamental Research Funds for the Central Universities (2015B14414), and the Innovation and Entrepreneurship Training Program for Hohai University (2017102941013).

Author Contributions: Chengpeng Lu and Gang Zhao conceived and designed the study; Ying Zhang performed the field experiments; Wei Qin and Wenpeng Wang performed the stochastic simulation and analyzed the data; Chengpeng Lu and Wei Qin wrote the paper.

Conflicts of Interest: The authors declare no conflict of interest. The founding sponsors had no role in the design of the study; in the collection, analyses, or interpretation of data; in the writing of the manuscript, and in the decision to publish the results.

References

1. Freeze, R.A.; Cherry, J.A. *Groundwater*; Prentice Hall: Englewood Cliffs, NJ, USA, 1979.
2. Kohlbecker, M.V.; Wheatcraft, S.W.; Meerschaert, M.M. Heavy-tailed log hydraulic conductivity distributions imply heavy-tailed log velocity distributions. *Water Resour. Res.* **2006**, *42*, W04411. [CrossRef]
3. Wang, K.; Huang, G. Effect of permeability variations on solute transport in highly heterogeneous porous media. *Adv. Water Resour.* **2011**, *34*, 671–683. [CrossRef]
4. Schertzer, D.; Lovejoy, S. Physical modeling and analysis of rain and clouds by anisotropic scaling multiplicative processes. *J. Geophys. Res.* **1987**, *92*, 9693–9714. [CrossRef]
5. Freeze, R.A. A stochastic-conceptual analysis of one-dimensional groundwater flow in nonuniform homogeneous media. *Water Resour. Res.* **1975**, *11*, 725–741. [CrossRef]
6. Bjerg, P.L.; Hinsby, K.; Christensen, T.H.; Gravesen, P. Spatial variability of hydraulic conductivity of an unconfined sandy aquifer determined by a mini slug test. *J. Hydrol.* **1992**, *136*, 107–122. [CrossRef]
7. Hess, K.M.; Wolf, S.H.; Celia, M.A. Large-scale natural gradient tracer test in sand and gravel, cape cod, massachusetts: 3. Hydraulic conductivity variability and calculated macrodispersivities. *Water Resour. Res.* **1992**, *28*, 2011–2027. [CrossRef]
8. Turcke, M.A.; Kueper, B.H. Geostatistical analysis of the borden aquifer hydraulic conductivity field. *J. Hydrol.* **1996**, *178*, 223–240. [CrossRef]
9. Sudicky, E.A. A natural gradient experiment on solute transport in a sand aquifer: Spatial variability of hydraulic conductivity and its role in the dispersion process. *Water Resour. Res.* **1986**, *22*, 2069–2082. [CrossRef]
10. Woodbury, A.D.; Sudicky, E.A. The geostatistical characteristics of the borden aquifer. *Water Resour. Res.* **1991**, *27*, 533–546. [CrossRef]
11. Cheng, C.; Song, J.; Chen, X.; Wang, D. Statistical distribution of streambed vertical hydraulic conductivity along the platte river, nebraska. *Water Resour. Manag.* **2011**, *25*, 265–285. [CrossRef]
12. Sánchez-Vila, X.; Carrera, J.; Girardi, J.P. Scale effects in transmissivity. *J. Hydrol.* **1996**, *183*, 1–22. [CrossRef]
13. Hyun, Y. Multiscale Anaylses of Permeability in Porous and Fractured Media. Ph.D. Thesis, The University of Arizona, Tucson, AZ, USA, 2002.
14. Rehfeldt, K.R.; Boggs, J.M.; Gelhar, L.W. Field study of dispersion in a heterogeneous aquifer: 3. Geostatistical analysis of hydraulic conductivity. *Water Resour. Res.* **1992**, *28*, 3309–3324. [CrossRef]
15. Vereecken, H.; Döring, U.; Hardelauf, H.; Jaekel, U.; Hashagen, U.; Neuendorf, O.; Schwarze, H.; Seidemann, R. Analysis of solute transport in a heterogeneous aquifer: The krauthausen field experiment. *J. Contam. Hydrol.* **2000**, *45*, 329–358. [CrossRef]
16. Bagarello, V.; Iovino, M.; Elrick, D. A simplified falling-head technique for rapid determination of field-saturated hydraulic conductivity. *Soil Sci. Soc. Am. J.* **2004**, *68*, 66–73. [CrossRef]
17. Dagan, G. *Flow and Transport in Porous Formations*; Springer: Berlin/Heidelberg, Germany, 1989.
18. Liang, J.; Zeng, G.; Guo, S.; Li, J.; Wei, A.; Shi, L.; Li, X. Uncertainty analysis of stochastic solute transport in a heterogeneous aquifer. *Environ. Eng. Sci.* **2009**, *26*, 359–368. [CrossRef]

19. Ye, B.; Zhang, Z.; Mao, T. Petroleum hydrocarbon in surficial sediment from rivers and canals in Tianjin, China. *Chemosphere* **2007**, *68*, 140–149. [CrossRef] [PubMed]
20. Cong, S. *Probability and Statistical Methods in Water Science and Technology*; Science Press: Beijing, China, 2010.
21. Painter, S. Flexible scaling model for use in random field simulation of hydraulic conductivity. *Water Resour. Res.* **2001**, *37*, 1155–1163. [CrossRef]
22. Yang, C.-Y.; Hsu, K.-C.; Chen, K.-C. The use of the levy-stable distribution for geophysical data analysis. *Hydrogeol. J.* **2009**, *17*, 1265–1273. [CrossRef]
23. Liang, Y.; Chen, W. A survey on computing lévy stable distributions and a new matlab toolbox. *Signal Process.* **2013**, *93*, 242–251. [CrossRef]
24. Nolan, J.P. Parameterizations and modes of stable distributions. *Stat. Probab. Lett.* **1998**, *38*, 187–195. [CrossRef]
25. Helsel, D.R.; Hirsch, R.M. *Statistical Methods in Water Resources Techniques of Water Resources Investigations*; U.S. Geological Survey: Reston, VA, USA, 2002; Volume 4, p. 522.
26. Arnold, T.B.; Emerson, J.W. Nonparametric goodness-of-fit tests for discrete null distributions. *R J.* **2011**, *3*, 34–39.
27. Stephens, M.A. Edf statistics for goodness of fit and some comparisons. *J. Am. Stat. Assoc.* **1974**, *69*, 730–737. [CrossRef]
28. De Jong, G.D.J. Longitudinal and transverse diffusion in granular deposits. In *Soil Mechanics and Transport in Porous Media: Selected works of g. De Josselin de Jong*; Schotting, R.J., van Duijn, H.C.J., Verruijt, A., Eds.; Springer: Dordrecht, The Netherlands, 2006; pp. 261–268.
29. Gégo, E.L.; Johnson, G.S.; Hankins, M. An evaluation of methodologies for the generation of stochastic hydraulic conductivity fields in highly heterogeneous aquifers. *Stoch. Environ. Res. Risk Assess.* **2001**, *15*, 47–64. [CrossRef]
30. Teatini, P.; Ferronato, M.; Gambolati, G.; Baù, D.; Putti, M. Anthropogenic venice uplift by seawater pumping into a heterogeneous aquifer system. *Water Resour. Res.* **2010**, *46*, W11547. [CrossRef]
31. Painter, S. Stochastic interpolation of aquifer properties using fractional lévy motion. *Water Resour. Res.* **1996**, *32*, 1323–1332. [CrossRef]
32. Tennekoon, L.; Boufadel, M.C.; Lavallee, D.; Weaver, J. Multifractal anisotropic scaling of the hydraulic conductivity. *Water Resour. Res.* **2003**, *39*, SBH 8-1. [CrossRef]
33. Boufadel, M.C.; Lu, S.L.; Molz, F.J.; Lavallee, D. Multifractal scaling of the intrinsic permeability. *Water Resour. Res.* **2000**, *36*, 3211–3222. [CrossRef]
34. Liu, H.H.; Molz, F.J. Multifractal analyses of hydraulic conductivity distributions. *Water Resour. Res.* **1997**, *33*, 2483–2488. [CrossRef]

water MDPI

Article

Regional Groundwater Flow Assessment in a Prospective High-Level Radioactive Waste Repository of China

Xiaoyuan Cao [1,2], Litang Hu [1,2,*], Jinsheng Wang [1,2] and Jingrui Wang [1,2]

[1]   College of Water Sciences, Beijing Normal University, Beijing 100875, China;
      caoxiaoyuan1988@126.com (X.C.); wangjs@bnu.edu.cn (J.W.); 201221470032@mail.bnu.edu.cn (J.W.)
[2]   Engineering Research Center of Groundwater Pollution Control and Remediation of Ministry of Education,
      Beijing Normal University, Beijing 100875, China
*   Correspondence: litanghu@bnu.edu.cn; Tel.: +86-138-111-63348

Received: 13 June 2017; Accepted: 17 July 2017; Published: 23 July 2017

Abstract: The production of nuclear energy will result in high-level radioactive waste (HLRW), which brings potential environmental dangers. Selecting a proper disposal repository is a crucial step in the development of nuclear energy. This paper introduces firstly the hydrogeological conditions of the Beishan area in China. Next, a regional groundwater model is constructed using a multiphase flow simulator to analyze the groundwater flow pattern in the Beishan area. Model calibration shows that the simulated and observed hydraulic heads match well, and the simulated regional groundwater flow pattern is similar to the surface flow pattern from the channel network, indicating that the groundwater flow is mainly dependent on the topography. In addition, the simulated groundwater storage over the period from 2003 to 2014 is similar to the trend derived from the Gravity Recovery and Climate Experiment satellite-derived results. Last, the established model is used to evaluate the influences of the extreme climate and regional faults on the groundwater flow pattern. It shows that they do not have a significant influence on the regional groundwater flow patterns. This study will provide a preliminary reference for the regional groundwater flow assessment in the site of the HLRW in China.

Keywords: nuclear waste disposal; the Beishan area; TOUGH2; groundwater flow; assessment

1. Introduction

Nuclear energy is considered as an additional dependable and clean energy source all over the world [1]. However, high-level radioactive waste (HLRW), which are the highly radioactive materials produced as a byproduct of the reactions that occur inside nuclear reactors, will probably bring potential dangers. Therefore, the proper disposal of nuclear waste is highly challenging.

The safety of high-level radioactive waste disposal has gained much attention in the world. In America, Yucca Mountain was regarded as a potential repository site of high-level nuclear waste in 2002, with many experimental and numerical studies performed on groundwater flow. The initial assessment of the regional groundwater system at Yucca Mountains was carried out in 1997 [2], and then the comprehensive evaluation and simulation of the system was performed in 2010 by using MODFLOW-2000 [3]. In addition, a site-scale model for fluid flow in the unsaturated zone of Yucca Mountain, including fracture flow [4–7] and parallel computing [8], was developed by Transport Of Unsaturated Groundwater and Heat 2 (TOUGH2) to reproduce the overall system behavior. Finland was the first country to propose an ultimate disposal way for HLRW. In 2001, Finland selected a high-level radioactive waste disposal site based on field investigation and numerical simulation [9]. A kind of crystalline rock, which is tectonically and geochemically stable, was selected as the ultimate rock of high-level radioactive waste disposal site in Finland. According to the regional groundwater assessment, the groundwater flow in fractures plays a significant role in the regional groundwater system. In Sweden, because one of the candidate sites is located on the coast of the Baltic Sea and composed of 26 catchment areas and 96 subcatchments in total, the integrated hydrogeological-hydrological analysis are required to assess groundwater flow paths in the bedrock for a reasonable repository site selection [10]. After some detailed tests and numerical studies, Forsmark in Sweden was considered as the final repository for short-lived radioactive waste in 2009. This method provides an optional method for safety assessment of regional groundwater flow in the regions with special topographic or climate conditions. In China, research on the disposal of HLRW has only been performed since 1985. The Beishan area was selected as the most prospective site for a disposal repository of HLRW in 1993 [11], with the tectonically stable rock and arid climate. A simplified numerical model of the regional groundwater flow in the Beishan area was developed [12]. In recent years, to trace the recharge source of groundwater and explore groundwater age in the Beishan area, hydrogeological and isotopic studies have been performed [13,14], which suggest that the main recharge source of groundwater is rainfall. However, the regional groundwater flow pattern in the Beishan area has not been fully discussed to date. Therefore, it is necessary to develop an understanding of a saturated-unsaturated groundwater model to assist site selection for the disposal of HLRW. In addition, the influences of climate change or faults on regional groundwater flow patterns needed to be further addressed.

The objective of this study includes two aspects. One is to analyze and evaluate the regional groundwater flow patterns by using an established saturated-unsaturated groundwater model. The model results then will be further used to subdivide groundwater flow systems, which is helpful to refine the radioactive waste disposal site selection in China. The other objective is to test how the flow in the proposed disposal site is influenced by the hydraulic properties of regional fault zones and by climate change. After the model construction, the changes of the groundwater level and the soil moisture will be discussed in detail. Next, the simulated variations of groundwater storage and soil water are compared with the Gravity Recovery and Climate Experiment (GRACE) satellite-derived results. Last, the groundwater budgets and the influences of extreme climate and regional faults on the groundwater flow system will be evaluated. This study will provide a preliminary reference for the regional groundwater flow assessment in a potential high-level radioactive waste repository of China.

2. Method

2.1. Study Area

The Beishan area is a typical inland arid region with an area of approximately 68,900 km^2. The area is located in Gansu Province, northwestern China (Figure 1) and is surrounded by Mongolia to the north, and Yumen City in the south. The area includes the Shu-le river basin in the west and the Heihe river basin in the east. The yearly average rainfall is 60–100 mm, and 60% of the total rainfall is from June to August. The mean annual potential evaporation is 2900–3200 mm, and the mean annual surface temperature is 4–5 °C.

The study area is mainly composed of flatter Gobi and small hills, and depression and valleys are distributed among them, forming an undulating topography. The elevation varies significantly from 873 m to 2556 m over the study area. And the higher mountains, with the elevation from 1700 m to 2500 m, are located in the middle east of the study area. The geological materials in the study area can be classified into Proterozoic metamorphic (schist), intrusive (granite) rocks, Quaternary clay and sandy loam weathered from schist and granite, with occasional alluvium deposits. Part of locations with rock represent locations without clay or sandy loam (scale is 1:500,000, Figure 1). The thicknesses of the clay and sandy loam and alluvium deposits are less than 50 m. Faults and fractures are distributed from Quaternary deposits to the intensely weathered granite in the study area, but most faults are tension faults and filled with quartz according to the investigation [15].

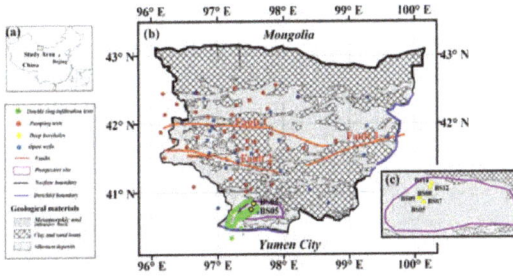

Figure 1. Sketch map of the study area: (**a**) map of China; (**b**) map of the study area; and (**c**) location of boreholes in the prospective site of the disposal repository.

2.1. Model Boundaries and Their Representations

To obtain the drainage divide, the USGS/NASA Shuttle Radar Topography Mission (SRTM) data in 3 arc seconds is used to extract channel network and watershed information. The main procedures in Arc Hydro Tools in ArcGIS software include Digital Elevation Model (DEM) pretreatment, determination of the direction of flow, and extraction of flow accumulation, channel network, watersheds and watershed division. The colormap of the DEM and generated channel network are shown in Figure 2. The study area is surrounded by a natural surface divide in the north, the west and the partial east boundaries (Figure 2). Considering human activities, such as groundwater pumping in Yumen City, the south boundary is set between Yumen City and Gobi. There are nearly no rivers in the study area. The top layer of this model is set as the atmosphere layer. The bottom layer is assumed to be a no-flow boundary. No long-term observation well data are available, and hydraulic heads at 33 open wells are collected from 2008 to 2009. The depth to groundwater in open wells almost varies from 5 m to 50 m, occasionally over hundreds of meters. Field investigation from 2012 to 2015 found that the inter-annual variation of water table in open wells is very small, usually less than 0.1 m, and yearly average of water table remain almost the same. According to the measured results [16], the depth to the groundwater level in BS09, BS12, BS14, BS05 are 59.5 m, 5.2 m, 36.6 m, 65 m, respectively. Vertical changes of soil moisture are not measured. Vertically, soil water usually changes from unsaturated state to fully-saturated state.

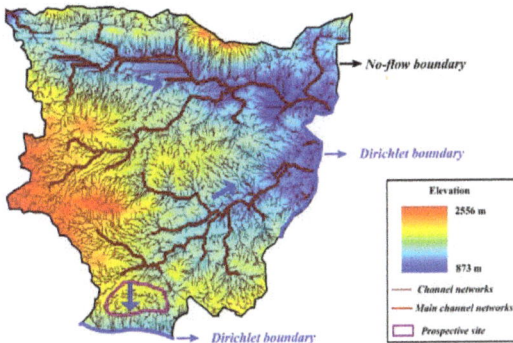

Figure 2. Schematic map of the channel network and model boundary.

As investigated by the Beijing Research Institute of Uranium Geology (2010) [16], the metamorphic rock and intrusion rock extend from the surface to the depth of over 600 m. The permeability of original metamorphic and intrusion rock is low. However, because faults and fractures are occasionally distributed in the study area, hydraulic conductivity at the intersections of faults and fractures is relatively high. The metamorphic rock and intrusion rock can be simplified into three types according to logs of deep drillings [16], i.e., intensely weathered, middle weathered, and weakly weathered or even unweathered layers, where the thicknesses of the three layers are 50, 100, and 500 m, respectively. To better present the change of groundwater level and soil moisture in the unsaturated zone, the intensely weathered layer is refined to 15 model layers, where the thickness of each layer is 0.1, 1.4, 0.5, 1.0, 1.0, 2.0, 2.0, 2.0, 2.5, 2.5, 3.0, 3.0, 9.0, 10.0 and 10.0 m. The middle weathered layer is refined to five layers with the same thickness (20 m). The weakly weathered or unweathered layer is refined to three layers, and the thicknesses of the three layers are 100, 200 and 200 m (Figure 3). In total, the model is divided into 23 model layers, so that the processes of the unsaturated-saturated interface can be identified by the transition from two-phase flow to single-phase flow in the numerical simulation.

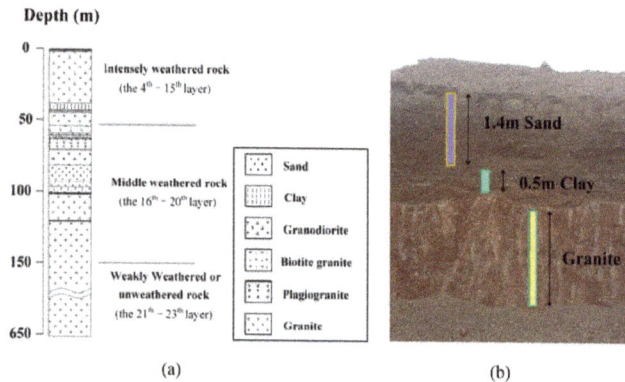

(a) (b)

Figure 3. Lithologic description of the study area: (**a**) lithological column; and (**b**) the lithology profile in the areas with of the Quaternary deposits.

2.2. Numerical Model

The parallel version of multiphase flow simulator is adopted in this study. TOUGH2 is a model based on the integral finite difference method, which could offer the advantage of being applicable to regular or irregular discretization in one, two and three dimensions [17]. The governing equations of TOUGH2 are established from mass and energy balance. Fluid flow is described with a multiphase extension of Darcy's law; in addition there is diffusive mass transport in all phases. Heat flow is governed by conduction and convection, also including sensible as well as latent heat effects. Time will be discretized fully implicitly as a first-order backward finite difference. The nonlinear equations in the residual form are solved by Newton/Raphson iteration [18]. The EOS3 module in TOUGH2 can simulate water and air movement in isothermal and nonisothermal conditions. The single and two-phase conditions can be easily known from the range of a primary variable [17]. This module was employed to simulate the regional groundwater flow in the saturated-unsaturated zone of the study area, and it could describe fluid and heat flow in multiphase, multi-component systems [18,19]. The mass and energy equations for the two-phase flow of two components (water, air) in the TOUGH2/EOS3 module are summarized in Table 1 [20]. The unsaturated-saturated interface can be judged by the transition from the state of two-phase flow to that of single-phase flow.

Table 1. Governing equations for TOUGH2/EOS3.

Description	Equation
Conservation of mass and energy	$\frac{d}{dt}\int_{v_n} M^k dv_n = \int_{\Gamma_n} F^k \times n d\Gamma_n + \int_{v_n} q^k dv_n, k = 1,2,3$
Mass accumulation	$M^K = \Phi\sum_{\beta} S_{\beta}\rho_{\beta}X_{\beta}^k, k = 1,2$
Mass flux	$F^K = \sum_{\beta} X_{\beta}^k \rho_{\beta} u_{\beta}, k = 1,2$
Energy flux	$F^3 = -\lambda\nabla T + \sum_{\beta} h_{\beta}\rho_{\beta}u_{\beta}$
Energy accumulation	$M^3 = (1-\phi)\rho_R C_R T + \phi\sum_{\beta}\rho_{\beta}S_{\beta}U_{\beta}$
	$u_{\beta} = -k\frac{k_{\gamma\beta}\rho_{\beta}}{\mu_{\beta}}(\nabla P_{\beta} - \rho_{\beta}g)$

where M^K are the accumulation terms of the components and energy, kg m^{-3} for components and J m^{-3} for energy; k is the index for the components, =1 (water), 2 (air), and 3 (temperature); v_n is the volume of the nth grid cell, m^3; F is the flux, g m^{-2} s^{-1}; n is the normal vector on the surface element, dimensionless; Γ_n is the area of closed surface, m^2; q^k is the source/sink terms for mass or energy components, kg m^{-3}s^{-1} or J m^{-3}s^{-1}; ϕ is the porosity, dimensionless; S_{β} is the saturation of phase β, dimensionless; ρ_{β} is the density of phase β, kg m^{-3}; $\beta = G$ for gas and = L for liquid; X_{β}^k is the mass fraction of component k in fluid phase β, dimensionless; u_{β} is the Darcy velocity in phase β, m s^{-1}; λ is the thermal conductivity, W K^{-1} m^{-1}; u_{β} is the viscosity, Pa s; h_{β} is the specific enthalpy in phase β, J kg^{-1}; C_R is the heat conductivity, W K^{-1} m^{-1}; T is temperature, °C or k; $k_{\gamma\beta}$ is the relative permeability to phase β; g is the gravitational acceleration constant, m s^{-2}; and k is the absolute permeability, m^2.

3. Data Preparation

3.1. Elevation and Model Discretization

The topography is important for both regional and local groundwater flow patterns [21]. The elevation is derived from the SRTM data in 3 arc seconds, where Reuter recreated products by adopting void-filling interpolation methods to obtain higher-quality SRTM data [22]. The elevation changes from over 2300 m in the west to 900 m in the eastern part. By using IGMESH [23], which is a convenient irregular-grid-based pre- and post-processing tool for the TOUGH2 simulator, the study area is discretized into 12,390 gridblocks in the plane and 23 model layers in the vertical direction. There are 284,970 gridblocks and 1,120,107 connections in total (Figure 4). The prospective site of the disposal repository is refined for better characterization of the groundwater flow.

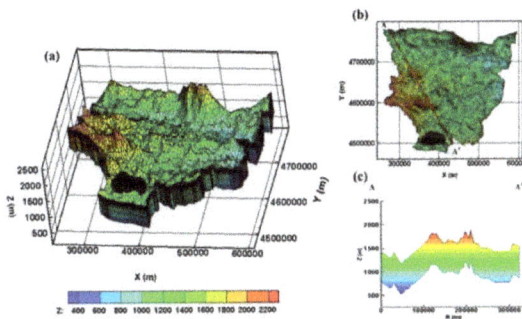

Figure 4. Mesh of the study area (scale ratio of x:y:z = 1:1:65): (**a**) three-dimensional mesh; (**b**) mesh of Horizontal plan; and (**c**) elevation of profile A-A′.

3.2. Hydraulic Properties

Since 2005, the Beijing Research Institute of Uranium Geology has performed many field investigations via borehole drilling as well as hydrogeochemical and isotope sampling in an effort to identify a feasible site for high level radioactive waste disposal. The institute summarized hydraulic conductivity data from many pumping tests [16]. Hydraulic tests can be divided into three types: pumping tests, double ring infiltration tests and double packer hydraulic tests. There are 37 shallow wells with traditional pumping tests, and the locations of these tests are shown in Figure 1, where the range of the estimated hydraulic conductivity is from 0.0015 to 3.82 m/day. Double ring infiltration tests were conducted to estimate the hydraulic conductivity of the shallow soil in 2013 and 2014, and there are 113 test data points in total. The range of hydraulic conductivity estimated by the double ring infiltration tests is 0.11–18.8 m/day, and the arithmetic mean value of hydraulic conductivity is approximately 4.27 m/day. Double packer hydraulic tests were performed at deep wells BS05 and BS06, where the depth of the drilling borehole is approximately 600 m below the surface. According to a large number of double packer hydraulic tests, the mean values of the hydraulic conductivities at BS05 and BS06 are 0.0088 and 0.00018 m/day, respectively. Considering all the test results, the change of permeability with the depth to the surface obtained from all test data [16] is summarized in Figure 5. The permeability is found to have an almost exponentially deceasing trend with the depth. The permeability obtained from these wells are in the range of 1×10^{-11} to 1×10^{-18} m^2.

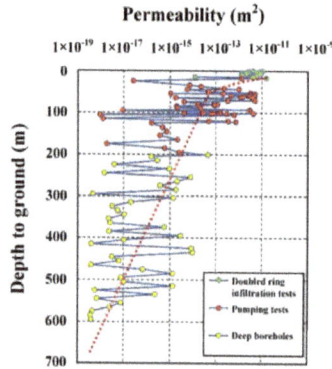

Figure 5. Change of permeability with depth to ground obtained from field tests.

Vertically, there are 23 model layers in total. Table 2 lists the vertical model layers and the zonation of hydraulic properties for each layer. The lithology zoning map at the 2nd layer and the 3rd layer (Figure 6) are made from the geological zones in Figure 1. The rock grain density, formation heat conductivity and pore compressibility are assumed to be uniform, at 2650 $kg \cdot m^{-3}$, 1 $W \cdot m^{-1} \cdot {}^{\circ}C^{-1}$ and 3.7×10^{-10} Pa^{-1}, respectively. The heat capacity of the top layer is set as 8.0×10^{99} $J \cdot kg^{-1} \cdot {}^{\circ}C^{-1}$, and the values of the other layers are set as 1000 $J \cdot kg^{-1} \cdot {}^{\circ}C^{-1}$. The relative permeability and capillary pressure were calculated by the van Genuchten–Mualem model [24]. In addition, related parameters and the pore compression coefficient of the model were obtained according to the empirical value of soil. Because the study area is almost composed of low-permeability media, such as metamorphic and intrusive rocks, parameters for estimating relative permeability in the van Genuchten-Mualem model remain the same for each type of rock. Among them, λ (a parameter in Van Genuchten–Mualem function) is equal to 0.492. Residual liquid saturation S_{lr} and residual gas saturation S_{gr} are both set as 0.01, and saturated liquid saturation S_{ls} in the Van Genuchten–Mualem Model is 1.0. Initial capillary pressure P_0 and maximum capillary pressure P_{max} are set as 1.0×10^3 Pa and 1.0×10^{12} Pa also based on the empirical value, respectively.

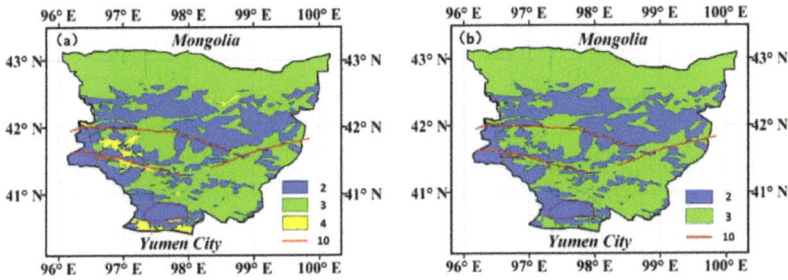

Figure 6. Lithology zoning map of the shallow layer: (**a**) zoning map of Layer 2; and (**b**) zoning map of Layer 3. The numbers are consistent with the Table 2.

Table 2. List of the hydrogeology zones for the 23 model layers.

Serial Number of Layer	Description	Zone	Permeability (m^2)		Porosity(−)
			kx = ky	kz	
1	Atmosphere layer	1	1.0×10^{-12}	1.0×10^{-12}	0.99
2	Quaternary deposits and weakly permeable strata (10 represents faults)	2	1.0×10^{-13}	5.0×10^{-14}	0.25
		3	1.0×10^{-13}	5.0×10^{-14}	0.05
		4	1.0×10^{-11}	5.0×10^{-12}	0.2
		10	1.0×10^{-13}	5.0×10^{-14}	0.9
3	Weakly permeable strata (10 represents faults)	2	1.0×10^{-13}	5.0×10^{-14}	0.25
		3	1.0×10^{-13}	5.0×10^{-14}	0.05
		10	1.0×10^{-13}	5.0×10^{-14}	0.9
4–15	Intensely weathered rock (10 represents faults)	5	2.0×10^{-15}	2.0×10^{-16}	0.05
		10	1.0×10^{-13}	5.0×10^{-14}	0.9
16–20		6	1.0×10^{-16}	1.0×10^{-17}	0.02
21	Weakly weathered rock or Unweathered rock	7	1.0×10^{-17}	1.0×10^{-18}	0.01
22		8	1.0×10^{-18}	1.0×10^{-19}	0.01
23		9	1.0×10^{-18}	1.0×10^{-19}	0.01

3.3. Regional Faults

Geological investigation [15] verified that fractures and regional faults are developed in the Beishan area (Figure 1). Usually, there are granite and quartz fractures with filling materials. The field investigation in 2013 and 2014 and previous geological studies (scale is 1:500,000) found that the regional faults extend along the south-eastern, north-eastern, and east-western directions, and the inclination of those faults is mainly between 60° and 70°. In the model, the regional faults are assumed to extend vertically to a depth of approximately 80 m. In addition, the widths of the faults are usually greater than 0.01 m on surface rocks, and then are reduced to 0.001–0.003 m in the depth of 650 m. However, the properties of faults are not fully recognized. In this model, three groups of regional faults (Figure 1) are modeled by the method of Equivalent Permeability Medium, and key parameters are listed in Table 2 (zone 10). They are considered to analyze their influences on the regional groundwater flow pattern.

3.4. Initial Conditions and Boundary Conditions

The initial pressure field is assumed to be under hydrostatic equilibrium by running the model over hundred thousands of years with yearly average rainfall infiltration (90 mm). The temperature field is determined from a surface temperature of 15 °C and a common geothermal gradient of 30 °C/km. This model is assumed to be isothermal. The hydraulic head can be converted from pressure according to Equation (1). The fluxes at the southern and partial eastern boundaries are difficult to ascertain. Thus, the model sets gridblocks at the boundary as constant pressure in TOUGH2, which is estimated by field investigations.

$$H = \frac{P - P_0}{\rho g} \tag{1}$$

where P is the water pressure, Pa; P_0 is the standard atmospheric pressure, Pa; ρ is the water density, kg/m^3; g is the gravitational acceleration, m/s^2; and H is the hydraulic head, m.

3.5. Source and Sinks

The main groundwater recharge item is precipitation infiltration. Groundwater discharge mainly includes evapotranspiration and lateral outflow. Monthly precipitation data over the study area are obtained from data of the Global Precipitation Climatology Project (GPCP), which is a section of the WCRP (the World Climate Research program). The GPCP dataset integrated infrared and microwave satellite observations, and 6000 global conventional observation data points are proven to be reliable precipitation data sources [25–27]. The change of monthly rainfall from 2003 to 2014 is shown in Figure 7, and the yearly average rainfall is approximately 90 mm. Most rainfall is from June to September.

The methods for recharge estimation include in situ experiments of tritium migration (TM) and chloride mass balance (CMB) [28]. According to the tests, the estimated infiltration recharge rates are approximately 2.55 to 6.57 mm/year based on in situ experiments of TM, 0.05 to 1.57 mm/year based on the CMB method in unsaturated zones, and 0.287 mm/year based on the CMB method in saturated zones [14]. So the infiltration rates from the precipitation on the top boundary are set as 0.01 and 0.004 for Quaternary and the weakly permeable strata respectively.

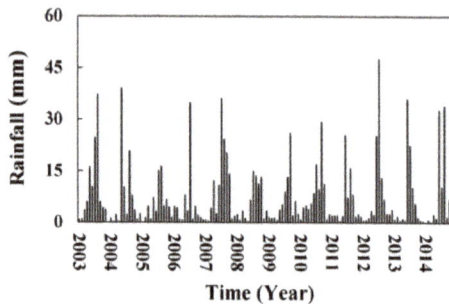

Figure 7. Monthly rainfall from 2003 to 2014 in the Beishan area.

4. Model Calibration

The time period for model calibration is set from January of 2003 to December of 2014. In this study, calibration targets include observed hydraulic heads, and GRACE (Gravity Recovery and Climate Experiment)-derived groundwater storage.

4.1. Comparison of Simulated and Observed Hydraulic Heads

The simulated pressure from TOUGH2 software is converted to the hydraulic head by Equation (1). A comparison of the simulated and observed hydraulic heads at 33 open wells is shown inFigure 8. Here, the relative error is defined as the ratio of the absolute error of observed and simulated values to the observed values. Of the simulated data, 91% are in the range of the relative error in 2%. Overall, the fit between the simulated values and observed values is reasonable. The poor match between the observed and simulated results may be caused by the representations in the models of coarse resolution of hydrogeology zonation, averagely allocation of sources and sinks over a region, and heterogeneity of rock.

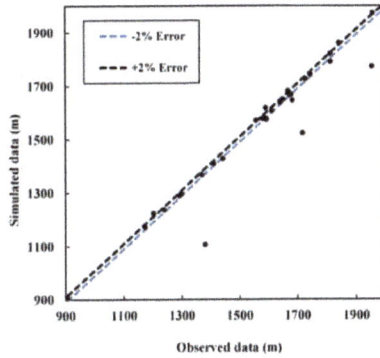

Figure 8. Fit between the observed data and the simulated data.

The change of soil moisture with depth at the end of 2014 is shown in Figure 9. From Figure 9, the depth of water table is basically in accordance with the measured depth to the groundwater level [16], which reveals that the simulated values fit the measured values well. In addition, some scholars performed tests on the soil permeability of the aeration zone in the study area [14], and they found that the soil permeability varied significantly at the spatial scale. As the graph shows, there is an obvious sudden change at the depth of 2 m for BS12 and approximately 10 m for the other boreholes. Because the shallow layer is composed of sand, clay and some intensely weathered granite and the permeability of these rocks are different, there is also some difference among the liquid saturation levels. For BS12, the depth of sand and clay is approximately 2 m; thus, a change occurs at that depth, and then, the liquid saturation reached saturation quickly. Regarding the other boreholes, the granite weathering degree is more serious than that below 10 m; thus, an abrupt change occurred at approximately 10 m below the surface.

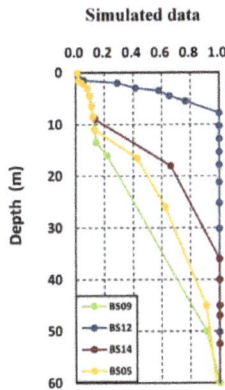

Figure 9. Changes of the liquid saturation with depth for four boreholes at the end of 2014.

4.2. Groundwater Flow Patterns

Figure 10 shows the groundwater flow pattern in the Beishan area. The groundwater level in Figure 10 ranges approximately from 700 m to 2300 m (a.s.l.), with the lowest in the northeast and the highest in the mid-west. Then, the groundwater system in the study area can be divided into three local groundwater systems. Local groundwater in unit I and II flow towards the southeast and the northeast respectively, and then flow into the downstream of the Heihe River Basin. Groundwater in unit III flows to the south boundary. Meanwhile, the prospective site for high-level radioactive waste disposal is located in the unit III, so the hydrostratigraphic units should be further investigated for finalizing site selection. Comparing Figures 2 and 10, the groundwater flow is found to have almost the same pattern as surface flow, suggesting that groundwater flow is mainly dependent on the topography.

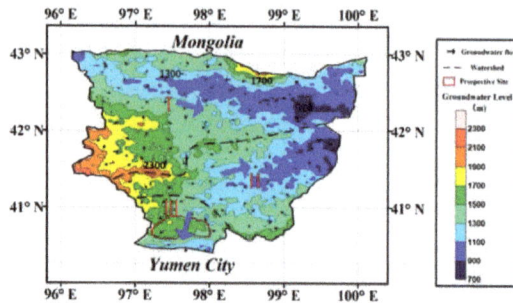

Figure 10. Groundwater flow levels in the Beishan area at the end of 2014.

4.3. Change of Groundwater Storage

Limited point-based observation well data make it difficult to sufficiently calibrate the numerical model in the study area. In this situation, the simulated groundwater storage changes are compared with the GRACE-derived results. The Gravity Recovery and Climate Experiment (GRACE) satellite can monitor the time-dependent earth's gravitational field and then obtain the groundwater storage variation by inversion calculation [29,30]. The GRACE-derived terrestrial water storage (TWS) includes surface water, groundwater, soil water and snow water. Recent studies verified that TWS can be directly estimated from GRACE data for certain large basins when appropriate smoothing is applied to reduce striping [31], and the TWS changes agree reasonably well, at least for long wavelengths, with those estimated from hydrology models and observation networks [29,32–34]. In data-poor regions of the world, such as desert and some remote mountainous regions, GRACE may represent the only option for regional groundwater assessments [35]. Because the study area is in an arid region, and the infiltration rate is so small that there are nearly no rivers; thus, the changes of both surface water and snow water are set as zero [36]. The GRACE-derived data in the study area represent the sum of groundwater and soil water. The numerical model in the saturated–unsaturated zone of the study area can output the changes of groundwater and soil water. A comparison of the groundwater and soil water storage changes with the GRACE-derived values is shown in Figure 11, which shows that the simulated storage trend is accordance with the GRACE-derived TWS variations.

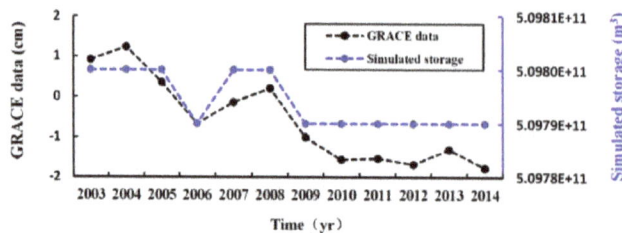

Figure 11. Comparison between the simulated and GRACE-derived TWS variations.

5. Discussions

5.1. Groundwater Budgets

Groundwater budgets play important roles in analyzing regional groundwater flow. Groundwater budgets can be used to evaluate the generation of the groundwater and guide the groundwater management, especially in a large-scale region. In the Beishan area, the main groundwater recharge item is precipitation infiltration, and groundwater discharge mainly includes evapotranspiration and lateral outflow. The Beijing Research Institute of Uranium Geology conducted some field tests to obtain the precipitation infiltration recharge rate of unsaturated soil in the Beishan area.

According to the principle of water balance, the water resources budget can be expressed in Equation (2).

$$P \times W = \Delta Q + O + S \qquad (2)$$

where P is the annual average precipitation, m; W is the study area, m^2; ΔQ is the total groundwater storage variation (including soil water and saturated groundwater), m^3; O is the outflow volume at the Dirichlet boundaries, m^3; and S denotes the surface runoff and evaporation, m^3.

In the numerical model, the yearly averaged precipitation is 90 mm. The yearly averaged outflow at the Dirichlet boundaries is approximately 1.03×10^6 m^3. Changes in the soil water and saturated groundwater are approximately -3.22×10^5 m^3/year. Thus, from Equation (2), surface runoff and evaporation account for approximately 99.99% of precipitation. Over the study area, approximately 0.01%–0.05% of the precipitation infiltrates into the groundwater.

For assessing the influence of the south boundary on groundwater flow in the Beishan area, different yearly annual precipitations (Table 3) are set to estimate the groundwater budgets. The results suggest the south boundary is not sensitive for the groundwater flow pattern in the Beishan area.

Table 3. The variation of groundwater budgets in five scenarios (m^3/year).

Precipitation (mm)		40	90	200	300	450
Recharge		2.76×10^9	6.55×10^9	1.38×10^{10}	2.07×10^{10}	3.10×10^{10}
Discharge	Evaporation and surface runoff	2.76×10^9	6.55×10^9	1.38×10^{10}	2.07×10^{10}	3.10×10^{10}
	O1 [1]	9.70×10^5	9.70×10^5	9.70×10^5	9.70×10^5	9.70×10^5
	O2	6.12×10^4	6.12×10^4	6.12×10^4	6.12×10^4	6.12×10^4
Total groundwater storage variation		-3.22×10^5	-3.22×10^5	-3.21×10^5	-3.20×10^5	-3.18×10^5

[1] O1 is the eastern Dirichlet boundary, O2 is the south Dirichlet boundary.

5.2. Influences of Climate Change on Groundwater Flow Pattern

The safety of the disposal repository in the Beishan area is the primary concern. As discussed earlier, the main recharge of groundwater is precipitation. As a result, the climate changes under extreme conditions, such as wet and dry conditions, should be considered to evaluate their influences on the regional groundwater flow [37]. A long-term calculation is thus required for the variation of climate; as a result, the prediction period is set as 10,000 years. Five precipitation scenarios are set with yearly value at 90, 40, 200, 300 and 450 mm for S1, S2, S3, S4 and S5 scenario, respectively. S1 is the base scenario. S2–S5 represent different annual mean precipitation scenarios.

Figure 12 reveals the liquid saturation at the depth of 2 m in two boreholes. The result suggests that liquid saturation causes a change in a small range for each scenario, and higher precipitation will incur an increase of the liquid saturation level. The highest difference of liquid saturation between S5 and S1 is 7.0×10^{-4}. The changes of liquid saturation with time at the two boreholes are very similar. From the pattern of the water table over the study area, the maximum difference of the water table relative to the base case is approximately 0.015 m for 10,000 years. The average groundwater velocity at the prospective site is approximately 4.66×10^{-13} m/s, and it shows no obvious change under five climate scenarios, suggesting that climate changes will have a limited effect on the groundwater flow in the Beishan area.

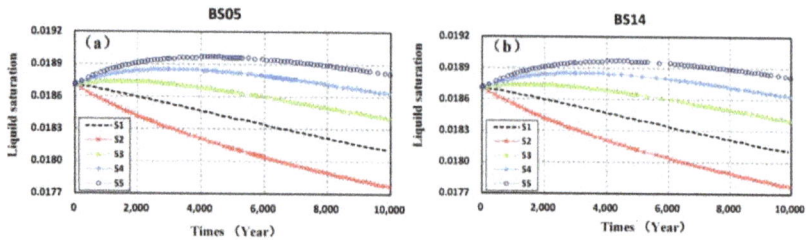

Figure 12. Change of liquid saturation under different precipitation conditions with time: (**a**) for the BS05 well; and (**b**) for the BS14 well.

5.3. Influences of Faults on Regional Groundwater Flow

Because the properties of three faults are not fully understood, the method of scenario analysis is used to ascertain the influences of regional faults on the groundwater flow patterns. Four scenarios (shown in Table 4) are set by changing the permeability of different faults (S2–S5), and S1 is set as the base scenario. All models are run for 10,000 years. The changes of liquid saturation with time at the BS05 and BS14 wells under the four scenarios are nearly the same. To determine the influence on the regional groundwater level, the predicted contour maps of the water table difference in these four scenarios relative to the base scenario are shown in Figure 13. The result reveals that Fault 3 plays a significant role in the groundwater flow pattern, with the maximum variation of 30 m, and S5 considering all three faults together is also similar with S4. However, the permeability values of Fault 1 and Fault 2 have a small influence on the water table, with the greatest variation of 6.5 m. According to the flow directions of groundwater system (Figure 10), faults 1 and 2 occupy the topographically high area between regions I and II such that groundwater drains away from them. Only fault 3 lies transverse to regional flow paths, and so the influence of fault 3 on the regional groundwater flow is significantly. Overall, the influences of the permeability of the three faults are mostly located at the region near the faults, and they will not change the pattern of regional groundwater flow.

Table 4. Permeability distribution under four scenarios (m^2).

Scenarios	Fault 1	Fault 2	Fault 3
Base scenario (S1)	1.0×10^{-13}	1.0×10^{-13}	1.0×10^{-13}
Scenario 2	1.0×10^{-9}	1.0×10^{-13}	1.0×10^{-13}
Scenario 3	1.0×10^{-13}	1.0×10^{-9}	1.0×10^{-13}
Scenario 4	1.0×10^{-13}	1.0×10^{-13}	1.0×10^{-9}
Scenario 5	1.0×10^{-9}	1.0×10^{-9}	1.0×10^{-9}

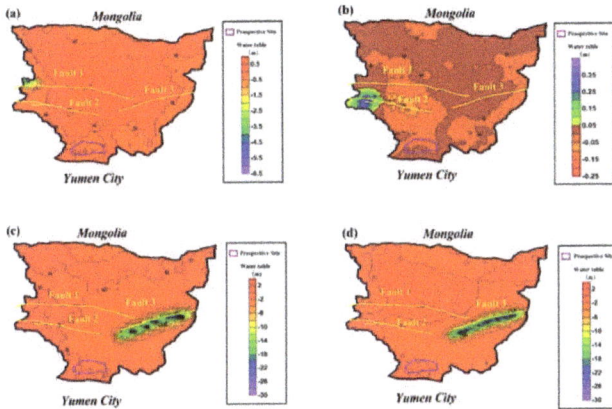

Figure 13. Contour map of the water table difference in the scenarios relative to the base case: (**a**) water table difference between S2 and S1; (**b**) water table difference between S3 and S1; (**c**) water table difference between S4 and S1; and (**d**) water table difference between S5 and S1.

6. Conclusions

The proper disposal of high-level radioactive wastes is highly challenging. Numerical simulation is a wise method to evaluate the safety of a high-level radioactive waste repository [38]. The numerical method is also an effective approach to ascertain the groundwater flow patterns and the influences of climate change on the groundwater flow. In this study, the regional groundwater flow model in the Beishan area, a prospective disposal site for HLRW, was constructed based on the TOUGH2-MP/EOS3 module, which could describe fluid flow in multiphase, multi-component systems. The arid study area covers approximately 68,900 km^2. This model was divided into 23 layers in the vertical direction to allow the unsaturated and saturated zones to be simulated accurately. The parameters of the model were identified by a series of investigations and statistical analysis. Initial pressure field was obtained by running the established model over 10,000 years after comparing a large period, and the differences of the pressure field between two periods were very small, so the initial pressure had been considered as a steady state.

The calibration targets included the hydraulic head from observation wells, and the GRACE-derived groundwater storage changes. The good match between the simulated and observed results proved the reliability of this model. In addition, the model revealed that the regional groundwater flow pattern is quite similar to the surface flow from the channel network given by the DEM model, suggesting that groundwater is mainly dependent on the topography. The study demonstrated that only 0.01%–0.05% of the precipitation infiltrates into the groundwater, which is mainly caused by the wide distribution of low permeability media. Model predictions showed that extreme climate and the permeability of three groups of faults will not have a noticeable effect on the regional groundwater flow in the Beishan area within a period of 10,000 years.

In this paper, field investigation data were integrated to establish the numerical model to fully analyze and evaluate the regional groundwater flow pattern under extreme climate and uncertain fault properties, although the information on hydrogeological conditions is limited. This study is focused on analyzing the regional groundwater flow patterns, refining the groundwater flow systems and testing the influences of hypothesis that both climate change and the fault zone hydraulic properties. Results show three local groundwater systems are subdivided from the regional groundwater flow. To finalize the site selection, a series of chemical and physical experiments, and much data of observation well are highly required. It is suggested that experiments and refined field investigations could be carried out in the region of local flow system, such as the unit III. The study also found that local groundwater flow is sensitive to faults developed in the study area. A detailed investigation on fracture distribution, and development of a refined model, such as dual porosity or dual permeability model in the local area, are required to be carried out.

Acknowledgments: This study was supported by the Research and Development Project on Geological Disposal of High Level Radioactive Waste by the State Administration of Science, Technology and Industry for National Defense (Grant Number: 2012-240) and the National Natural Science Foundation of China (Grant number: 41572220). We thank the Beijing Research Institute of Uranium Geology for supporting the field test data. We would also like to thank the China Scholarship Council for their financial support and constructive comments and suggestions from two anonymous reviewers.

Author Contributions: Xiaoyuan Cao, Litang Hu, Jinsheng Wang and Jingrui Wang. Litang Hu conceived the study; Xiaoyuan Cao and Jingrui Wang analyzed the data. Xiaoyuan Cao conducted the research work under the supervision of Litang Hu and Jinsheng Wang. Xiaoyuan Cao and Litang Hu analyzed and discussed the results. All co-authors contributed to the discussion of the results. Xiaoyuan Cao wrote the first draft of the paper. Litang Hu and Xiaoyuan Cao corrected the first draft of the paper. All co-authors revised and commented the final draft of the paper.

Conflicts of Interest: The authors declare no conflict of interest.

References

1. Rosner, R.; Goldberg, S.M. A practical, regional approach to nuclear waste storage. *Bull. At. Sci.* **2013**, *69*, 58–66. [CrossRef]

2. D'Agnese, F.A.; Faunt, C.C.; Turner, A.K.; Hill, M.C. *Hydrogeologic Evaluation and Numerical Simulation of the Death Valley Regional Ground—Water Flow System*; 4300; U.S Geological Survey Water—Resources Investigations Report: Nevada, CA, USA, 1997; p. 96.

3. Belcher, W.R.; Sweetkind, D.S. *Death Valley Regional Groundwater Flow System, Nevada and California—Hydrogeologic Framework and Transient Groundwater Flow Model*; No.1711; U.S. Geological Survey: Nevada, CA, USA, 2010.

4. Thoma, S.G.; Gallegos, D.P.; Smith, D.M. Impact of fracture coatings on fracture/matrix flow interactions in unsaturated, porous media. *Water Resour. Res.* **1992**, *28*, 1357–1367. [CrossRef]

5. Liu, H.H.; Bodvarsson, G.S.; Doughty, C. *An Active Fracture Model for Unsaturated Flow and Transport*; No. MOL. 20000125.0485; Yucca Mountain Project: Las Vegas, NV, USA, 1999. [CrossRef]

6. Wu, Y.S.; Haukwa, C.; Bodvarsson, G.S. A site-scale model for fluid and heat flow in the unsaturated zone of Yucca Mountain, Nevada. *J. Contam. Hydrol.* **1999**, *38*, 185–215. [CrossRef]

7. Bodvarsson, G.S.; Wu, Y.S.; Zhang, K. Development of discrete flow paths in unsaturated fractures at Yucca Mountain. *J. Contam. Hydrol.* **2003**, *62*, 23–42. [CrossRef]

8. Zhang, K.; Wu, Y.S.; Bodvarsson, G.S. Parallel computing simulation of fluid flow in the unsaturated zone of Yucca Mountain, Nevada. *J. Contam. Hydrol.* **2003**, *62*, 381–399. [CrossRef]

9. Tim, M. *The Site Selection Process for a Spent Fuel Repository in Finland-Summary Report*; Posiva Oy: Helsinki, Finland, 2000.

10. Werner, K.; Bosson, E.; Berglund, S. Analysis of water flow paths: Methodology and example calculations for a potential geological repository in Sweden. *AMBIO J. Hum. Environ.* **2006**, *35*, 425–434. [CrossRef]

11. Wang, J.; Chen, W.; Su, R.; Guo, Y.; Jin, Y. Geological disposal of high-level radioactive waste and its key scientific issues. *J. Rock Mech. Eng.* **2006**, *25*, 801–812. (In Chinese)

12. Dong, Y.H.; Li, G.M.; Li, M. Numerical modeling of the regional ground water flow in Beishan area, Gansu Province. *Chin. Sci. Bull.* **2009**, *54*, 3112–3115. [CrossRef]

13. Xiao, F.; Guo, Y.H.; Wang, Z.M. Outlook of Isotopic Hydrologic Study in Beishan Preselected Area of High Level Waste Repository. *Uranium Geol.* **2012**, *2*, 83–114.

14. Li, J.B.; Su, R.; Zhou, Z.C.; Guo, Y.; Ji, R.; Zhang, M. Soil permeability of aeration zone in Xinchang-Xiangyangshan—A preselected site for high level radioactive waste disposal. *Acta Pedol. Sin.* **2015**, *52*, 1412–1421. [CrossRef]

15. Beijing Research Institute of Uranium Geology. *Research on the Fault Activity in the Beishan Area, Gansu, China*; Science Report; Beijing Research Institute of Uranium Geology: Beijing, China, 2011. (In Chinese)

16. Beijing Research Institute of Uranium Geology. *Researches on Regional Hydrogeological Characterization in the Beishan Area, Gansu, China*; Science Report; Beijing Research Institute of Uranium Geology: Beijing, China, 2010. (In Chinese)

17. Pruess, K.; Oldenburg, C.M.; Moridis, G.J. *TOUGH2 User's Guide*; Version 2; Lawrence Berkeley National Laboratory: Berkeley, CA, USA, 1999.

18. Zhang, K.; Wu, Y.S.; Pruess, K. *User's Guide for TOUGH2-MP-a Massively Parallel Vrsion of the TOUGH2 Code*; Report LBNL-315; Lawrence Berkeley National Laboratory: Berkeley, CA, USA, 2008.

19. Pruess, K. A mechanistic model for water seepage through thick unsaturated zones in fractured rocks of low matrix permeability. *Water Resour. Res.* **1999**, 35, 1039–1051. [CrossRef]

20. Hu, L.T.; Winterfeld, P.H.; Fakcharoenphol, P.; Wu, Y.S. A novel fully-coupled flow and geomechanics model in enhanced geothermal reservoirs. *J. Petrol. Sci. Eng.* **2013**, 107, 1–11. [CrossRef]

21. Wang, X.S.; Jiang, X.W.; Wan, L.; Ge, S.; Li, H. A new analytical solution of topography—Driven flow in a drainage basin with depth-dependent anisotropy of permeability. *Water Resour. Res.* **2011**, 47. [CrossRef]

22. Reuter, H.I.; Nelson, A.; Jarvis, A. An evaluation of void filling interpolation methods for SRTM data. International. *Int. J. Geogr. Inf. Sci.* **2007**, 21, 983–1008. [CrossRef]

23. Hu, L.T.; Zhang, K.N.; Cao, X.Y.; Li, Y.; Guo, C. IGMESH: A convenient irregular-grid-based pre-and post-processing tool for TOUGH2 simulator. *Comput. Geosci.* **2016**, 95, 11–17. [CrossRef]

24. Van Genuchten, M.T. A closed-form equation for predicting the hydraulic conductivity of unsaturated soils. *Soil Sci. Soc. Am. J.* **1980**, 44, 892–898. [CrossRef]

25. Skomorowski, P.; Rubel, F.; Rudolf, B. Verification of GPCP-1DD global satellite precipitation products using MAP surface observations. *Phys. Chem. Earth* **2001**, 26, 403–409. [CrossRef]

26. Huffman, G.J.; Adler, R.F.; Bolvin, D.T.; Gu, G. Improving the global precipitation record: GPCP version 2.1. *Geophys. Res. Lett.* **2009**, 36. [CrossRef]

27. Wang, K.; Niu, J.; Chen, J. Features of global hydrological processes using the Variable Infiltration Capacity Model simulation: Focusing on five major river basins. AGU Fall Meeting Abstracts. 2013. Available online: http://adsabs.harvard.edu/abs/2013AGUFM.H33B1347W (accessed on 20 July 2017).

28. Gee, G.W.; Zhang, Z.F.; Tyler, S.W.; Albright, W.H.; Singleton, M.J. Chloride mass balance. *Vadose Zone J.* **2005**, 4, 72–78. [CrossRef]

29. Schmidt, R.; Schwintzer, P.; Flechtner, F.; Reigber, C.; Güntner, A.; Döll, P.; Ramillien, G.; Cazenave, A.; Petrovic, S.; Jochmann, H.; Wünsch, J. GRACE observations of changes in continental water storage. *Glob. Planet. Chang.* **2006**, 50, 112–126. [CrossRef]

30. Gonzalez, R.; Ouarda, T.B.; Marpu, P.R.; Allam, M.M.; Eltahir, E.A.; Pearson, S. Water Budget Analysis in Arid Regions, Application to the United Arab Emirates. *Water* **2016**, 8, 415. [CrossRef]

31. Chen, J.L.; Rodell, M.; Wilson, C.R.; Famiglietti, J.S. Low degree spherical harmonic influences on Gravity Recovery and Climate Experiment (GRACE) water storage estimates. *Geophys. Res. Lett.* **2005**, 32. [CrossRef]

32. Swenson, S.; Yeh, P.J.F.; Wahr, J.; Famiglietti, J. A comparison of terrestrial water storage variations from GRACE with in situ measurements from Illinois. *Geophys. Res. Lett.* **2006**, 33. [CrossRef]

33. Syed, T.; Famiglietti, J.; Rodell, M.; Chen, J.; Wilson, C.R. Analysis of terrestrial water storage changes from GRACE and GLDAS. *Water Resour. Res.* **2008**, 44, 57–79. [CrossRef]

34. Tiwari, V.M.; Wahr, J.; Swenson, S. Dwindling groundwater resources in northern India, from satellite gravity observations. *Geophys. Res. Lett.* **2009**, 36. [CrossRef]

35. Rodell, M.; Chen, J.L.; Kato, H.; Famiglietti, J.S.; Nigro, J.; Wilson, C.R. Estimating groundwater storage changes in the Mississippi River basin (USA) using GRACE. *Hydrogeol. J.* **2007**, 15, 159–166. [CrossRef]

36. Hu, L.T.; Jiao, J.J. An innovative method to estimate regional-scale hydraulic diffusivity using GRACE data. *Hydrol. Sci. J.* **2016**. [CrossRef]

37. Van Roosmalen, L.; Christensen, B.S.; Sonnenborg, T.O. Regional differences in climate change impacts on groundwater and stream discharge in Denmark. *Vadose Zone J.* **2007**, 6, 554–571. [CrossRef]

38. Qiu, S.; Liang, X.; Xiao, C.; Huang, H.; Fang, Z.; Lv, F. Numerical simulation of groundwater flow in a river valley basin in Jilin Urban area, China. *Water* **2015**, 7, 5768–5787. [CrossRef]

water

MDPI

Article

Effect of Roughness on Conservative Solute Transport through Synthetic Rough Single Fractures

Zhou Chen [1,2,*], Hongbin Zhan [3], Guiqing Zhao [1], Yong Huang [1] and Yefei Tan [4]

1 School of Earth Science and Engineering, Hohai University, Nanjing 210098, China; zhaogq@hhu.edu.cn (G.Z.); hyong38@gmail.com (Y.H.)
2 Department of Civil Engineering, University of Toronto, 35 St. George Street, Toronto, ON M5S 1A4, Canada
3 Department of Geology and Geophysics, Texas A&M University, College Station, TX 77843, USA; zhan@geos.tamu.edu
4 Nanjing Hydraulic Research Institute, 225 Guangzhou road, Nanjing 210029, China; tan8112@gmail.com
* Correspondence: chenzhoucly@hhu.edu.cn; Tel.: +86-138-1541-8015; Fax: +86-025-8378-7234

Received: 31 July 2017; Accepted: 29 August 2017; Published: 1 September 2017

Abstract: Understanding solute transport in fractured rocks is of particular importance in many applications. Aperture values ranging from 4.7 to 8.7 mm and Reynolds number (R_e) values at 9.38~1743.8 were set for investigating fluid flow through synthetic horizontal single smooth and rough fractures. The Brilliant Blue FCF dye was chosen as the tracer to visualize the transport process. This paper focuses on the dispersion process in rough single fractures under non-Darcian flow conditions. Non-Darcian flow existed in both smooth and rough single fractures and the average flow velocity–hydraulic gradient (V–J) relationships were best described by the Forchheimer equation. The main objectives were to check the existing flow and transport models and to study possible correlations between fitting parameters and heterogeneities. The classical advection dispersion equation (ADE) model failed to capture the long-tailing of breakthrough curves (BTCs). Instead, the continuous time random walk (CTRW) model was better at explaining BTCs in both smooth and rough fractures, especially in capturing the long-tailing feature. The non-Darcian coefficient β_c in the Forchheimer equation and the coefficient β in the CTRW model appeared to be most relevant for characterizing the heterogeneity of the rough single fractures.

Keywords: rough single fracture; solute transport; non-Darcian; non-Fickian; heterogeneity

1. Introduction

The management of groundwater resources and control of contaminated aquifers require an understanding of the processes of flow and transport in porous or fractured rocks [1–5]. Fractures and bedding planes or faults give rise to preferential flow paths for groundwater, pollution in dilution, and free product, and thus are of great concern [6].

A single fracture has been traditionally idealized as parallel plates to obtain a tractable description of fluid flow and solute transport, and the model used to describe fluid flow in such an idealized single fracture is the local cubic law (LCL), which is essentially the expression of Darcy's law for a single fracture [7]. Also, Fickian transport is believed to be the "right" form of governing law for transport in the subsurface, where the dispersive mass flux is assumed to be proportional to the first derivative of the resident solute concentration [8]. However, real fractures have rough walls with points of contact, in which the transport is found to be non-Fickian on many occasions [1,2,7]. In general, there are two distinctive features of non-Fickian transport. First, the peak value of a breakthrough curve (BTC) originated from a pulse input of a contaminant or tracer arrives earlier than expected from the advection dispersion equation (ADE) (early arrival). The early arrival often suggests one or multiple preferential paths for the solute transport. Second, the tailing of the BTC lasts much longer than expected (the long tail) [2]. The long tail phenomena can possibly be interpreted in terms of: a multi-rates mass exchange process between mobile and immobile domains [9,10]; fracture wall roughness [11]; the absorption of solute on fracture walls [12,13]; a permeable rock matrix [14]; and the entrapment of solute in eddies inside the fracture [15–17].

There have been several attempts to seek an explanation for non-Fickian transport in single fractures. These include the continuous time random walk (CTRW) [18–22]; the fractional advection dispersion equation (FADE) [23]; the mobile–immobile approach (MIM) [7,24,25]; and the boundary layer dispersion [26]; among others.

Bauget and Fourar [1] and Nowamooz et al. [27] studied the solute transport in a transparent replica of a real single fracture and found that the Fickian ADE was incapable of modeling the heavy (or long) tails' behavior. Alternatively, CTRW was able to describe the non-Fickian dispersion. The influence of surface roughness on nonlinear flow behaviors in three-dimensional (3D) self-affine rough single fractures was determined by Wang et al. [28] using Lattice Boltzmann simulations. Non-Fickian transport through two-dimensional (2D) rough single fractures was numerically studied by Wang and Cardenas [29], and their results clearly showed early arrival and heavy tailing in BTCs. The degree of deviation of transport from Fickian to non-Fickian is captured by the parameter β of the truncated power law used in CTRW. Specifically, β was found to be proportional to fracture heterogeneity [1,24].

On the other hand, non-Darcian flow, which can arise in a number of different ways, may have a great impact on solute transport. Non-Darcian flow is particularly prone to occuring in heterogeneous geological formations such as in a single fracture [11,30–32] or in a fractured network [33] due to relatively high speeds of flow.

To investigate solute transport under non-Darcian flow conditions, Qian et al. [8] established a well-controlled physical model with a vertical smooth parallel fracture, and the non-Fickian transport process was identified with early arrival and long tails. A MIM model fitted both peak and tails of the observed BTCs better than the ADE model. On the basis of this experience, with the purpose of describing solute transport under different flow velocities and fracture apertures, Chen et al. [34] carried out a series of flow and tracer test experiments on an artificial channeled single fracture—a single fracture with contact in certain areas—constructed in the laboratory. The flow condition showed a non-Darcian feature (best described by the nonlinear Forchheimer equation) and BTCs showed a non-Fickian feature such as early arrival of the peak value, long tailing and multi-peak phenomena. The results of this study [34] showed that ADE was not adequate to describe BTCs.

In a similar way to that undertaken by the study of Chen et al. [34], laboratory flow and tracer tests have been carried out on an artificially created fractured rock sample by Cherubini et al. [7], which showed a Forchheimer type of non-Darcian flow, and profound mobile–immobile mass exchange in terms of transport.

The main goals here were to test different models on the solute transport through a rough single fracture under non-Darcian flow conditions and to study possible correlations between fitting parameters and fracture heterogeneities. With these objectives in mind, the following tasks were carried out in sequence. First, synthetic rough single fractures with different roughness elements were designed. Then a series of flow and tracer test experiments were performed with a Reynolds number (*Re*) ranging from 9.38 to 1743.8. An imaging process was introduced to obtain a precise 2D concentration distribution for visual inspection. Subsequently, the characteristics of BTCs were analyzed and two models (ADE and CTRW) tested for their capacity to characterize the non-Fickian transport behavior under non-Darcian flow conditions.

2. Theory

2.1. Flow Model

Generally the model used to describe fluid flow in fractured media is LCL, which adapts Darcy's law [35]:

$$J = -\frac{\mu}{K\rho g}V = -kV,$$

(1)

where J is the hydraulic gradient [dimensionless], K is the hydraulic conductivity [m/s], μ is dynamic viscosity [mPa s], ρ is water density [kg/m^3], g is the gravitational acceleration [m/s^2], V is flow velocity [m/s] and k is the Darcian coefficient. A minor note is that porosity in a single fracture is expected to be 1, meaning that there is no filling materials inside the single fracture.

When a nonlinear flow feature occurs, the flow models commonly used to represent non-Darcian flow behavior are the Forchheimer law and the Izbash law. The former includes a quadratic term of velocity to represent the inertial effect [36,37]:

$$J = -(\frac{\mu}{K\rho g}V + \frac{\beta_c}{g}V^2) = -(aV + b_c V^2)$$

(2)

where β_c [m^{-1}] is the inertial term friction coefficient or non-Darcian coefficient, and a [s/m] and b_c [s^2/m^2] are the linear and inertial coefficients, respectively. If $b_c = 0$, flow becomes Darcian. If $a = 0$, flow becomes fully developed turbulent [30].

The Izbash law, or the power-law, equation is as follows [35]:

$$J = -cV^m$$

(3)

where c [s^2/m^2] and m [dimensionless] are two constant coefficients. The Forchheimer and Izbash equations are equivalent when $a = 0$ in Equation (1) and $m = 2$ in Equation (2).

Another dimensionless parameter to analyze the inertial versus viscous forces is the Reynolds number (R_e), which is defined as [30]:

$$R_e = \frac{2Vb}{\nu}$$

(4)

where b is the fracture aperture [mm], and ν is the kinematic viscosity of water [mm^2/s] (here $\nu = 1.308$ mm^2/s at 10 °C).

The V–J relationships obtained by experiments were fitted by the utility Cftool in Matlab. The quality of the fitting was discussed based on the decision coefficient r^2, the error squared SSE, and the root mean square error (RMSE), as:

$$r^2 = 1 - \frac{\sum\limits_{i=1}^{N}(J_{i0} - J_{ie})^2}{\sum\limits_{i=1}^{N}(J_{i0} - \bar{J}_{i0})^2}$$

(5)

$$SSE = \frac{1}{N}\sum_{i=1}^{N}(J_i - \bar{J}_i)^2$$

(6)

$$RMSE = \sqrt{\frac{1}{N}\sum_{i=1}^{N}(J_{ie} - \bar{J}_{i0})^2}$$

(7)

where J_i is the ith hydraulic gradient, $\bar{J}_i$ is the average hydraulic gradient, J_{ie} is fitted hydraulic gradient, J_{i0} is the test hydraulic gradient, $\bar{J}_{i0}$ is the average test hydraulic gradient, N is the number of the test hydraulic gradients.

2.2. The Advection Dispersion Approach

The general formation of the well-known ADE for a conservative solute without any sink/sources in a one-dimensional framework is given as,

$$\frac{\partial C}{\partial t} + V\frac{\partial C}{\partial x} = D\frac{\partial^2 C}{\partial x^2} \tag{8}$$

where C is the average solute concentration, D is the longitudinal hydrodynamic dispersion coefficient, t is time, and x is distance from the injected position along the flow direction. An analytical solution of Equation (8) with an instantaneous injection of tracer is [38]:

$$C(x,t) = \frac{M/A}{2\sqrt{\pi Dt}} e^{-\frac{(x-Vt)^2}{4Dt}} \tag{9}$$

where M is the injected mass, A is the cross-sectional area of the fracture over which injection occurs. The tracer is assumed to be injected uniformly over the entire saturated thickness of the fracture.

2.3. Continuous Time Random Walk

The continuous time random walk theory (CTRW) has been developed specifically to model the non-Fickian tracer transport [1,39–42]. The CTRW transport equation is based on the Fokker-Planck equation with a memory equation (FPME) which originally describes the temporal evolution of the probability density function of the velocity of a particle under the influence of drag and random forces. In CTRW, the Laplace transformed concentration function $\tilde{C}(x,w)$, is given by the one-dimensional form of FPME,

$$w\tilde{C}(x,w) - C_0(x) = -\tilde{M}(w)[u_\psi \frac{\partial}{\partial x}\tilde{C}(x,w) - D_\psi \frac{\partial^2}{\partial x^2}\tilde{C}(x,w)] \tag{10}$$

where w is the Laplace variable, and u_ψ and D_ψ are transport velocity and dispersion in the framework of CTRW respectively. In Equation (10), the memory function $\tilde{M}(w)$ captures the anomalous or non-Fickian transport induced by local heterogeneity or process and the corresponding formulation is given by

$$\tilde{M}(w) = \bar{t}w\frac{\tilde{\psi}(w)}{1 - \tilde{\psi}(w)} \tag{11}$$

where $\bar{t}$ is some characteristic time and $\tilde{\psi}(w)$ is the transition rate probability, which is the core of the CTRW model. In general, there are three different models for $\tilde{\psi}(w)$, the asymptotic model, the truncated power law (TPL) model, and the modified exponential model. Details for the three different models of $\tilde{\psi}(w)$ can be found in the reference Cortis and Berkowitz [18]. Here, the most-used TPL model was adopted in the framework of CTRW,

$$\tilde{\psi}(w) = (1 + \tau_2 w t_1)^\beta exp(t_1 w)\frac{\Gamma(-\beta, \tau_2^{-1} + t_1 w)}{\Gamma(-\beta, \tau_2^{-1})} \tag{12}$$

where t_1 represents the lower limit time when the power law behavior begins, t_2 is the cut-off time describing the Fickian behavior, and $\Gamma()$ represents the incomplete Gamma function. The parameter β indicates different types of anomalous transport. Following Berkowitz et al. [40], $\tilde{\psi}(w)$ may be approximated by a power-law decay function, $t^{-1-\beta}$, where β is a parameter that characterizes dispersion. For $\beta > 2$, the process is Fickian and CTRW is equivalent to the classical ADE model. If $2 > \beta > 1$, the process is no longer Fickian and shows moderately anomalous transport; If $1 > \beta > 0$, the process indicates highly anomalous transport.

3. Experimental Results

3.1. Experiment Setup and Test Process

Artificial horizontal single fractures were constructed in the laboratory using organic glass plates, as per the schematic shown in Figure 1a. Although such artificial single fractures may not represent all the features of natural fractures, excellent control of flow field can be imposed on the apparatus, and the influence of roughness on water flow through single fractures can be studied in great detail. A detailed setup can be found in Chen et al. [11]. Three different kinds of roughness were used for the upper wall (Figure 1b): a smooth parallel plate; a rough plate with rectangular rough elements; and a rough plate with trapezoidal rough elements. For each type of rough single fractures, four types of asperity heights (Δ), including 2 mm, 4 mm, 6 mm and 8 mm, were used. The corresponding relative roughness (ε), which equals the ratio of Δ and the maximum aperture b_{max} (10 mm), were 0.2, 0.4, 0.6 and 0.8 respectively. Therefore, the average apertures (b_a), which equal the average asperity heights of the rough single fracture (calculation shown in Figure 1b), were obtained and shown in Table 1.

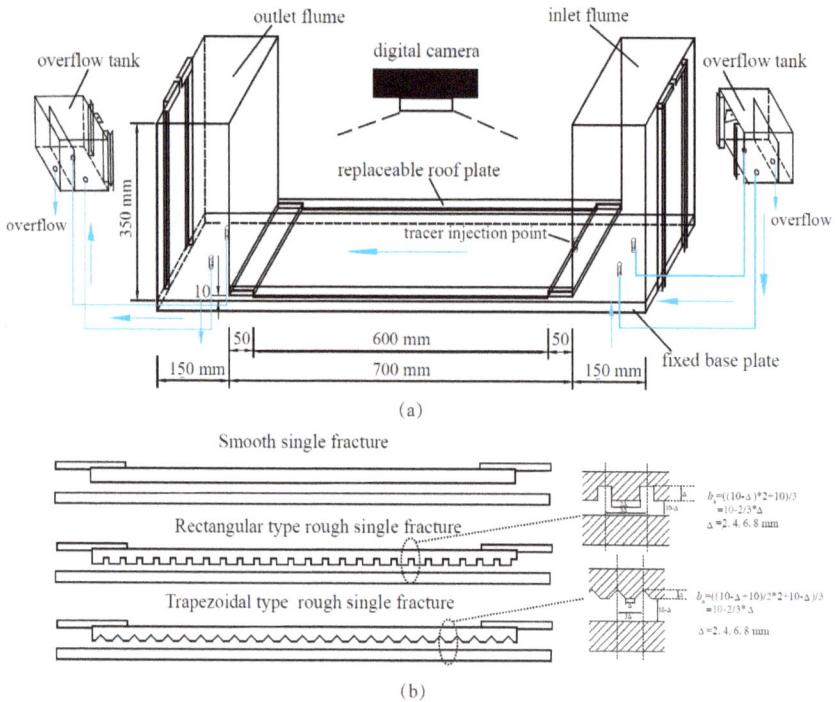

(a)

(b)

Figure 1. Schematic figure of the experimental setup.

Table 1. Average aperture and relative roughness for different single fractures.

Roughness Element	Average Aperture b_a (mm)				Relative Roughness ($\varepsilon = \Delta/b_{max}$)			
	A	B	C	D	A	B	C	D
Smooth	4.7	6.0	7.3	8.7			0	
Trapezoidal	4.7	6.0	7.3	8.7	0.8	0.6	0.4	0.2
Rectangular	4.7	6.0	7.3	8.7	0.8	0.6	0.4	0.2

The water levels at the recharge and discharge reservoirs were measured using piezometers with errors less than 0.5 mm. The hydraulic gradient through the single fracture was controlled through adjusted water levels between the inflow and the outflow reservoirs. The average flow rate was monitored using a calibrated flow meter LZB-25 (glass Rotameter made by Chengfeng Flowmeter Co. Ltd., Changzhou, Jiangsu, China) with a maximum permissible error of 1.5%. The average velocity was calculated by the average flow rate divided by the average aperture and width of the fracture. The flow experiments under different hydraulic gradients were carried out for a given single fracture. After that, the upper plate was replaced by a plate with a different type of roughness, and the entire flow experiment was repeated.

The dye Brilliant Blue FCF F (bis {4-(*N*-ethyl-N-3-sulfophenylmethyl) aminophenyl}-2-sulfophenyl methylium disodium salt) was chosen as the tracer here for visual inspection and image process. It is important to choose an appropriate tracer concentration. If the selected concentration is too high, the gravity and density of the tracer may be a concern. If the concentration is too low, it may not generate images of tracer transport that are of sufficiently high quality for visual inspection. Concentration of the dye Brilliant Blue FCF was set as 1200 mg/L after a series of preliminary tests. Instantaneous tracer was adopted using an injection syringe and 1 mL tracer was injected in each test. The tracer injection point was illustrated in Figure 1a.

The digital camera (Canon EOS 500D) was fixed on a tripod. The length of the camera lens from the main body of the single fracture was set about 650 mm. A light source was placed under the fracture and the light emitted by the lamp was evenly distributed over the transparent plexiglas plate of the main fracture. All tracer tests were carried out in a home-made digital darkroom and the only light source during the test was provided by the lamp.

3.2. Image Process

Image analysis was improved to the extent that estimation of dye concentration from soil color was possible [43]. The main purpose here was to analyze the color information in the image to obtain the mathematical relationship between the color and concentration, and to further analyze the temporal and spatial variation of the concentration. The main factors that affect the color of the image during the test are the intensity of the light source, the color temperature, and the camera's parameter settings (mainly the exposure and white balance). In order to ensure the accuracy of the digital image's color, color temperature adjustment is needed for each photo.

In this study, the water-filled fracture image was used as the background of each tracer test. The background image was subtracted from the image containing the tracer using the image subtract code by Matlab software. The result is shown in Figure 2. The residual color of the fracture surface caused by the dye tracer was reduced.

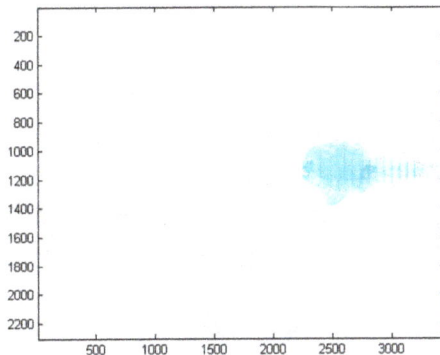

Figure 2. Difference in value between the adjusted image and the background image.

The RGB color model is the most commonly used on digital cameras. We tested the relationship of red (R), green (G), and blue(B) values and the concentration and finally found that the R value in the RGB model had the strongest correlation with the concentration [44].

The polynomial fitting of the relationship between the R value of the image and the concentration C was performed with a coefficient of determination of $r^2 = 0.990$ and RMSE = 0.757. The high r^2 value and the relatively low RMSE value suggested that the image processing method was acceptable.

Different fracture thickness and roughness for each test meant that the color of the same concentration solution in the medium was different. Therefore, the C–R relationships for each roughness type were obtained and the results were shown in Table 2, where the concentration has a unit of mg/L.

Table 2. The concentration C–R value equations for different single fractures (unit of C is mg/L).

Roughness Element	C–R Relation Equation	r^2
Rectangular A	$C = 0.0010 \times R^2 - 0.6442 \times R + 97.31$	0.995
Rectangular B	$C = 0.0007 \times R^2 - 0.3693 \times R + 50.92$	0.998
Rectangular C	$C = 0.0012 \times R^2 - 0.8386 \times R + 133.6$	0.990
Rectangular D	$C = 0.0007 \times R^2 - 0.3693 \times R + 50.92$	0.998
Trapezoidal A	$C = 0.003 \times R^2 - 1.670 \times R + 230.9$	0.992
Trapezoidal B	$C = 0.0014 \times R^2 - 0.9311 \times R + 144.2$	0.993
Trapezoidal C	$C = 0.0003 \times R^2 - 0.3572 \times R + 67.70$	0.992
Trapezoidal D	$C = -0.0006 \times R^2 - 0.101 \times R + 67.78$	0.995
Smooth A	$C = -0.0008 \times R^2 + 0.0506 \times R + 34.98$	0.990
Smooth B	$C = 0.0007 \times R^2 - 0.6243 \times R + 107.4$	0.991
Smooth C	$C = 0.0014 \times R^2 - 0.7719 \times R + 106.6$	0.992
Smooth D	$C = 0.0008 \times R^2 - 0.489 \times R + 74.03$	0.991

3.3. Flow and Tracer Experiments

The present works focused on relatively low velocity and small aperture. In order to investigate the tracer test in non-Darcian flow conditions, we chose a medium aperture ranging from 0.0047 to 0.0087 m and the Reynolds number (R_e) at 9.38~1743.8. The evidence of non-Darcian flow is reflected in the nonlinear relationship between the hydraulic gradient and the flow rate at the given range of R_e.

To analyze the characteristics of solute transport through single fractures more accurately, instantaneous tracer tests using the dye Brilliant Blue FCF were performed. The horizontal fractures with smooth parallel plates, rectangular cross-section rough fractures, and trapezoidal cross-section rough fractures, were chosen. For demonstration purposes, concentrations at position $x = 555$ mm downstream from the point of injection, which is located at the middle elevation at $x = 0$ mm, were calculated under flow velocity of 9.7 mm/s and 51.7 mm/s for smooth parallel plates; 44.3 mm/s and 71.0 mm/s for rectangular cross-section rough fractures; and 20.5 mm/s and 51.2 mm/s for trapezoidal cross-section rough fractures; and the results can be found in Figures 3–5 respectively.

A few observations can be made about Figure 3. For the smooth parallel plates shown in Figure 3, when flow velocity V was as small as 9.7 mm/s, the solute plume moved forward as a lump, but the color inside the plume for each point was not the same, indicating that the solute had not yet fully mixed. The front of the plume showed a typical Gaussian distribution. Lateral dispersion was also clearly visible and irregular movement of the solute with a long tail occurred. When the flow velocity V reached a relatively higher value of 55.1 mm/s, the front of the plume also showed a Gaussian distribution but the lateral dispersion was reduced. Similarly, with that for V of 9.7 mm/s, the concentration inside the plume showed apparent difference and a distinctive long tail also appeared in this case.

Figure 4 showed the solute plume of Brilliant Blue FCF through a rough single fracture with trapezoidal roughness. The differences from that shown in Figure 3 were as follows: the front of the plume was irregular, rather than smooth, which was obviously caused by the heterogeneity of fracture wall roughness. Figure 4 also indicates that the solute concentration inside the plume was different and the solute migration rate was not the same for each point. In addition, retention of solute in some "dead-ends" caused by fracture wall roughness was clearly visible. Specifically, the retained solute was trapped for a long time in those dead-ends, leading to a long tail.

Figure 5 shows the solute plume through a rough single fracture with rectangular roughness. It is similar to Figure 4, with the exception that the irregularity degrees of plume were greater and the long tailing and solute-retention capacity were more pronounced.

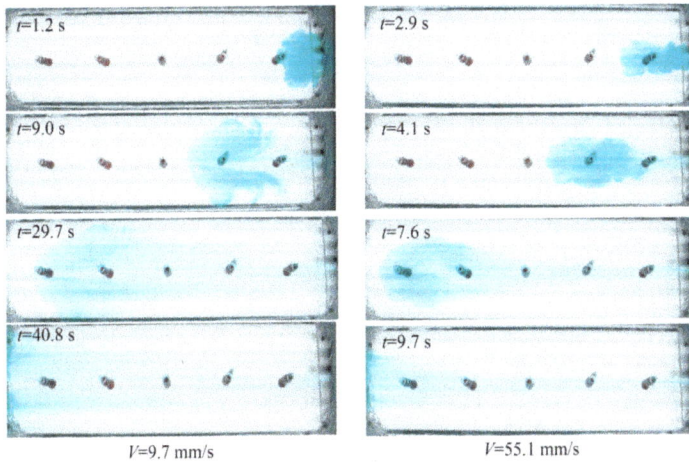

Figure 3. Solute plume of the tracer test in single fractures with smooth parallel plates.

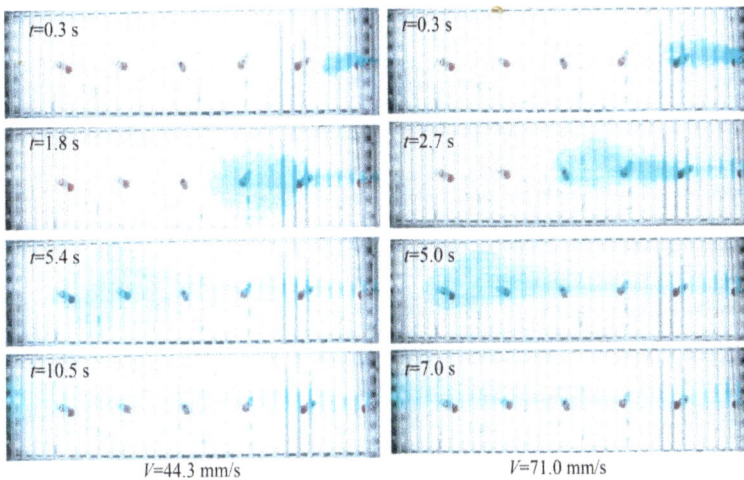

Figure 4. Solute plume of the tracer test in rough single fractures with rectangular roughness.

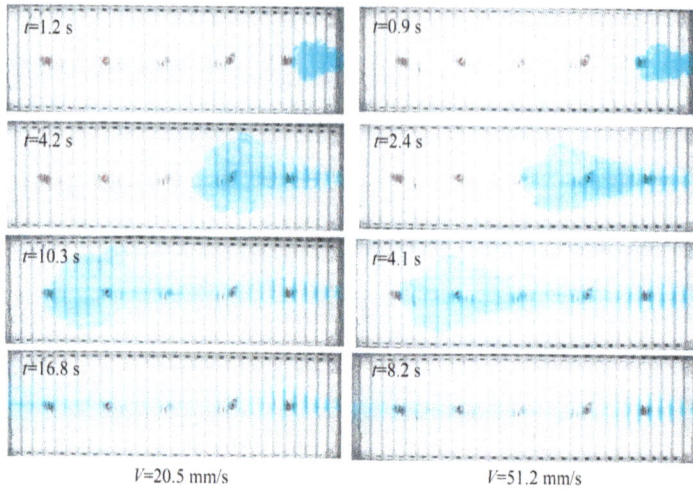

Figure 5. Solute plume of the tracer test in rough single fractures with trapezoidal roughness.

4. Interpretation of Experiments

In the following section, we seek to interpret the obvious non-Fickian transport phenomenon observed in Figures 3–5 using an appropriate theoretical model. Before discussing the transport process, it is necessary to address the flow process first as it was the fundamental driving force for hydrodynamic dispersion.

4.1. The Non-Darcian Equations

The experimental flow velocity–hydraulic gradient (V–J) relationships for different rough single fractures under different flow conditions illustrated in Figures 6–8 and the fitting results, using linear Equation (1) and nonlinear Equations (2) and (3), and goodness of fit were also shown in Tables 3–5. Fitting parameters are listed in Table 6.

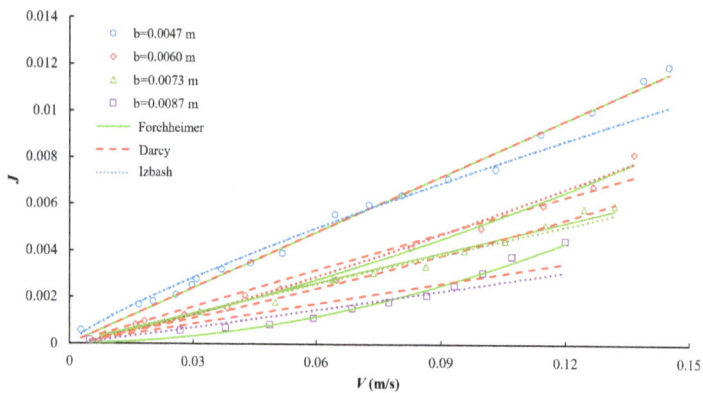

Figure 6. The V–J relationship and fitting results for flow through single fractures with smooth parallel plates.

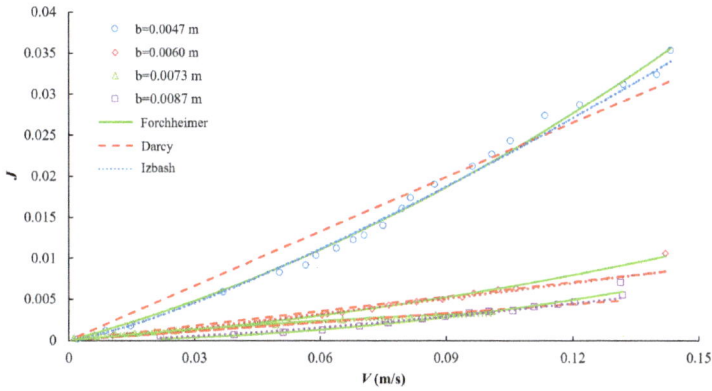

Figure 7. The *V–J* relationship and fitting results for flow through single fractures with rectangular roughness.

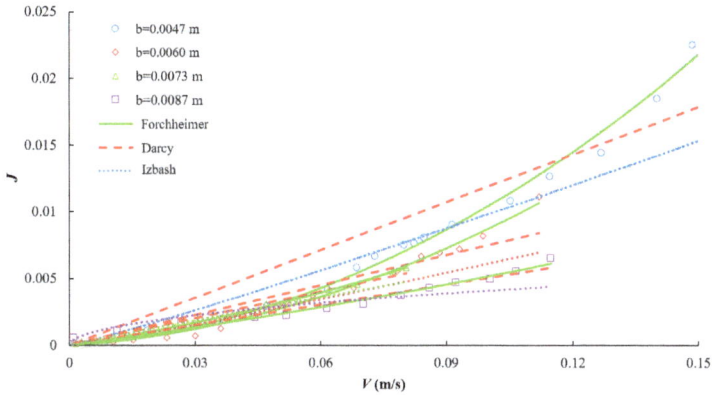

Figure 8. The *V–J* relationship and fitting results for flow through single fractures with trapezoidal roughness.

Table 3. Fitting results and goodness of fit for the *V–J* relationship for flow through single fractures with smooth parallel plates.

b (mm)	*Re*	Fitting Equations	r^2
4.7	20.0~1043.2	$J = 0.081V$	0.993
		$J = 0.049V^{0.813}$	0.977
		$J = 0.0036V^2 + 0.079V$	0.993
6.0	67.2~1253.7	$J = 0.052V$	0.974
		$J = 0.087V^{1.21}$	0.963
		$J = 0.14V^2 + 0.037V$	0.993
7.3	95.6~1472.4	$J = 0.043V$	0.981
		$J = 0.04V^{0.975}$	0.988
		$J = 0.084^2 + 0.034V$	0.991
8.7	15.0~1596.6	$J = 0.028V$	0.852
		$J = 0.031V^{1.07}$	0.937
		$J = 0.28^2 + 0.03V$	0.985

Table 4. Fitting results and goodness of fit for the *V–J* relationship for flow through single fractures with rectangular roughness.

b (mm)	*Re*	Fitting Equations	r^2
4.7	15.1~1021.1	$J = 0.22V$	0.96
		$J = 0.41V^{1.286}$	0.997
		$J = 0.78V^2 + 0.14V$	0.993
6.0	96.2~1300.0	$J = 0.059V$	0.925
		$J = 0.074V^{1.113}$	0.979
		$J = 0.33^2 + 0.0018V$	0.963
7.3	12.7~1131.1	$J = 0.037V$	0.948
		$J = 0.012V^{0.596}$	0.958
		$J = 0.27V^2 + 0.034V$	0.994
8.7	291.8~1743.8	$J = 0.037V$	0.809
		$J = 0.114V^{1.526}$	0.946
		$J = 0.16V^2 + 0.049V$	0.982

Table 5. Fitting results and goodness of fit for the *V–J* relationship for flow through single fractures with trapezoidal roughness.

b (mm)	*Re*	Fitting Equations	r^2
4.7	9.38~1097.4	$J = 0.12V$	0.864
		$J = 0.12V^{1.097}$	0.99
		$J = 0.82V^2 + 0.023V$	0.992
6.0	12.3~1025.6	$J = 0.075V$	0.89
		$J = 0.082V^{1.127}$	0.866
		$J = 0.67V^2 + 0.021V$	0.991
7.3	16.5~1114.2	$J = 0.067V$	0.981
		$J = 0.052V^{0.948}$	0.961
		$J = 0.31^2 + 0.047V$	0.996
8.7	12.1~1519.0	$J = 0.051V$	0.956
		$J = 0.013V^{0.499}$	0.869
		$J = 0.11^2 + 0.041V$	0.966

Table 6. Fitting parameters of non-Darcian equations.

Roughness Element	*b* (mm)	Darcy	Izbash		Forchheimer		Non-Darcian Coefficient
		k	*c*	*m*	*a*	b_c	β_c
Smooth parallel plates	4.7	0.081	0.049	0.813	0.079	0.0036	0.04
	6.0	0.052	0.087	1.21	0.037	0.14	1.37
	7.3	0.043	0.04	0.975	0.034	0.084	0.82
	8.7	0.028	0.031	1.07	0.03	0.28	2.74
Rectangular roughness	4.7	0.22	0.41	1.286	0.14	0.78	7.64
	6.0	0.059	0.074	1.113	0.0018	0.33	3.23
	7.3	0.037	0.012	0.596	0.034	0.27	2.65
	8.7	0.037	0.114	1.526	0.049	0.16	1.57
Trapezoidal roughness	4.7	0.12	0.12	1.097	0.023	0.82	8.04
	6.0	0.075	0.082	1.127	0.021	0.67	6.57
	7.3	0.067	0.052	0.948	0.047	0.31	3.04
	8.7	0.051	0.013	0.499	0.041	0.11	1.08

It was found that non-Darcian flow existed in both smooth and rough single fractures. The nonlinear phenomenon was more obvious in rough single fractures with high asperity heights, which are the heights of roughness inside the fracture. The non-Darcian *V*–*J* relationships were best described by the Forchheimer equation.

The fitting parameters in Table 6 exhibited the following patterns. First, with the decrease of the asperity heights of the rough elements from 8 mm to 2 mm, the relative roughness (ε) decreased from 0.8 to 0.2 and the average aperture increased from 4.7 mm to 8.7 mm, the value of *k* in the Darcian equation decreased, indicating an increasing of permeability for the fracture. The value of *c* in the Izbash equation also decreased with the average aperture, but the value of *m* in the Izbash equation did not show any obvious trend. Second, fitting parameters showed different trends for smooth and rough single fractures. For smooth single fractures, the value of *a* in the Forchheimer equation decreased with the average aperture but the value of b_c and the value of the non-Darcian coefficient β_c increased with the average aperture. Third, the nonlinear phenomenon was more striking in smooth single fractures with larger average apertures. However, for rough single fractures in which flow can be described by the Forchheimer equation, *a* increased with the average aperture but b_c and β_c decreased with the average aperture. This was probably because the increase of asperity heights of rough elements in a rough single fracture would create greater fluid–solid contact surface, and enhance the frictional force for flow. Therefore, the relative importance of viscous flow versus inertial flow for a greater relative roughness is increased, reducing the flow nonlinearity (which is primarily caused by the inertial effect of flow).

4.2. The ADE and CTRW Models

The molecular diffusion coefficient D_m is set as 5.0×10^{-11} m²·s⁻¹ based on the Taylor–Aris experiment [1]. The Peclet numbers ($P_e = bV/D_m$) in our experiment were greater than 5×10^6. As pointed out by Bauget and Fourar [1], when the Peclet number is greater than 4000, transverse dispersion can be ignored. The P_e numbers of our experiments were more than 100 times greater than 4000, which warranted the use of the one-dimensional approach. The CTRW Matlab Toolbox V3.1 [20] was employed for the CTRW TPL model.

BTCs at position *x* = 555 mm in single fractures with smooth parallel plates and rectangular rough single fractures were illustrated and fitted by ADE and CTRW TPL models. The results are shown in Figures 9 and 10 and Tables 7 and 8.

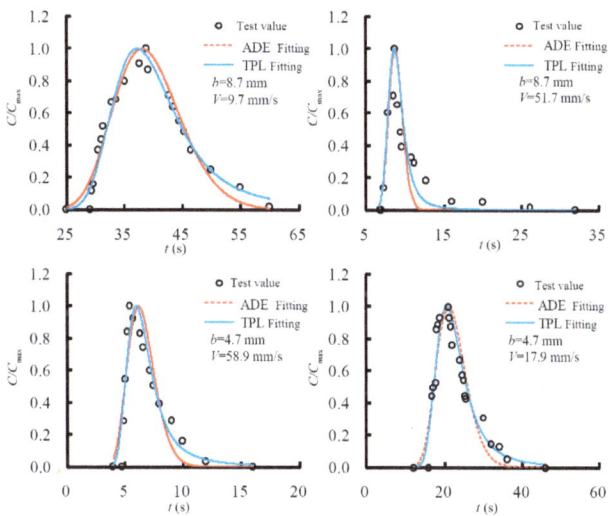

Figure 9. BTCs and fitting results in single fractures with smooth parallel plates (*x* = 555 mm).

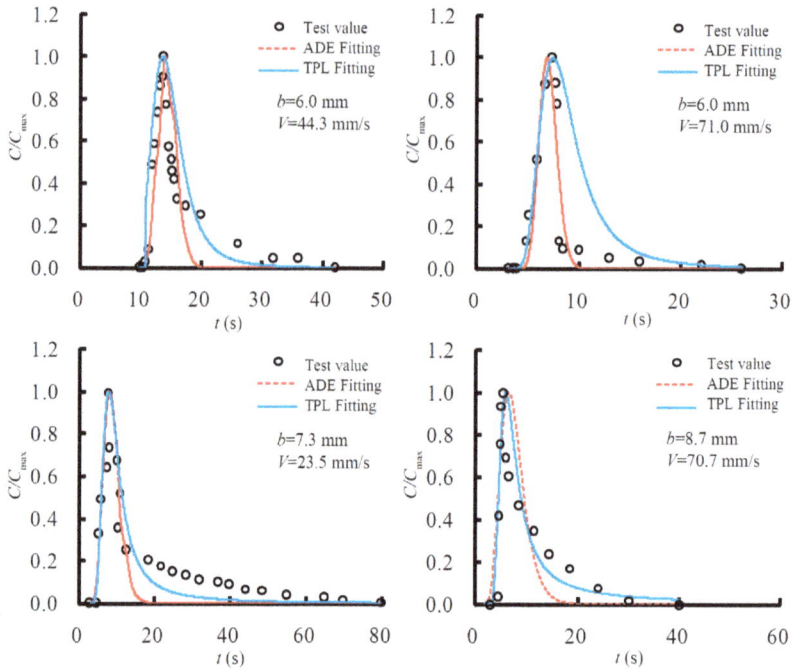

Figure 10. BTCs and fitting results in single fractures with rectangular rough fractures (x = 555 mm).

Table 7. Fitting parameters of BTC in single fractures with smooth parallel plates.

b mm	V mm/s	ADE						TPL				
		V_A mm/s	$D_A \times 10^{-4}$ m²/s	r^2	RMSE	$Lg(t_1)$	$Lg(t_2)$	v_ψ mm/s	$D_\psi \times 10^{-4}$ m²/s	β	r^2	RMSE
4.7	17.9	25.2	2.16	0.932	0.165	−1.26	9.92	38.9	0.39	1.6	0.963	0.0895
	58.9	84.8	8.0	0.891	0.172	−2.15	8.63	178.4	1.79	1.43	0.951	0.103
8.7	9.7	14.0	0.86	0.943	0.133	−1.5	18.7	25.9	0.262	1.51	0.974	0.0537
	51.7	62.5	1.72	0.924	0.172	−3.04	7.83	141.2	1.43	1.41	0.988	0.0428

Table 8. Fitting parameters of BTC in single fractures with rectangular rough fractures.

b mm	V mm/s	ADE						TPL				
		V_A mm/s	$D_A \times 10^{-4}$ m²/s	r^2	RMSE	$Lg(t_1)$	$Lg(t_2)$	v_ψ mm/s	$D_\psi \times 10^{-4}$ m²/s	β	r^2	RMSE
6.0	44.3	61.0	0.12	0.881	0.221	−2.62	8.6	365.2	3.72	1.05	0.962	0.121
	71.0	71.9	0.21	0.904	0.157	−3.24	3.26	631.6	6.38	1.00	0.966	0.0895
7.3	23.5	38.0	1.2	0.912	0.164	−1.01	9.75	41.3	0.43	1.87	0.957	0.185
8.7	70.7	78.5	3.39	0.942	0.112	−1.71	4.34	112.3	1.45	1.7	0.985	0.0562

We can make several interesting observations from Figures 9 and 10. First, BTCs followed the normal distribution in smooth single fractures, especially when the flow velocity was small. Second, as the flow velocity V increased, BTCs appeared to exhibit non-normal distributions with long tails. These non-normal distributions could be found ubiquitously in rough single fractures. Third, the classical ADE model was incapable of capturing the long-tailing of BTCs either in smooth or rough single fractures. However, the CTRW TPL model could better explain BTCs in both smooth and rough single fractures, especially in capturing the long-tailing phenomenon.

Similar conclusions can be obtained from Tables 7 and 8, as demonstrated by relatively smaller values of R^2 and larger RMSE values for the ADE model compared with the CTRW TPL model. The R^2 values were greater than 0.95 and the RMSE values were less than 0.1 with the CTRW TPL model, indicating satisfactory goodness of fit. It could be seen that the t_1 value in the CTRW TPL model was less than the observation time of the test and the t_2 value in the same model was much larger than the observation time of the test, indicating that the solute transport through single fractures under experimental flow conditions was far from reaching Fickian transport.

It can be seen from Tables 7 and 8 that $1 < \beta < 2$, reflecting how the solute transport in both smooth and rough single fractures under non-Darcian flow conditions had not reached Fickian migration. The β value decreased with the flow velocity and the relative roughness (ε), indicating that the degree of nonlinearity was stronger with larger flow velocity; and the degree of deviation from Fickian transport was also higher. Meanwhile, an increase of the asperity height would further aggravate the irregularity of solute transport. The above observations were made based on a β value of 1.05 and a ε value of 0.6 in single fractures with rectangular roughness.

In addition, the parameter v_ψ, which is the average velocity of solute particles in the CTRW TPL model, was much larger than the actual average velocity of V. This implied that the average velocity of solute particles was greater than the average velocity of flow.

To further study the effect of relative roughness (ε) on solute transport, BTCs at position $x = 355$ mm downstream, V of 71.0 m/s, b of 8.7 mm and 6.0 mm, and ε of 0.2 and 0.6 in single fractures with rectangular roughness were fitted by the CTRW TPL model; while BTCs at position $x = 355$ mm downstream, V of 49.5 m/s, b of 8.7 mm and 7.3 mm, and ε of 0.2 and 0.4 in single fractures with trapezoidal roughness were also fitted by the CTRW TPL model. The fitting results can be seen in Table 9. As a reference for comparison, BTCs at position $x = 355$ mm downstream, V of 59.0 m/s, and b of 4.7 mm and 8.7 mm in single fractures with smooth parallel plates were also fitted by the CTRW TPL and the results are also listed in Table 9.

Table 9. Fitting results of TPL for single fractures with different ε and b.

Roughness Element Structure	V mm/s	b mm	ε	TPL Fitting Parameters				
				v_ψ mm/s	$D_\psi \times 10^{-4}$ m²/s	β	$\log_{10}(t_1)$	$\log_{10}(t_2)$
Rectangular	71.0	6.0	0.6	183.9	0.81	1.28	−0.746	11.257
		8.7	0.2	135.9	0.13	1.89	−1.291	1.210
Trapezoidal	49.5	7.3	0.4	256.4	0.26	1.19	−2.536	3.162
		8.7	0.2	98.6	0.57	1.67	−0.909	9.294
Smooth parallel plates	59.0	4.7	0	59.7	0.36	1.69	0.534	5.160
		8.7	0	131.2	0.13	1.70	−1.318	11.415

From Table 9, one can see that an increase of ε resulted in a significant reduction in the value of β; that is, the degree of non-Fickian transport (or the deviation from Fickian transport) was affected significantly by ε in rough single fractures. However, the change of the b values in single smooth fractures had little effect on the value of β.

Furthermore, to analyze the effect of fracture roughness type (rectangular or trapezoidal) on solute transport, BTCs from rough single fractures with rectangular and trapezoidal roughness under the same flow condition ($V = 36.0$ mm/s) and the same ε value of 0.4 were fitted by the CTRW TPL model, and the results are shown in Table 10. As a reference for comparison, BTCs from smooth single fractures and rough single fractures with trapezoidal roughness under the same flow condition ($V = 51.5$ mm/s) and the same b value of 6 mm were fitted by the CTRW TPL model and the results are also listed in Table 10.

Table 10. Fitting results of the CTRW TPL model for single fractures with the same b with different roughness structure.

V mm/s	b mm	ε	Roughness Element Structure	TPL Fitting Parameters				
				v_ψ mm/s	$D_\psi \times 10^{-4}$ m^2/s	β	$\log_{10}(t_1)$	$\log_{10}(t_2)$
36.0	7.3	0.4	Rectangular	51.1	0.18	1.329	0.748	7.45
			Trapezoidal	110.8	0.11	1.469	−1.293	8.03
51.5	6.0	0.6	Trapezoidal	144.3	0.180	1.346	−1.334	5.26
		0	Smooth parallel plates	39.0	0.14	1.513	1.132	4.32

We can conclude from Table 10 that the β value under the same flow condition in a rough single fracture with a rectangular-type of roughness was smaller than that with a trapezoidal-type of roughness. As shown in Chen et al. [11], recirculation of flow in the eddies and curved streamline in a rough single fracture with the rectangular-type of roughness caused greater energy loss than that in the rough single fracture with a trapezoidal-type roughness. Such additional energy loss caused a greater degree of flow nonlinearity or stronger non-Darcian flow, leading to stronger non-Fickian transport and a smaller β value.

In summary, the heterogeneity of the fractured media, which is reflected in the relative roughness and roughness structure, was primarily responsible for the non-Fickian transport through rough single fractures. Also, the increase of flow nonlinearity of non-Darcian flow can exacerbate the non-Fickian transport phenomenon.

5. Conclusions

Artificial smooth and rough single fractures were constructed in the laboratory using organic glass plates. Compared with previous works focused on relatively low velocity and small apertures, medium aperture values ranging from 4.7 to 8.7 mm and the Reynolds numbers of 9.38~1743.8 were set to characterize the tracer transport through smooth and rough single fractures under non-Darcian flow conditions. The dye Brilliant Blue FCF was chosen as the tracer to facilitate the image process, enabling the solute transport process to be directly observed more intuitively.

It was found that non-Darcian flow existed in both smooth and rough single fractures under these experimental conditions. The nonlinear flow phenomenon was more obvious in rough single fractures, especially with large asperity heights. The non-Darcian flow velocity and hydraulic gradient (V–J) relationships were best described by the Forchheimer equation.

The classical ADE model was incapable of capturing the long-tailing of BTCs either in smooth or rough single fractures; the CTRW TPL model, with more adjustable parameters, can explain BTCs both in smooth and rough single fractures better, especially when it comes to capturing the long-tailing phenomenon.

The heterogeneity of the fractured media, which was manifested through the relative roughness and roughness structure, had a significant effect on the degree of deviation from Fickian transport (or the degree of non-Fickian transport) through rough single fractures. Also, the increase in the flow nonlinearity can further exacerbate the non-Fickian transport phenomenon. The non-Darcian coefficient β_c and the coefficient β in the CTRW TPL model seem to be good parameters for characterizing the heterogeneity of the rough fracture. The results may enhance our understanding of solute transport under non-Darcian flow conditions through fractured media. The study of solute transport and reactive transport through a real single fracture under non-Darcian conditions needs to be studied further.

Acknowledgments: The study is financially supported by the National Natural Science Foundation of China (Grant Nos. 41402197 and 41472233), the national key research and development plan (2016YFC0402807), and the Central University Fundamental Research Project of China (No. 26120132013B02914).

Author Contributions: Z.C. and H.Z. conceived and designed the experiments; G.Z. performed the experiments; Z.C. and H.Z. analyzed the data; Y.H. and Y.T. contributed reagents, materials and analysis tools; Z.C. wrote the paper and H.Z. revised the paper.

Conflicts of Interest: The authors declare no conflict of interest.

References

1. Bauget, F.; Fourar, M. Non-fickian dispersion in a single fracture. *J. Contam. Hydrol.* **2008**, *100*, 137–148. [CrossRef] [PubMed]
2. Berkowitz, B. Characterizing flow and transport in fractured geological media: A review. *Adv. Water Resour.* **2002**, *25*, 861–884. [CrossRef]
3. Dou, Z.; Zhou, Z.; Sleep, B.E. Influence of wettability on interfacial area during immiscible liquid invasion into a 3d self-affine rough fracture: Lattice boltzmann simulations. *Adv. Water Resour.* **2013**, *61*, 1–11. [CrossRef]
4. Dou, Z.; Zhou, Z.-F. Numerical study of non-uniqueness of the factors influencing relative permeability in heterogeneous porous media by lattice boltzmann method. *Int. J. Heat Fluid Flow* **2013**, *42*, 23–32. [CrossRef]
5. Wang, L.; Bayani Cardenas, M. Transition from non-fickian to fickian longitudinal transport through 3-d rough fractures: Scale-(in)sensitivity and roughness dependence. *J. Contam. Hydrol.* **2017**, *198*, 1–10. [CrossRef] [PubMed]
6. Becker, M.W.; Shapiro, A.M. Tracer transport in fractured crystalline rock: Evidence of nondiffusive breakthrough tailing. *Water Resour. Res.* **2000**, *36*, 1677–1686. [CrossRef]
7. Cherubini, C.; Giasi, C.I.; Pastore, N. Evidence of non-darcy flow and non-fickian transport in fractured media at laboratory scale. *Hydrol. Earth Syst. Sci.* **2013**, *17*, 2599–2611. [CrossRef]
8. Qian, J.; Zhan, H.; Chen, Z.; Ye, H. Experimental study of solute transport under non-darcian flow in a single fracture. *J. Hydrol.* **2011**, *399*, 246–254. [CrossRef]
9. Cvetkovic, V.; Haggerty, R. Transport with multiple-rate exchange in disordered media. *Phys. Rev. E* **2002**, *65*, 051308. [CrossRef] [PubMed]
10. Haggerty, R.; Gorelick, S.M. Multiple-rate mass transfer for modeling diffusion and. *Water Resour. Res.* **1995**, *31*, 2383–2400. [CrossRef]
11. Chen, Z.; Qian, J.; Zhan, H.; Zhou, Z.; Wang, J.; Tan, Y. Effect of roughness on water flow through a synthetic single rough fracture. *Environ. Earth Sci.* **2017**, *76*, 186. [CrossRef]
12. Bodin, J.; Delay, F.; De Marsily, G. Solute transport in a single fracture with negligible matrix permeability: 1. Fundamental mechanisms. *Hydrogeol. J.* **2003**, *11*, 418–433. [CrossRef]
13. Zimmerman, M.D.; Bennett, P.C.; Sharp, J.M., Jr.; Choi, W.-J. Experimental determination of sorption in fractured flow systems. *J. Contam. Hydrol.* **2002**, *58*, 51–77. [CrossRef]
14. Zhou, Q.; Liu, H.-H.; Bodvarsson, G.S.; Molz, F.J. Evidence of multi-process matrix diffusion in a single fracture from a field tracer test. *Transp. Porous Media* **2006**, *63*, 473–487. [CrossRef]
15. Cardenas, M.B.; Slottke, D.T.; Ketcham, R.A.; Sharp, J.M. Navier-stokes flow and transport simulations using real fractures shows heavy tailing due to eddies. *Geophys. Res. Lett.* **2007**, *34*. [CrossRef]
16. Cardenas, M.B.; Slottke, D.T.; Ketcham, R.A.; Sharp, J.M. Effects of inertia and directionality on flow and transport in a rough asymmetric fracture. *J. Geophys. Res.* **2009**, *114*, 258–266. [CrossRef]
17. Lee, S.H.; Yeo, I.W.; Lee, K.K.; Detwiler, R.L. Tail shortening with developing eddies in a rough-walled rock fracture. *Geophys. Res. Lett.* **2015**, *42*, 6340–6347. [CrossRef]
18. Cortis, A.; Berkowitz, B. Anomalous transport in "classical" soil and sand columns. *Soil Sci. Soc. Am. J.* **2004**, *68*, 1539–1548. [CrossRef]
19. Cortis, A.; Berkowitz, B. Computing "anomalous" contaminant transport in porous media: The ctrw matlab toolbox. *Ground Water* **2005**, *43*, 947–950. [CrossRef] [PubMed]
20. Dentz, M.; Cortis, A.; Scher, H.; Berkowitz, B. Time behavior of solute transport in heterogeneous media: Transition from anomalous to normal transport. *Adv. Water Resour.* **2004**, *27*, 155–173. [CrossRef]
21. Geiger, S.; Cortis, A.; Birkholzer, J.T. Upscaling solute transport in naturally fractured porous media with the continuous time random walk method. *Water Resour. Res.* **2010**, *46*. [CrossRef]
22. Jiménez-Hornero, F.; Giráldez, J.; Laguna, A.; Pachepsky, Y. Continuous time random walks for analyzing the transport of a passive tracer in a single fissure. *Water Resour. Res.* **2005**, *41*, 1325–1327. [CrossRef]

23. Benson, D.A. The Fractional Advection—Dispersion Equation: Development and Application. Ph.D. Thesis, University of Nevada Reno, Reno, NV, USA, 1998.

24. Cherubini, C.; Giasi, C.I.; Pastore, N. On the reliability of analytical models to predict solute transport in a fracture network. *Hydrol. Earth Syst. Sci.* **2014**, *18*, 2359. [CrossRef]

25. Van Genuchten, M.T.; Wierenga, P.J. Mass transfer studies in sorbing porous media I. Analytical solutions1. *Soil Sci. Soc. Am. J.* **1976**, *40*, 473–480. [CrossRef]

26. Moutsopoulos, K.N.; Koch, D.L. Hydrodynamic and boundary-layer dispersion in bidisperse porous media. *J. Fluid Mech.* **1999**, *385*, 359–379. [CrossRef]

27. Nowamooz, A.; Radilla, G.; Fourar, M.; Berkowitz, B. Non-fickian transport in transparent replicas of rough-walled rock fractures. *Transp. Porous Media* **2013**, *98*, 651–682. [CrossRef]

28. Wang, M.; Chen, Y.-F.; Ma, G.-W.; Zhou, J.-Q.; Zhou, C.-B. Influence of surface roughness on nonlinear flow behaviors in 3d self-affine rough fractures: Lattice boltzmann simulations. *Adv. Water Resour.* **2016**, *96*, 373–388. [CrossRef]

29. Wang, L.; Cardenas, M.B. Non-fickian transport through two-dimensional rough fractures: Assessment and prediction. *Water Resour. Res.* **2014**, *50*, 871–884. [CrossRef]

30. Iwai, K. Fundamental Studies of Fluid Flow through a Single Fracture. Ph.D. Thesis, University of California, Berkeley, CA, USA, 1976.

31. Lomize, G.M. *Flow in Fractured Rocks*; Gosenergoizdat: Moscow, Russia, 1951; pp. 127–129. (In Russian)

32. Qian, J.; Zhan, H.; Zhao, W.; Sun, F. Experimental study of turbulent unconfined groundwater flow in a single fracture. *J. Hydrol.* **2005**, *311*, 134–142. [CrossRef]

33. Novakowski, K.S.; Lapcevic, P.A.; Voralek, J.; Bickerton, G. Preliminary interpretation of tracer experiments conducted in a discrete rock fracture under conditions of natural flow. *Geophys. Res. Lett.* **1995**, *22*, 1417–1420. [CrossRef]

34. Chen, Z.; Qian, J.-Z.; Qin, H. Experimental study of the non-darcy flow and solute transport in a channeled single fracture. *J. Hydrodyn.* **2011**, *23*, 745–751. [CrossRef]

35. Qian, J.Z.; Chen, Z.; Zhan, H.B.; Luo, S.H. Solute transport in a filled single fracture under non-darcian flow. *Int. J. Rock Mech. Min.* **2011**, *48*, 132–140. [CrossRef]

36. Bear, J. *Dynamics of Fluids in Porous Media*; Courier Corporation: North Chelmsford, MA, USA, 2013.

37. Ranjith, P.G.; Darlington, W. Nonlinear single-phase flow in real rock joints. *Water Resour. Res.* **2007**, *43*. [CrossRef]

38. Bear, J. *Dynamics of Fluids in Porous Materials*; Society of Petroleum Engineers: Dallas, TX, USA, 1972.

39. Berkowitz, B.; Cortis, A.; Dentz, M.; Scher, H. Modeling non-fickian transport in geological formations as a continuous time random walk. *Rev. Geophys.* **2006**, *44*. [CrossRef]

40. Berkowitz, B.; Kosakowski, G.; Margolin, G.; Scher, H. Application of continuous time random walk theory to tracer test measurements in fractured and heterogeneous porous media. *Ground Water* **2001**, *39*, 593–604. [CrossRef] [PubMed]

41. Berkowitz, B.; Scher, H. Theory of anomalous chemical transport in random fracture networks. *Phys. Rev. E* **1998**, *57*, 5858–5869. [CrossRef]

42. Berkowitz, Y.; Edery, Y.; Scher, H.; Berkowitz, B. Fickian and non-fickian diffusion with bimolecular reactions. *Phys. Rev. E* **2013**, *87*. [CrossRef]

43. Persson, M. Accurate dye tracer concentration estimations using image analysis. *Soil Sci. Soc. Am. J.* **2005**, *69*, 967. [CrossRef]

44. Forrer, I.; Papritz, A.; Kasteel, R.; Flühler, H.; Luca, D. Quantifying dye tracers in soil profiles by image processing. *Eur. J. Soil Sci.* **2000**, *51*, 313–322. [CrossRef]

water

MDPI

Article

Laboratory Investigation of the Effect of Slenderness Effect on the Non-Darcy Groundwater Flow Characteristics in Bimsoils

Yu Wang *, Changhong Li *, Xiaoming Wei and Zhiqiang Hou

Beijing Key Laboratory of Urban Underground Space Engineering, Department of Civil Engineering, School of Civil & Resource Engineering, University of Science & Technology Beijing, Beijing 100083, China; wxming_ustb@163.com (X.W.); houzq_ustb@126.com (Z.H.)
* Correspondence: wyzhou@ustb.edu.cn (Y.W.); lch@ustb.edu.cn (C.L.); Tel.: +86-10-6233-3745 (Y.W. & C.L.)

Received: 1 June 2017; Accepted: 5 September 2017; Published: 7 September 2017

Abstract: A series of experimental flow tests for artificial block-in-matrix-soils (bimsoils) samples with various slenderness ratios were performed to study the Non-Darcy groundwater flow characteristics. The variations of seepage velocity, permeability coefficient, critical sample height, and non-Darcy flow factor for samples against slenderness ratios were investigated. A servo-controlled flow testing system that was developed by the authors was adopted to conduct the flow test. Cylindrical bimsoil samples (50 mm diameter and various heights) with staggered rock block percentages (RBPs) (30, 40, 50, and 60%, by mass) were prepared by compaction tests to roughly insure the same porosity of the soil matrix. The testing results show that flow the distance has a strong influence on the flow characteristics of bimsoil, and the relationship between the permeability coefficient and slenderness ratio is proposed. In addition, the critical sample height to eliminate the slenderness effect was determined, and the relationship between the critical sample height and RBP was established. Moreover, the responses of non-Darcy flow were studied by using an index of non-Darcy βfactor, which reveals the internal mechanism of the effect of flow distance on groundwater flow characteristics. The research results can be useful to the prediction of flow piping disaster for geological body made up of bimsoils.

Keywords: bimsoils; water flow; slenderness effect; permeability coefficient; non-Darcy flow

1. Introduction

Geological formations are generally classified as either soils or rocks from an engineering point of view. In the geomechanical literature, the term block-in-matrix-soils (bimsoils) is used to describe the structurally chaotic geomaterials characterized by structurally complex formations composed of a variety of stronger rock block inclusions with various sizes, strengths and different lithologies embedded in a pervasively fine-grained weaker matrix (soil) [1–9]. In the literature, despite some researchers having focused on this special inhomogeneous and loose geomaterial, different terms have been used to describe those mixed geomaterials similar to bimsoil, such as mélange [10–15], bimrocks [16–21], SRM [5–8], rock and soil aggregate [3,22], conglomerates [23], coarse-grain alluviums and colluviums [20], to name a few. These complex mixtures occur globally and originated by several geological processes (fault rocks, mélanges, olistostromes, breccias, weathering eluvia deposit, etc.) [1]. The mechanical and physical properties of bimsoils are characterized by the extreme inhomogeneity, looseness and environmental sensitivity, and the mechanical and physical properties of bimsoil are controlled by the interactions between rock blocks and the soil matrix [1,2,4,7].

A number of studies have been done to investigate the effect of sample sizes on the mechanical properties of bimsoils subject to internal and external loading, namely the scale effect. Medley [13]

considered that bimsoils have the same general appearance regardless of scale, and he pointed out that bimsoils were independent of the RBP in bimsoils. Bagnold and Barndorff-Nielsen [24] conducted in-situ measurement of the size of blocks in bimsoil, and studied the log-histogram relationship of bimsoil with different scales. They found that for the measured area, the rock blocks show very similar shapes, and the fractal dimensions are almost the same [10]. Based on this study, other researchers assumed that the geomechanical behaviors of bimsoil are also scale-independent [1,23,25]. Xu et al. [25] conducted a series of in-situ shear tests, and pointed out that the sample height should be five times larger than the maximum rock block diameter. Co li et al. [1] conducted in-situ shear tests, and pointed out that a maximum dimension of the rock block should be 0.1 times the bimsoil sample. To study the deformability and failure process, Zhang et al. [9] have performed numerical simulations to conduct uniaxial compression for bimsoil samples with different RBP, size, and slenderness ratios. They found the geomechanical behavior of bimsoil is scale-dependent; changing the ratio of height and diameter alters the shear strain band, and the associated peak strength. They also found that he slenderness effect of ail rezone formation for bimsoil was not obvious for bimsoil samples with lower RBP, but became appreciable for samples with higher RBP. The permeable properties of bimsoil have equal importance forits strength and deformation characteristics in soil and rock mechanics. The issues on the permeability of bimsoil have been deeply studied by many scholars and engineers, as the permeability of bimsoil is directly related to the stability of geological bodies [5,26–29]. Bimsoilis a kind of typical porous medium, and its flow characteristics are closely related to the content of rock blocks, soil matrix properties, random distribution of blocks, the size of the blocks, etc. One of the special characteristics of bimsoilis its sensitivity to water. To study the flow rule of bimsoil, it is important to understand the deformation and failure mechanism under a stress–flow coupled environment. To study the seepage characteristics of bimsoil, different testing methods have been used, such as the conventional seepage test (e.g., constant head laboratory test) [27,30,31], servo-controlled laboratory seepage tests [5], in-situ seepage tests [31–33], and numerical simulations [25,34]. Physical experimental approaches are essential to studying the flow behavior for geomaterials. Direct observations by means of in-situ flowtests and laboratory experiments can provide plenty of insights into the complicated flow behaviors of bimsoil. In summary, the RBP has the most remarkable influence on the flow properties of bimsoil. When the physical and mechanical properties of the soil matrix is roughly the same, adding rock blocks to the soil matrix causes the permeability coefficient of bimsoil to first increase and then decrease, with increasing RBP. The influence of rock block content on the permeability of bimsoil has been widely studied. However, the influences of flow distance on the permeability mechanisms of bimsoil (e.g., the relationship between the hydraulic gradient and seepage velocity, the link among the non-Darcy flow factor and flow distance, etc.) have not been involved up to now.

A review of the literature shows that the study of the slenderness effect on the permeability properties for bimsoil materials is very limited. In addition, the critical H/D (ratio of sample height to diameter) for bimsoil samples with different RBP has not been published. It is clear that the slenderness effect is an important aspect of the scale effect, and the study of it can reveal the flow characteristics of bimsoil along different flow distances, the relationship between the permeability coefficient and flow distance, and point out the mechanism that influences the flow distance on the non-Darcy flow rule. The basic purpose of this study is to investigate the flow slenderness effect for artificial bimsoils, with different RBP. The authors carried out a series of systematic testing on bimsoil samples with RBP of 30%, 40%, 50%, and 60%,with various slenderness ratios (i.e., H/D = 40/50, 60/50, 80/50, 100/50, 120/50, 140/50, 160/50, 180/50, and 200 mm/50 mm, respectively). The water was injected into the samples by using a self-developed servo-controlled permeability testing system. The newest test results presented here show that the permeability coefficient of bimsoil is strongly related to flow distance, and the permeability coefficient of bimsoil was almost kept constant after a critical flow distance. The critical flow distance is also different for bimsoils with different RBP. Inaddition, by introducing the Forchheimer non-Darcy flow law, the effect of flow distance on the degree of non-Darcy flow properties was first discussed in this work.

2. Experimental Methods

2.1. Experimental Setup

This testing setup was previously reported by Wang et al. [5]. Figure 1 shows the layout of the flow test system made up of rigid sample holder, the servo pressurized water-supply system, and the sample chamber system.

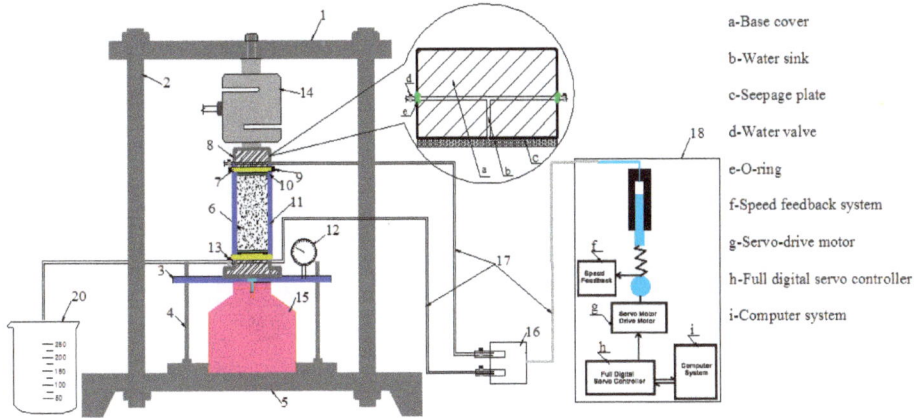

a-Base cover

b-Water sink

c-Seepage plate

d-Water valve

e-O-ring

f-Speed feedback system

g-Servo-drive motor

h-Full digital servo controller

i-Computer system

Figure 1. Schematic diagram of the flow testing system [5], which is composed of the rigid sample holder, the servo pressurized water-supply system, and the sample chamber system.

The rigid sample holder is composed of the beams, rigid column, rigid platform, guide bar, etc. Its purpose is to keep the sample chamber system steady on the platform during the flow test.

The servo pressurized water-supply system includes the main parts of the speed feedback component, servo and drive motor, full digital servo controller, and the computer. It is the core component of the overall setup. The Doli servo controller made in German is used to control the servo pressurized water-supply system, a ball screw stepping servo motor is used to drive it to work. According to the operating principle of a piston, the fluid is pressurized and supplied to the bimsoil sample. We can control the servo pressurized water supply system by computer, the injected fluid into the samples can be supplied at a constant pressure or rate. Before the flow test, water is returned to the piston from water tank, and then the piston is driven in servo controlled mode to control the injected speed of water at constant hydraulic pressure or constant flow rate, into the sample chamber.

The sample chamber system is composed of two metal seepage plates, two metal cushions (upper one and lower one), two hose clamps, and a length of heat shrink tubing accommodating the bimsoil sample. The metal permeability cushions are specially designed for the flow test; they contain the inlet valves, outlet valves, and some grooves. The diameters of the inlet and outlet are 3 mm. The heat shrinks tube and metal cushion is connected with self-adhesive type and hose clamps, purpose of the self-adhesive type is to prevent leakage, and its sealing hydraulic pressure can reach 1 MPa. The detailed dimensions and structure of the metal cushion, and locations of the inlet and outlet valve are shown in Figure 2. For samples with the same RBP of 30%, the tested sample height varies from 40 mm to 200 mm, and the diameter was kept constant at 50 mm, as shown in Figure 3.

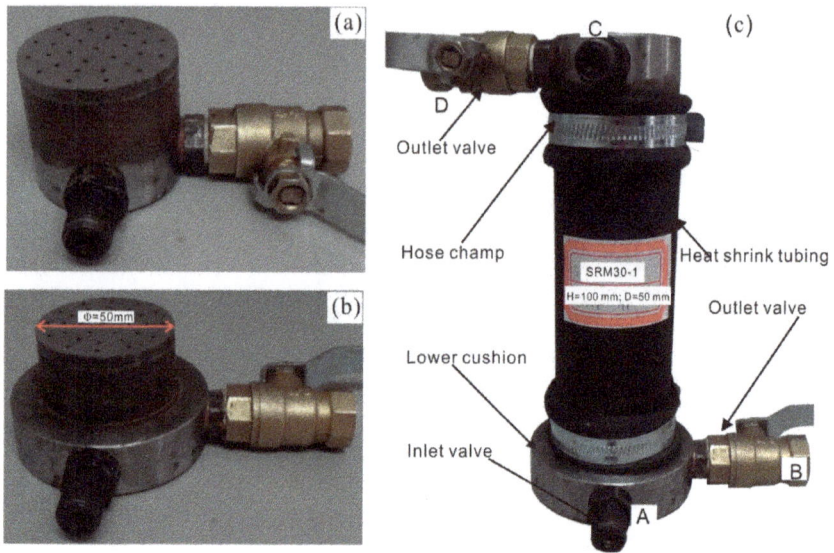

Figure 2. Photograph of the sample chamber structure and metal permeability cushions [5]: (**a**) the upper cushion; (**b**) the lower cushion; and (**c**) structure of the sample chamber.

Figure 3. The sample chamber system for bimsoil samples with different height, taking samples with RBP (rock block percentages) of 30% as an example.

2.2. Material Properties

2.2.1. Soil Matrix

The soil used in this experimental work was a kind of clay soil. The gradation curve of this soil is shown in Figure 4a. The physical and mechanical properties of the soil are summarized in Table 1. Scanning Electron Microscope (SEM) and X-Ray diffraction (XRD) tests were performed to obtain the mineral content and composition to the typical soil matrix. Many irregular and rod like quartz grains surrounded by clay minerals can be observed under SEM (scanning electron microscope) scanning, the grain size is between 0.001 and 0.003 mm, as shown in Figure 5a–b. The detailed mineral composition was acquired by XRD (X-ray diffraction) tests and listed in Table 2. From the analysis results, it can be seen that plenty of clay minerals (e.g., montmorillonite, kaolinite, and illite) exist in soil matrix.

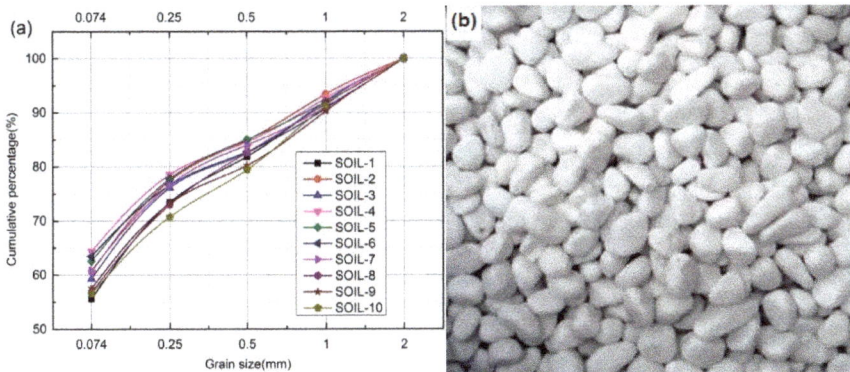

Figure 4. The soil and rock blocks used in the preparation of bimsoil samples [5]: (**a**) grain size distribution of soil matrix, with particle size larger than 0.074; and (**b**) rock blocks used in bimsoil samples, size range between 2 and 5 mm.

Table 1. Basic physical and mechanical properties of the used soil matrix and rock blocks for bimsoil samples.

Index	Soil Matrix	Rock Block
Bulk density (g/cm^3)	1.64	2.53
Dryweight density (g/cm^3)	2.06	/
Optimum water content (%)	9.5	/
Specific gravity (G$_S$)	2.73	/
Effective particle size, D10 (mm)	0.01	/
Coefficient of uniformity, Cu	4.2	/
Coefficient of curvature, Cc	1.32	/
Liquid limit (%)	64	/
plastic limit (%)	36	/
plasticity index	28	/
liquidity index	0.121	/
Wet compressive strength (MPa)	0.57	43.21
Dry compressive strength (MPa)	2.27	80.75

Note: for soil matrix, the wet state corresponds to natural state, and, for rock block, the wet state corresponds to saturation state.

Table 2. Mineralogical composition of soil sample obtained from XRD (X-ray diffraction).

Mineral	Soil Matrix 1	Soil Matrix 2
Montmorillonite	61.52	63.28
Kaolinite	26.73	24.66
Illite	6.25	6.58
chlorite	5.5	5.48

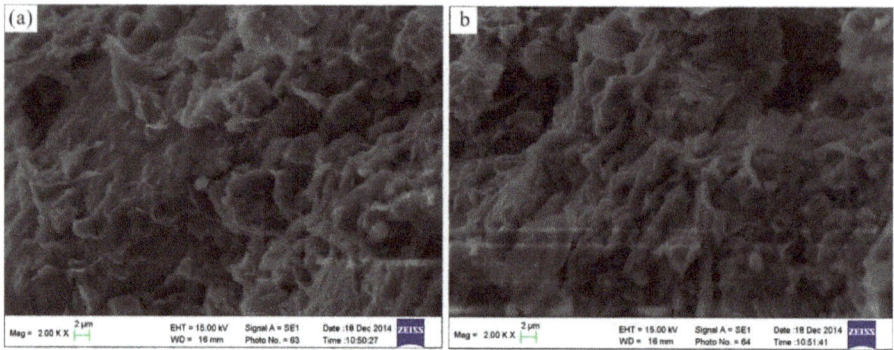

Figure 5. SEM (scanning electron microscope) results for soil Matrix 1 and Matrix 2 used for the preparation of bimsoil sample. (**a**) SEM result for matrix 1; (**b**) SEM result for matrix 2.

2.2.2. Rock Blocks

According to the geotechnical test standards [35,36] and the preparation of bimsoil samples, the threshold value for soil particle and rock block is determined as 2 mm. This is to say, when the grain size exceeds 2 mm, it is treated as block, while it is treated as soil matrix if the grain size is below 2 mm. Lithology of rock blocks used in the flow test was white marble (Figure 4b), the size of rock blocks ranges between 2 and 5 mm. The physical and mechanical properties of rock block are also listed in Table 1. Generally, the morphological characteristics of the rock block have great effect on the geomechanical properties of bimsoil. Therefore, quantitative morphological feature of the rock blocks with weighted average indices are obtained by the digital image process [3], as follows: (1) outline indices: flakiness is 0.954, elongation is 1.343, sphericity is 0.845, and shape factor is 0.943; and (2) angularity indices: angularity (Gradient Method) is 0.917, and convexity ratio is 0.902.

2.3. Remolded Sample Preparation

Because of the special geological and structural characteristics of the bimsoil, obtaining the undisturbed samples is very difficult; therefore, using the remolded sample to conduct experiments is necessary [37]. Many researchers [6,37,38] have adopted a hand mixing method to mix rock blocks uniformly within the soil matrix. To ensure homogeneity of the samples, the rock blocks were mixed by hand into the soil with 10 min. Hand mixing method can better avoid damage occurring in the soil matrix and rock blocks compared to other methods, e.g., machine mixing. Machine mixing may affect the permeability characteristics of the tested material. According to the study results of Wang et al. [5], the permeability coefficient of soil matrix can change after loading and unloading confining pressure, and soil matrix damage occurs in this case.

Then compaction tests were conducted to produce samples similar to that used for natural soil. The maximum dry unit weight and optimum water content for all tests was determined. The following procedure was adopted when preparing the samples. In the preparation of bimsoil samples, the soil was mixed with an amount of water corresponding to the optimum water content. Moist soil was kept in a closed plastic bag and allowed to cure for 24 h. All mixing was conducted by hand and proper care was taken to prepare homogenous mixtures at each stage of mixing. For the dynamic compaction, the relationship between hammer count and soil density was studied, and the appropriate optimal hammer count was finally determined. Compaction was done in a split mold by applying a dynamic pressure, using a compaction test apparatus. Owing to the high difference of elastic modulus between the soil matrix and the rock block, the compactness of the bimsoil is actually the compactness of the soil matrix. Soil matrix density is a very important factor affecting the permeability of the bimsoil [28]. As a result, how to control the hammer count at different values is crucial to the analysis result. In this work, hammer count to produce specimens with different soil density is determined from the relationship between the soil density and the optimal hammer count, as shown in Figure 6a. In Figure 6a, the density of the soil matrix in bimsoil samples with a RBP of 30–60% increases with an increase of hammer count. To roughly keep the same soil density (i.e., void ratio) in the bimsoil samples, we draw a dot dash line to intersect with the curves in Figure 6a, and the value of abscissa is determined as the optimal hammer count. In Figure 6a, the optimal hammer count was determined as 3, 4, 5, and 11 counts for bimsoil samples with RBP of 30%, 40%, 50%, and 60%, respectively. According to the value of RBP (i.e., 30%, 40%, 50% and 60%), combined with the density of soil and rock blocks, which are already known in bimsoil sample, the total volume of the prepared sample is also known in advance. Therefore, the required amount of soil and rock blocks can be calculated for bimsoil samples with a certain RBP and a certain height. The number of compaction layers can be determined according to the specific prepared sample. Taking the sample with a height of 20 mm and 100 mm as examples, we can produce the samples with one layer and three layers, respectively. For convenience and to keep the produced sample entire, only samples with length of 20 mm, 40 mm, 60 mm, 80 mm, and 100 mm were produced. If we want to obtain the sample with height of 120 mm, we grouped the samples with height of 20 mm and 100 mm together, and so on. The samples were cylinder-shaped, and all the tested samples were sealed with heat shrink tubing to prevent water volatilization.

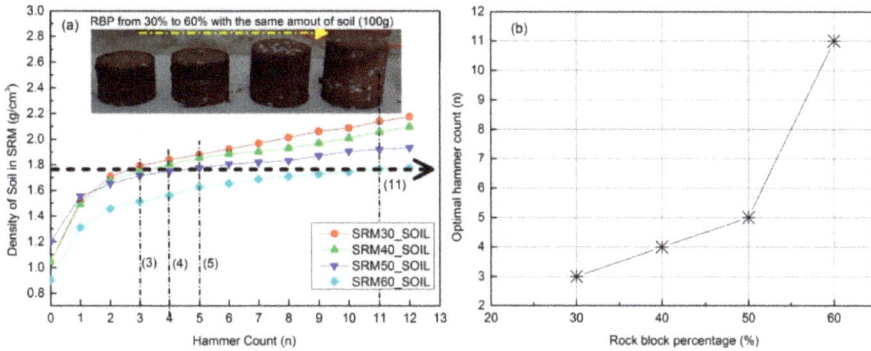

Figure 6. The methods to keep the same soil density in bimsoil samples: (**a**) relationship between the density of soil matrix and hammer count, for samples with RBP of 30–60%, and 100 g soil matrix; and (**b**) determination of the optimal hammer count for bimsoil sample.

2.4. Test Procedure

To investigate the effect of the slenderness ratio on the groundwater flow characteristics in bimsoil and to get some important conclusions from the flow test, the detailed technical flowchart is shown in Figure 7. First, the bimsoil sample was installed on the chamber system, and then the water was injected into the chamber at a constant rate until saturation of the bimsoil sample; at this moment, the seepage fluid in bimsoil reaches a steady state. When the sample reached saturation state, the hydraulic gradient was kept constant, and the flow test started. During the flow test, we monitored the variation of the hydraulic gradient and the flow water volume while collecting the corresponding experimental data. After analyzing the data, we obtained the permeability coefficient of bimsoil with a different heights with a certain RBP. The influence of the flow path on the permeability mechanism was analyzed.

The permeability coefficient of bimsoil was obtained when water flow reaches steady state. The water-outflow volume, hydraulic pressure and flow time at each of the injection steps were automatically recorded by a computer, and then we can calculate the hydraulic gradient, permeability coefficient based on Darcy law, as shown below [5,6]:

$$k = \frac{QL}{At(P_1 - P_2)} \frac{\eta_T}{\eta_{20}}$$

(1)

where Q is the total amount of water flow; A is the sample cross-section area; t is the flow time; L is flow distance (i.e., length of sample); P_1 and P_2 are the hydraulic pressure of the inlet valve and outlet valve, respectively; and η_T and η_{20} are the coefficient of water kinematic viscosity at $T\,°C$ and 20 °C, respectively.

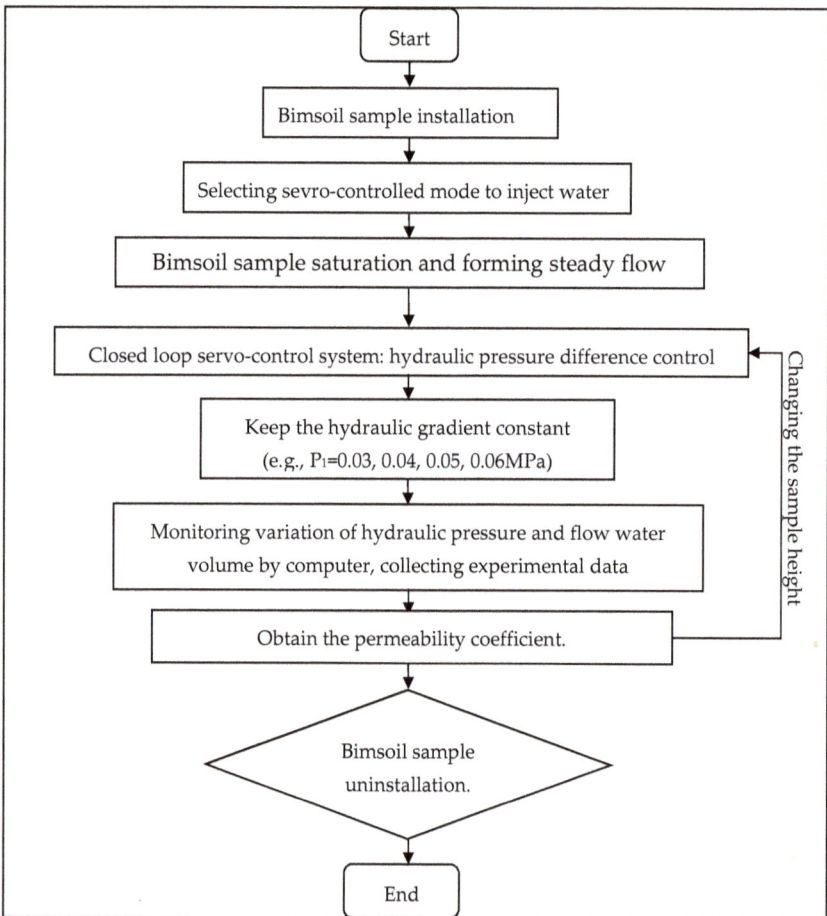

Start

Bimsoil sample installation

Selecting sevro-controlled mode to inject water

Bimsoil sample saturation and forming steady flow

Closed loop servo-control system: hydraulic pressure difference control

Keep the hydraulic gradient constant
(e.g., P_1=0.03, 0.04, 0.05, 0.06MPa)

Monitoring variation of hydraulic pressure and flow water
volume by computer, collecting experimental data

Obtain the permeability coefficient.

Bimsoil sample
uninstallation.

End

Changing the sample height

Figure 7. Technical flowchart for study on the slenderness ratio flow tests for bimsoil samples.

3. Experimental Results and Discussion

3.1. General Descriptions

The relationships between hydraulic gradient and seepage velocity for bimsoil samples with different RBPs are shown in Figure 8. As shown in Figure 8, samples with various heights were tested from 40 mm to 200 mm. It can be seen that the seepage velocity increases with increasing hydraulic gradient, and the increment rate for bimsoil samples with a RBP of 60% is the most evident. In addition, for the samples with various RBPs, the seepage velocity decreased with the increase of sample height. These results implied that the permeability coefficient of bimsoils is variational and not constant, and it depends on the hydraulic gradient. This result is consistent with the study of Wang et al. [5]: the permeability law of bimsoils does not comply with Darcy's law. With the increase of sample height, the curves tended to be stable. This indicates that the seepage field in inhomogeneous bimsoil becomes steady gradually after a certain flow distance; the flow distance is an important factor influencing the flow characteristics. Moreover, the critical sample height is different for samples with various RBPs.

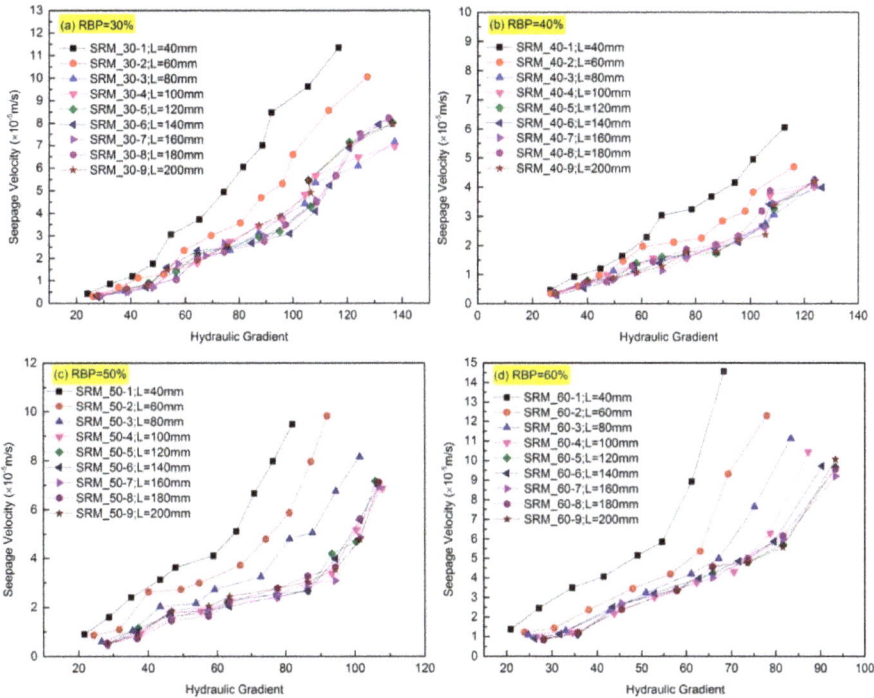

Figure 8. The relationship between hydraulic gradient and seepage velocity for samples with height from 40 mm to 200 mm: (**a–d**) the RBP is 30%, 40%, 50%, and 60%, respectively.

3.2. The Non-Darcy's Flow of Bimsoils

When water flows in bimsoils, with the increase of seepage velocity, the flow characteristic becomes non-Darcian, and the link between the seepage velocity and pressure gradient of Darcy's Law becomes nonlinear (i.e., the bimsoil permeability depends on the seepage velocity). This dependency is influenced by the randomly distributed rock blocks in the bimsoil samples. To interpret this phenomenon well, an empirical equation was proposed by Forchheimer [39] to correct for the nonlinearity of Darcy's Law. During the flow process in bimsoil, each stage of the stable value of the hydraulic gradient and seepage velocity is performed with a binomial fitting, and a modified version of Darcy's Law can be obtained as below (Wang et al. 2015a):

$$\rho C_a \frac{\partial V}{\partial t} = -\frac{\partial p}{\partial x} - \frac{\mu}{k} V + \rho \beta V^2 + f \qquad (2)$$

where ρ is the fluid mass density; C_a is the acceleration coefficient; f is the volume force of unit mass; V is seepage velocity; $\xi = \partial p/\partial x$ is pressure gradient; and coefficient β is termed the non-Darcy flow factor, m^{-1}, also known as the inertial coefficient, inertial resistance, or turbulence factor. Both β and k are regarded as material constants of the Forchheimer's equation in the range of its validity.

When the duration of the path of the fluid flow in bimsoil samples is long enough, the flow reaches stability, thus $\partial V/\partial t = 0$. Theoretical analysis shows that when ignoring the compressibility of the liquid, the pressure gradient presents a uniform distribution [39]. As a result, we rewrite the expression of the pressure difference as follows:

$$\xi = \frac{\partial p}{\partial x} = \frac{p_d}{H} = \frac{p_{base} - p_{top}}{H} \tag{3}$$

where p_{base} and p_{top} are the hydraulic pressure at the outlet and inlet of the bimsoil sample, respectively; and H is the height of the sample, which is the length of the flow path.

Neglecting the mass force, in the case the sample is not very large, the expression for Equation (3) can be written as follows:

$$\frac{p_d}{H} = -\frac{\mu}{k}V + \rho\beta V^2 \tag{4}$$

Using the experimental data above, the polynomial fitting equations for the typical specimens with different specimen height are listed in Tables 3–6. Figure 9 plots the curve fitting results of sample height from 40 mm to 200 mm, with RBP of 30%, 40%, 50%, and 60%, respectively. The correlation coefficient of all equations is good with a correlation coefficient larger than 0.9. From Equation (4), we can obtain the non-Darcy permeability coefficient and the non-Darcy flow β factor accordingly.

Table 3. The non-Darcy's flow equations for typical specimens with rock block percentage of 30%, using Forchheimer equation.

Specimen No.	$-J = -aV + bV^2$ (Equation (4))		K ($\times 10^{-5}$ m/s)	R^2
	a	b		
Bimsoil_30-1(H = 40 mm)	25.4676	57.28886	0.03966	0.975
Bimsoil_30-1(H = 60 mm)	27.3673	46.06427	0.03691	0.977
Bimsoil_30-1(H = 80 mm)	37.5564	36.6585	0.02689	0.990
Bimsoil_30-1(H = 100 mm)	37.7689	25.86399	0.02674	0.977
Bimsoil_30-1(H = 120 mm)	37.2458	25.04659	0.02712	0.980
Bimsoil_30-1(H = 140 mm)	38.612	26.41917	0.02616	0.981
Bimsoil_30-1(H = 160 mm)	38.827	27.59595	0.02601	0.979
Bimsoil_30-1(H = 180 mm)	37.0617	27.39719	0.02725	0.980
Bimsoil_30-1(H = 200 mm)	37.6457	26.13581	0.02683	0.976

Table 4. The non-Darcy's flow equations for typical specimens with rock block percentage of 40%, using Forchheimer equation.

Specimen No.	$-J = -aV + bV^2$ (Equation (4))		K ($\times 10^{-5}$ m/s)	R^2
	a	b		
Bimsoil_40-1(H = 40 mm)	33.21934	79.41739	0.0304	0.991
Bimsoil_40-1(H = 60 mm)	44.2489	54.85734	0.02283	0.957
Bimsoil_40-1(H = 80 mm)	55.9207	46.08953	0.01806	0.978
Bimsoil_40-1(H = 100 mm)	56.86151	38.04205	0.01776	0.986
Bimsoil_40-1(H = 120 mm)	56.6892	29.67069	0.01782	0.976
Bimsoil_40-1(H = 140 mm)	58.2059	30.95381	0.01735	0.990
Bimsoil_40-1(H = 160 mm)	58.0889	30.68529	0.01739	0.989
Bimsoil_40-1(H = 180 mm)	56.72037	30.81588	0.01781	0.957
Bimsoil_40-1(H = 200 mm)	57.1004	28.84796	0.01769	0.977

Table 5. The non-Darcy's flow equations for typical specimens with rock block percentage of 50%, using Forchheimer equation.

Specimen No.	$-J = -aV + bV^2$ (Equation (4))		K ($\times 10^{-5}$ m/s)	R^2
	a	b		
Bimsoil_50-1(H = 40 mm)	17.45548	48.85914	0.05786	0.996
Bimsoil_50-2(H = 60 mm)	21.70822	36.53996	0.04653	0.992
Bimsoil_50-3(H = 80 mm)	27.27899	25.15432	0.03702	0.991
Bimsoil_50-4(H = 100 mm)	33.96639	27.1969	0.02974	0.991
Bimsoil_50-5(H = 120 mm)	34.12356	26.77677	0.0296	0.992
Bimsoil_50-6(H = 140 mm)	34.85317	26.23295	0.02898	0.990
Bimsoil_50-7(H = 160 mm)	33.23728	27.15362	0.03039	0.992
Bimsoil_50-8(H = 180 mm)	34.50031	26.9832	0.02928	0.992
Bimsoil_50-9(H = 200 mm)	33.54192	26.56346	0.03011	0.993

Table 6. The non-Darcy's flow equations for typical specimens with rock block percentage of 60%, using Forchheimer equation.

Specimen No.	$-J = -aV + bV^2$ (Equation (4))		K ($\times 10^{-5}$ m/s)	R^2
	a	b		
Bimsoil_60-1(H = 40 mm)	12.5917	30.10298	0.08021	0.997
Bimsoil_60-2(H = 60 mm)	16.2290	22.59943	0.06223	0.993
Bimsoil_60-3(H = 80 mm)	18.3969	17.88247	0.0549	0.988
Bimsoil_60-4(H = 100 mm)	19.2867	17.64004	0.05237	0.990
Bimsoil_60-5(H = 120 mm)	21.6305	17.5256	0.04669	0.987
Bimsoil_60-6(H = 140 mm)	20.9707	17.21499	0.04816	0.989
Bimsoil_60-7(H = 160 mm)	21.9001	16.46893	0.04612	0.988

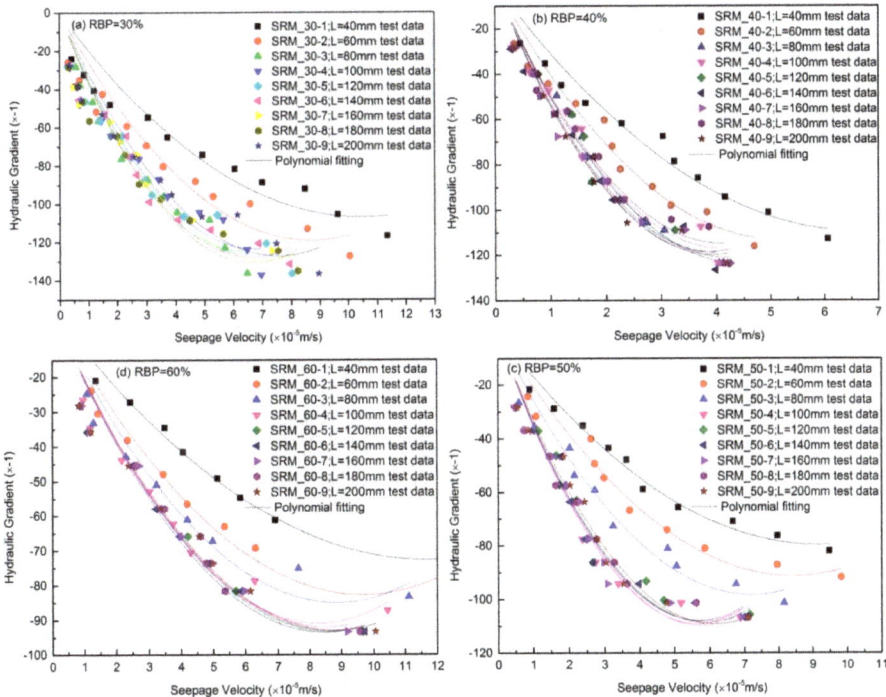

Figure 9. The curves of negative hydraulic gradient and seepage velocity for samples with height from 40 m to 200 mm: (a–d) the RBP is 30%, 40%, 50%, and 60%, respectively.

3.3. Slenderness Effect on Flow Characteristics

From Equation (4), we can obtain the non-Darcy permeability coefficient and the non-Darcy flow β factor. The plots of the permeability coefficient against sample height, for samples with different RBP, are shown in Figure 10. It can be seen that there exists an inflection on the curves; the permeability coefficient is prone to a stable value as the height increases. For the samples with RBP of 30% and 40%, the critical height is 80 mm; for the samples with RBP of 50%, the critical height is 100 mm; and for the samples with RBP of 60%, the critical height is 120 mm. These results indicate that when greater than the critical height, the seepage field in bimsoil tends to be stable. According to the results of Wang et al. [5], they conducted the flow tests five times for the bimsoil sample with the same RBP, and the results show that the permeability coefficient of the sample with the same RBP is almost same when using roughly the same morphology of the rock block in bimsoil samples. Therefore, distribution of rock blocks in samples with the same height may not be a primary factor influencing the flow characteristics. In this work, the tested samples have different height, the length of the sample may affect the distribution of rock blocks, which may further affect the heterogeneity and the associated permeability coefficient. The experimental results further imply that the flow distance is a crucial factor controlling the flow characteristics of bimsoil. We use a quadratic polynomial to fit the relationship between critical height and RBP, as shown in Equation (5). To eliminate the influence of the slenderness effect on the result of the flow test for bimsoil with various BRPs, using the equation fitting method to study the relationship between RBP and the critical sample height, it can be obtained from the equation as below (Figure 11):

$$C_H = 0.05 \times RBP^2 - 3.1 \times RBP + 127 \ (x > 25; D = 50 \text{ mm}) \ R^2 = 0.945 \tag{5}$$

where C_H is the critical sample height, and RBP is the rock block content of the bimsoil sample.

In this work, the slenderness effect has been studied by laboratory tests for samples with the same diameter but different heights. The change of the permeability coefficient with the sample slenderness ratio (H/D) is shown in Figure 12. It is noted that the permeability coefficient nonlinearly decreased with the H/D in an exponential fit, and the regression equations were listed in Table 7, which can be expressed as:

$$K = c + b \exp[a(H/D)] \tag{6}$$

where a, b, and c are the coefficients related to the RBP of bimsoil.

Figure 13 shows plots of the non-Darcy flow factor versus sample height. The degree of non-Darcy flow decreases with the increase of sample height. This result indicates that with the increase of flow distance, the non-Darcy degree becomes weaker, and the effect of the rock–soil interface flow becomes stable. Content of the rock blocks in bimsoil samples controls the orientation and tortuosity degree of flow and the associated non-Darcy degree. With the increase of flow distance, the incidence of rock blocks also decreases. In Figure 13, it can also be seen that the non-Darcy factor of bimsoil samples with RBP of 40% is larger than these samples with the RBP of 30%, 50%, and 60%. The non-Darcy flow characteristics for samples with RBP of 40% is very obvious, which indicates that the interaction between rock blocks and the soil matrix is great. From the results of Wang et al. [5], with the increase of the RBP, the average permeability coefficient decreases to a minimum at a RBP of 40%.As the RBP value continues to increase above 40%, the permeability increases again. The variation of permeability for bimsoil samples is a result of soil matrix properties combined with rock blocks and rock–soil interfaces. The results in this work further prove this phenomenon; for sample with RBP of 40%, interaction between soil matrix and rock blocks is stronger when the water flows in bimsoil.

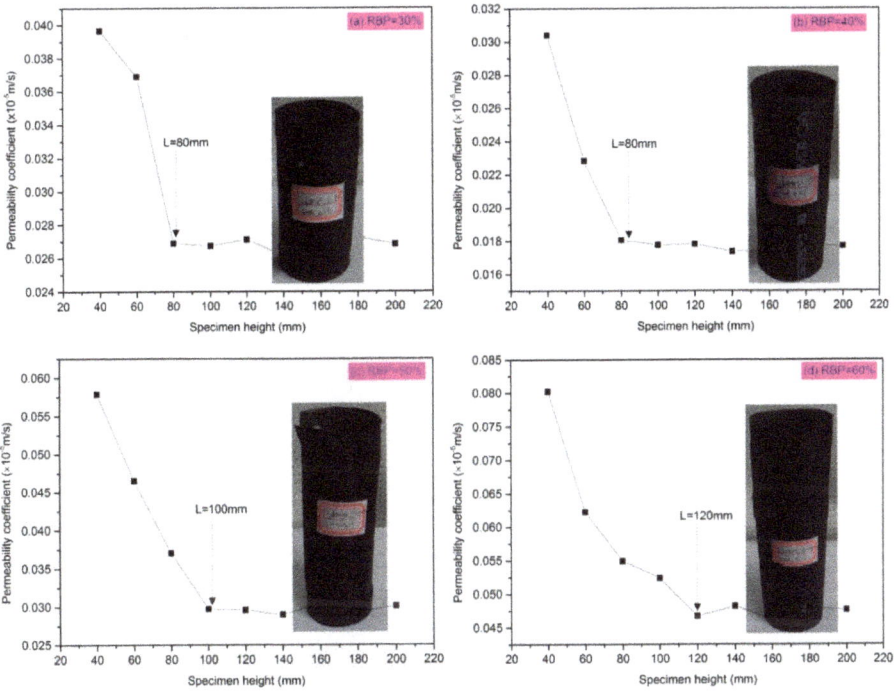

Figure 10. The plots of permeability coefficient against sample height for typical bimsoil samples:(**a–d**) the RBP is 30%, 40%, 50%, and 60%, respectively.

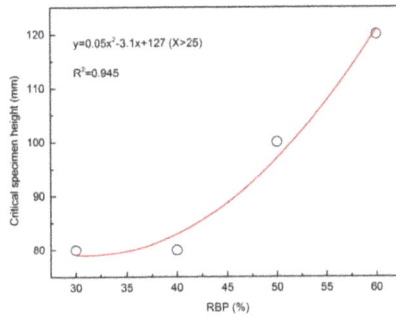

Figure 11. The relationship between critical sample height and RBP.

Table 7. Fitting regression functions of permeability coefficient with slenderness ratios (H/D). The R^2 is the correlation coefficient.

RBP (%)	Regression Function of Permeability Coefficient ($k \times 10^{-5}$)	R^2
30	$K = 0.02683 + 0.13314e^{-2.92448(H/D)}$	0.8455
40	$K = 0.01769 + 0.11061e^{-2.70369(H/D)}$	0.9831
50	$K = 0.03007 + 0.14022e^{-2.02304(H/D)}$	0.9663
60	$K = 0.04756 + 0.18436e^{-2.16395(H/D)}$	0.9862

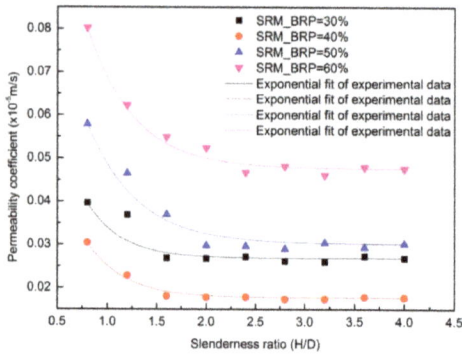

Figure 12. Evolution diagram of the permeability for samples with different slenderness ratio (H/D).

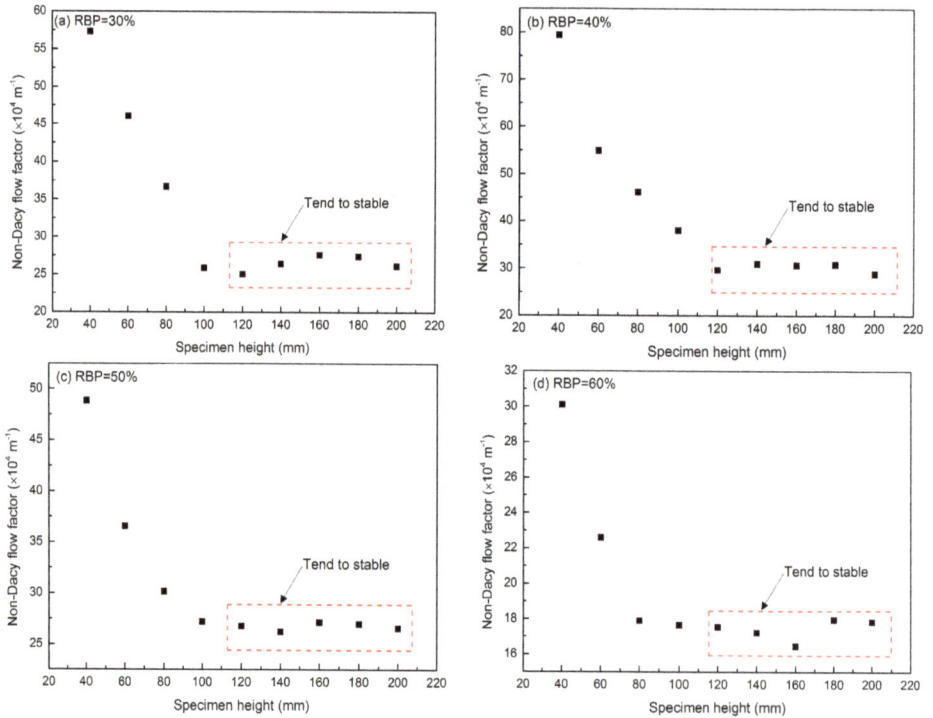

Figure 13. The plots of non-Darcy flow factor versus sample height against height for samples with RBPof30%, 40%, 50%, and 60%, respectively.

3.4. Discussions

The scale effect is a widespread phenomenon for various geomaterials (e.g., soil, rock, etc.), especially for the extreme soil and rock mixtures. Plenty of studies about the slenderness effect of bimsoil are mainly focused on the strength and deformation properties, and they think that the block size and distribution results in the scale effect of bimsoil. In our work, the slenderness effect also exists for the flow characteristic of bimsoil. We speculated that the reason may result from the seepage tortuosity along the flow direction. When water transports in bimsoil, the soil matrix combined with rock blocks and rock–soil interfaces simultaneously controls the flow characteristics. Flow tortuosity alters the seepage field and flow direction, and the water is prone to flowing along the path with least tortuosity, and the sample with larger RBP has the most tortuous flow path, therefore, the critical flow distance is larger than sample with low RBP. Specifically, the flow path tends to be along the direction with the least tortuosity, and the sample with more slender height, provides more possible flow paths along the flow direction. Therefore, as the sample slenderness increases, the tortuosity is reduced and the non-Darcy flow characteristic becomes not asobvious.

4. Conclusions

In this work, we focus on the slenderness effect of artificial bimsoil under unconfined flow test using a self-developed servo-controlled permeability testing system. Up to now, however, few reports have been published about this issue. The flow slenderness effect was studied according to the relationships among the seepage velocity, permeability coefficient, non-Darcy flow factor, and the critical sample height. Some main conclusions can be drawn as below:

(1) Flow distance is a key factor influencing the non-Darcy flow characteristics of bimsoil. With the increase flow distance, the seepage field in bimsoil tends to stabilize, and the seepage velocity is prone to be a constant value at the same hydraulic gradient. The permeability of bimsoil is influenced by the soil matrix properties combined with rock blocks and rock–soil interfaces.

(2) The permeability coefficient of bimsoil samples with different slenderness ratios is different. The value of the permeability coefficient decreases with the increase of flow distance. At the same sample diameter, the critical height varies for samples with different RBP, and its value increases with increasing RBP. The permeability coefficient varies in a monotonously decreasing nonlinear correlation with the sample slenderness. The slenderness effect for high RBP scenario is much more obvious.

(3) The degree of non-Darcy flow in bimsoil decreases with the increase of sample height. The flow path is prone to be along the channel with the least tortuosity, and a more slender sample can provide more possible flow paths along the flow direction.

Acknowledgments: The authors would like to thank the editors and the anonymous reviewers for their helpful and constructive comments. This work was supported by the Beijing Natural Science Foundation of China (Grants No. 8164070), the Fundamental Research Funds for the Central Universities, and National Natural Science Foundation of China (Grant No. 41502294).

Author Contributions: Y. Wang and C. Li designed the theoretical framework; Y. Wang conducted the experiment, analyzed the experimental date and wrote the manuscript; and X. Wei and Z. Hou corrected the figures.

Conflicts of Interest: The authors declare no conflict of interest.

References

1. Coli, N.; Berry, P.; Boldini, D. In situ non-conventional shear tests for the mechanical characterisation of a bimrock. *Int. J. Rock Mech. Min.* **2011**, *48*, 95–102. [CrossRef]
2. Wang, Y.; Li, X. Experimental study on cracking damage characteristics of a soil and rock mixture by UPV testing. *Bull. Eng. Geol. Environ.* **2015**, *74*, 775–788. [CrossRef]
3. Wang, Y.; Li, X.; Wu, Y.F.; Lin, C.; Zhang, B. Experimental study on meso-damage cracking characteristics of RSA by CT test. *Environ. Earth Sci.* **2015**, *73*, 5545–5558. [CrossRef]
4. Wang, Y.; Li, X.; Zheng, B.; Zhang, B.; Wang, J.B. Real-time ultrasonic experiments and mechanical properties of soil and rock mixture during triaxial deformation. *Géotech. Lett.* **2015**, *5*, 281–286. [CrossRef]
5. Wang, Y.; Li, X.; Zheng, B.; Zhang, Y.X.; Li, G.F. Experimental study on the non-Darcy flow characteristics of soil–rock mixture. *Environ. Earth Sci.* **2016**, *75*, 1–18. [CrossRef]
6. Wang, Y.; Li, X.; Zheng, B.; Li, S.D.; Duan, Y.T. A laboratory study of the effect of confining pressure on permeable property in soil-rock mixture. *Environ. Earth Sci.* **2016**, *75*, 1–16. [CrossRef]
7. Wang, Y.; Li, X.; Zheng, B.; Ma, C.F. An Experimental Investigation of the Flow–Stress Coupling Characteristics of Soil–Rock Mixture Under Compression. *Transp. Porous Med.* **2016**, *112*, 429–450. [CrossRef]

8. Wang, Y.; Li, X.; Zheng, B.; He, J.M.; Li, S.D. Macro–meso failure mechanism of soil–rock mixture at medium strain rates. *Géotech. Lett.* **2016**, *6*, 28–33. [CrossRef]
9. Zhang, S.; Tang, H.M.; Zhan, H.B.; Lei, G.P.; Cheng, H. Investigation of scale effect of numerical unconfined compression strengths of virtual colluvial–deluvial soil–rock mixture. *Int. J. Rock Mech. Min.* **2015**, *77*, 208–219. [CrossRef]
10. Lindquist, E.S. The Strength and Deformation Properties of Melange. Ph.D. Thesis, Department of Civil Engineering, University of California, Berkeley, CA, USA, June 1994.
11. Goodman, R.E.; Ahlgren, C.S. Evaluating safety of concrete gravity dam on weak rock. *J. Geotech. Geoenviron. Eng.* **2000**, *126*, 429–442. [CrossRef]
12. Medley, E.; Lindquist, E.S. The engineering significance of the scale-independence of some Franciscan Melanges in California, USA. In *Proceedings of the 35th US Rock Mechanics Symposium, Reno, NV, USA, 5–7 June 1995*; Daemen, J.K., Schultz, R.A., Eds.; Balkema: Rotterdam, The Netherlands, 1995; pp. 907–914.
13. Medley, E.W. Systematic characterization of mélange bimrocks and other chaotic soil/rock mixtures. *Felsbau-Rock Soil Eng.* **1999**, *17*, 152–162.
14. Medley, E.W. Observations on tortuous failure surfaces in Bimrocks. *Felsbau Rock Soil Eng.* **2004**, *5*, 35–43.
15. Medley, E.W.; Wakabayashi, J. Geological characterization of mélange for practitioners. *Felsbau Rock Soil Eng. J. Eng. Geol. Geomech. Tunn.* **2004**, *2*, 10–18.
16. Sonmez, H.; Kasapoglu, K.E.; Coskun, A.; Tunusluoglu, C.; Medley, E.W.; Zimmerman, R.W. A conceptional empirical approach for the overall strength of unwelded bimrocks. In *Rock Engineering in Difficult Ground Conditions, Karst and Soft Rock, Proceeding of the International Society Rock Mechanics Regional Symposium, Dubrovnik, Croatia, 29–31 October 2009*; CRC Press: Leiden, The Netherlands, 2009; pp. 29–31.
17. Coli, N.; Boldin, I.D.; Bandini, A.; Lopes, D.S. Modeling of complex geological rock mixtures under triaxial testing conditions. In Proceedings of the International Symposium on Rock Engineering & Technology for Sustainable Underground Construction (Eurock), Stockholm, Sweden, 28–30 May 2012.
18. Kahraman, S.; Alber, M. Estimating unconned compressive strength and elastic modulus of a fault breccias mixture of weak block sand strong matrix. *Int. J. Rock Mech. Min. Sci.* **2006**, *43*, 1277–1287. [CrossRef]
19. Kahraman, S.; Albe, R.M. Triaxial strength of a fault breccias of weak rocks in a strong matrix. *Bull. Eng. Geol. Environ.* **2008**, *67*, 435–441. [CrossRef]
20. Afifipour, M.; Moarefvand, P. Failure patterns of geomaterials with block-in-matrix texture: Experimental and numerical evaluation. *Arab. J. Geosci.* **2014**, *7*, 2781–2792. [CrossRef]
21. Medley, E.W. Uncertainty in estimates of block volumetric proportions in melange bimrock. In *Proceedings of the International Symposium, 22nd Engineering Geology and the Environment, Athens, Greece, 23–27 June 1997*; Marinos, P.G., Koukis, G.C., Tsiambaos, G.C., Stournaras, G.C., Eds.; Balkema: Rotterdam, The Netherlands, 1997; pp. 267–272.
22. Li, X.; Liao, Q.L.; He, J.M. In-situ tests and astochastic structural model of rock and soil aggregate in the three Gorges Reservoir Area, China. *Int. J. Rock Mech. Min. Sci.* **2004**, *41*, 702–727. [CrossRef]
23. Akram, M.S. Physical and Numerical Investigation of Conglomeratic Rocks. Ph.D. Thesis, University of New South Wales, Sydney, Australia, June 2010.
24. Bagnold, R.A.; Barndorff-Nielsen, O. The pattern of natural size distribution. *Sedimentology* **1980**, *27*, 199–207. [CrossRef]
25. Xu, W.J.; Wang, Y.G. Meso-structural permeability of S-RM based on numerical tests. *Chin. J. Geotech. Eng.* **2010**, *32*, 543–550. (In Chinese)
26. Shelley, T.L.; Daniel, D.E. Effect of gravel on hydraulic conductivity of compacted soil liners. *J. Geotech. Eng. ASCE* **1993**, *119*, 54–68. [CrossRef]
27. Shafiee, A. Permeability of compacted granule-clay mixtures. *Eng. Geol.* **2008**, *97*, 199–208. [CrossRef]
28. Vallejo, L.E.; Zhou, Y. The mechanical properties of simulated soil–rock mixtures. In Proceedings of the 13th International Conference on Soil Mechanics and Foundation Engineering, New Delhi, India, 5–10 January 1994; pp. 365–368.
29. Gutierrez, J.J.; Vallejo, L.E. Laboratory Experiments on the Hydraulic Conductivity of Sands with Dispersed Rock Particles. *Geotech. Geol. Eng.* **2013**, *31*, 1405–1410. [CrossRef]
30. Chen, X.B.; Li, Z.Y.; Zhang, J.S. Effect of granite gravel content on improved granular mixtures as railway subgrade fillings. *J. Cent. South Univ.* **2014**, *21*, 3361–3369. (In Chinese) [CrossRef]

31. Chen, Z.H.; Chen, S.J.; Chen, J.; Sheng, Q.; Yan, H.; Hu, W. In-situ Double-Ring Infiltration Test of Soil-Rock Mixture. *J. Yangtze River Sci. Res. Inst.* **2012**, *29*, 52–56. (In Chinese)

32. Gao, Q.; Liu, Z.H.; Li, X.; Li, J.H. Permeability Characteristics of Rock and Soil Aggregate of Backfilling Open-Pit And Particle Element Numerical Analysis. *Chin. J. Rock Mech. Eng.* **2009**, *28*, 2342–2348. (In Chinese)

33. Wu, J.H.; Chen, J.H.; Lu, C.W. Investigation of the Hsien-du-Shan rock avalanche caused by typhoon Morakot in 2009 at Kaohsiung county, Taiwan. *Int. J. Rock Mech. Min.* **2013**, *60*, 148–159. [CrossRef]

34. Liao, Q.L. Geological Origin and Structure Model of Rock and Soil Aggregate and Study on Its Mechanical and MH Coupled Properties. Ph.D. Thesis, Institute of Geology and Geophysics, Chinese Academy of Science, Beijing, China, July 2004. (In Chinese)

35. *British Standard Methods of Test for Soils for Civil Engineering Purposes-Part 1: General Requirements and Sample Preparation*; BS1377-1; BSI: London, UK, 1990.

36. *Standard for Soil Test Method, GB/T 50123-1999*; Ministry of Water Resources of the People's Republic of China: Beijing, China, 1999.

37. Donaghe, R.T.; Torrey, V.H. Proposed New Standard Test Method For Laboratory Compaction Testing of Soil-Rock Mixtures Using Standard Effort. *Geotech. Test. J.* **1994**, *3*, 387–392.

38. Rücknagel, J.; Götze, P.; Hofmann, B.; Christen, O.; Marschall, K. The influence of soil gravel content on compaction behaviour and pre-compression stress. *Geoderma* **2013**, *209–210*, 226–232. [CrossRef]

39. Forchheimer, P. Wasserbewegung durch Boden. *Zeitz. Ver. Dtsch Ing.* **1901**, *45*, 1782–1788.

water

MDPI

Article

A New Approach to Quantify Shallow Water Hydrologic Exchanges in a Large Regulated River Reach

Tian Zhou [1], Maoyi Huang [1,*], Jie Bao [1], Zhangshuan Hou [1], Evan Arntzen [1], Robert Mackley [2], Alex Crump [1,3], Amy E. Goldman [1], Xuehang Song [1], Yi Xu [1,4] and John Zachara [1]

[1] Pacific Northwest National Laboratory, Richland, WA 99352, USA; tian.zhou@pnnl.gov (T.Z.); jie.bao@pnnl.gov (J.B.); zhangshuan.hou@pnnl.gov (Z.H.); evan.arntzen@pnnl.gov (E.A.); acrump@uidaho.edu (A.C.); amy.goldman@pnnl.gov (A.E.G.); xuehang.song@pnnl.gov (X.S.); xuyee02@163.com (Y.X.); john.zachara@pnnl.gov (J.Z.)
[2] CH2M HILL Plateau Remediation Company, Richland, WA 99352, USA; Rob_D_Mackley@rl.gov
[3] Department of Soil and Water Systems, University of Idaho, Moscow, ID 84844, USA
[4] School of Geography Science, Nanjing Normal University, Nanjing 210023, China
* Correspondence: maoyi.huang@pnnl.gov

Received: 10 May 2017; Accepted: 11 September 2017; Published: 15 September 2017

Abstract: Hydrologic exchange is a crucial component of the water cycle. The strength of the exchange directly affects the biogeochemical and ecological processes that occur in the hyporheic zone and aquifer from micro to reach scales. Hydrologic exchange fluxes (HEFs) can be quantified using many field measurement approaches, however, in a relatively large river (scale > 10^3 m), these approaches are limited by site accessibility, the difficulty of performing representative sampling, and the complexity of geomorphologic features and subsurface properties. In rivers regulated by hydroelectric dams, quantifying HEF rates becomes more challenging because of frequent hydropeaking events, featuring hourly to daily variations in flow and river stages created by dam operations. In this study, we developed and validated a new approach based on field measurements to estimate shallow water HEF rates across the river bed along the shoreline of the Columbia River, USA. Vertical thermal profiles measured by self-recording thermistors were combined with time series of hydraulic gradients derived from river stages and inland water levels to estimate the HEF rates. The results suggest that the HEF rates had high spatial and temporal heterogeneities over the riverbed, with predicted flux rates varied from $+1 \times 10^{-6}$ m s^{-1} to -1.5×10^{-6} m s^{-1} under different flow conditions.

Keywords: hydrologic exchange; SW–GW interaction; field measurements; Columbia River

1. Introduction

1.1. Hydrologic Exchange and River Regulations

Hydrologic exchange is a concept introduced by Harvey and Gooseff [1] which combines surface water–groundwater interaction processes along river corridors at multiple spatiotemporal scales, including hyporheic exchange, bank storage, and regional groundwater discharge and recharge. River water interacts with subsurface water through the hydrologic exchange fluxes (HEFs), which facilitate the nutrient and carbon cycling, organic biodegradation, fish spawning, metal transport, and other key biogeochemical and hydroecological processes in the subsurface region of the river corridor [2–7].

Hydrologic exchange dynamics, including the direction, path, magnitude, and residence time of the HEFs, define where and when these aforementioned processes occur in the river bed and bank [8–10]. The HEF dynamics are controlled by channel geometry, catchment geology, and hydrology in both space and time [11–13], and the HEF rates are essentially governed by the permeability of the sediments and local hydraulic gradients. While the overall permeability of the river bed sediment is a physical property that is relatively stable (if not considering the sediment clogging/colmation at the river bed surface), the hydraulic gradient across the river bed is a time-varying variable that is highly dependent on the river stage variations. In natural rivers, the river stage reflects the hydrometeorological and drainage characteristics in the upstream watershed. However, in a dam-regulated river, variations in stage are dramatically altered by upstream dam operations on a wide spectrum of timescales from hourly to monthly [14], and these variations extend to a few hundred kilometers downstream of the dam [15]. Globally, over 30,000 large dams were built in the past 50 years, which considerably affected terrestrial water resources [16–18] and thus changed the HEF directions and rates as well as the biogeochemical processes in downstream reaches [19]. A number of studies have investigated HEFs in regulated river systems based on measurements or modeling at either the point [15] or transect scale [20–22]. However, monitoring HEFs in a large regulated river reach (>10³ m spatial scale) with high spatiotemporal resolutions remain less explored.

1.2. Approaches to Measuring HEF Rates

The challenges of monitoring the HEFs in a regulated large river reach arise from the size of the domain and dam-induced rapid hydraulic gradient variations that complicate the system dynamics. Current field measurement methods used in a large river system usually include mass balance approaches that integrate the groundwater discharge or river loss over the entire reach. These approaches include longitudinal flow gauging, longitudinal chemistry sampling, or hydraulic head measurements [23]. Alternatively, interpolation over a number of intensively measured points (<1 m spatial scale) is a widely used approach to establish spatial distributions across the river bed for a given river reach. This type of measurement can be categorized into three groups: direct measurements, indirect measurements based on Darcy's Law, and indirect measurements based on heat transport equations [24]. Direct measurements monitor the water volume flow across the surface water–groundwater interface using seepage meters, which are chambers with the bottom open to the sediment [25–28]. This type of measurement can only estimate the flux at the interface, and the seepage meters themselves disturb the surrounding flow field and therefore affect the flux measurements.

The method based on Darcy's Law estimates HEF rates by solving Darcy's equation:

$$q = -K\frac{dh}{dl} \tag{1}$$

where q is the flux, K is the hydraulic conductivity of the sediment, and dh/dl is the hydraulic gradient. K can be estimated from various experiments such as pumping tests, slug tests, and permeameter tests, while dh/dl can be derived from water level measurements in piezometers at different depths in the riverbed (for vertical flux) or at different locations at the same level beneath the riverbed (for horizontal flux) [29]. The concept of the Darcy's Law-based method is simple and equipment installation is straightforward, which makes this method popular in some small-scale applications. However, uncertainties could be introduced in K estimations due to the heterogeneity and variations of the hydraulic conductivity; and the approach may not be suitable for large river systems because of its intensive labor requirements [24].

Methods that use heat as a naturally occurring tracer to estimate HEFs have been developed, widely adopted, and improved since the 1960s [4]. The underlying concept of such methods is that the temperature distribution in the subsurface zone is determined by the surface water temperature with strong diurnal fluctuations and relatively stable temperature in deep groundwater. Heat transports in the subsurface region through thermal convection and conduction processes, which modifies the distribution and dynamics of temperature following the 3D heat transport equation. The one-dimensional thermal transport equation along the vertical direction can be written as:

$$\frac{\partial T}{\partial t} = -q_z \frac{\rho_w c_w}{\rho c} \frac{\partial T}{\partial z} + D \frac{\partial^2 T}{\partial z^2}$$ (2)

where T is temperature; t is time; q_z is the vertical flux; z is the depth in sediment; ρ_w and ρ are the density of water and water-sediment matrix; c_w and c are the specific heat of water and river-sediment matrix, respectively; and D is the thermal diffusion coefficient. A number of analytical solutions to the heat transport equation have been provided and applied successfully in specific cases by assuming the system is in steady state with vertical fluxes only, in a homogeneous semi-infinite sediment domain [30–32]. Transient state HEF rate can be estimated by applying numerical modeling tools (e.g., VS2DH [33]) based on the heat transport equations, with thermal boundary conditions and initial states defined at the river bed and in deep sediment [34–36]. Another method for transient state estimation is analyzing the amplitudes and phase shift of the damping diurnal temperature signals at different depths in the sediment [37–40]). Computational codes, such as Ex-Stream and VFLUX, were also developed to facilitate the application of these models [41,42]. One of the great advantages of these approaches is that temperature is a reliable tracer and can be relatively easily measured at different depths using one single measuring rod installed in the river bed compared to a piezometer nest for the water head measurements. However, it might be a challenge to apply these approaches in a highly regulated river system, because the temperature signals might be significantly dampened or distorted by the fluxes with rapidly changing directions and magnitudes.

1.3. Study Objectives

In this study, we aimed to synthesize the HEF measuring methods reviewed above and develop an approach that is suitable for a large river with highly regulated discharge. Such an approach has to fulfill the following criteria: (1) using less, easily accessible data to infer HEF rates in a relatively large domain; and (2) providing continuous HEF rates at high temporal resolutions (e.g., daily or sub-daily). Note that the river bed in our study reach has small slopes at both longitudinal (1/4000) and lateral (1/40) directions, so that the major HEF direction that penetrates the river bed is vertical, which is used to represent the total HEF rate in our analysis. By revisiting monitoring data from previous studies in our study reach [15,43], we developed and validated our approach by empirically relating the point vertical HEF rates with river hydrologic conditions and inland water table levels. Such an approach could provide point HEF rates for large scale groundwater modeling evaluations [44,45], and provide guidance for the field studies to identify strong HEF hotspots and active biogeochemical reaction areas at the river bed [46]. Although hydrologic exchanges occur almost everywhere at the river bed, numerical simulations and field measurements indicated that the exchange fluxes in the shallow water near the river banks were stronger than those in the center of the channel because of the greater pressure gradient between the inland water table and the river stage [21,47]. Therefore, in this study, we only focus on the shallow water area near the bank along the river reach.

To summarize, the objectives of this study are two-fold: (1) to demonstrate a new approach that combines field measurement data and regression analysis to infer long term river bed HEF rates in a highly dynamic river reach; and (2) to examine the spatial and temporal distributions of HEFs in the shallow water area along the river reach.

2. Method

2.1. Study Reach

Our study reach is a 5 km long, 800 m wide, nearly straight river segment of the Columbia River near the 300A Area of the U.S. Department of Energy Hanford Site in southeastern Washington State (Figure 1). The Columbia River flows from north to south through the study reach over an unconfined aquifer on top of the impermeable Columbia River Basalt Group. The total thickness of the aquifer ranges from 40 m to 60 m in this area, with a highly permeable Pleistocene flood gravel layer of the Hanford Formation on top and a more consolidated and less permeable Ringold Formation at the bottom. At the top of the river bed, there is a thin layer of alluvium with varying thicknesses ranging from one to three meters [48]. Measurements used in this study were all collected from sensors installed in the alluvial layer. Several sandbar islands were deposited in the center of the river, forming a deep primary channel with a meandering thalweg on one side of the islands and a relatively shallower secondary channel on the opposite side of the islands (Figure 1). Priest Rapids Dam, a hydroelectric dam located ~80 km upstream of the study reach, controls the stage for hydropower generation. Based on historical records over the past 40 years, the range of stage spans 105 m to 109 m based on the North American Vertical Datum of 1988 (NAD88), with a maximum daily fluctuation of 2 m. In this study, the river stage observations were recorded by a pressure transducer logger (SWS-1) installed inside our study domain since the year 2001. Nearly 100 monitoring wells were constructed over the past 40 years around the 300A Area to record the inland water levels and solute concentrations for environmental monitoring purposes [49]. In this study, one monitoring well (Well 2–3), about 150 m from the river, was selected as the reference point for the inland water level. Here we define "inland" as a contrast of the "river", where the water table does not show sub-daily variations in response to the dam regulations. According to the Koppen climate classification, our study area is located in a semi-arid—desert catchment in the Columbia River basin. The annual precipitation in this region is less than 200 mm, and the evapotranspiration to precipitation ratio is about 1–1.09 [50]. Under such conditions, recharge to the groundwater is close to ~2 mm year^{-1} [51] and precipitation over the river surface can be ignored compared to the volume of water in the river channel. Therefore, the fluctuations observed in the river stage time series are mainly caused by the dam operations.

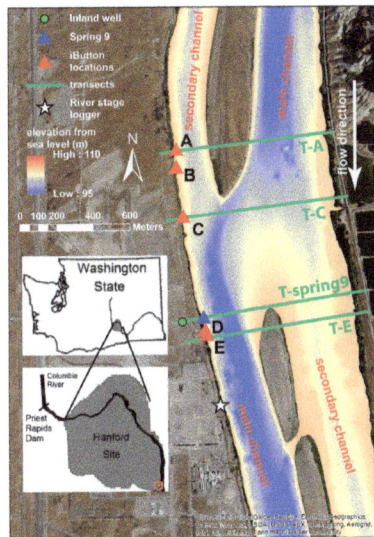

Figure 1. The study river reach. The iButton sites A–E are marked as red triangles. The Spring 9 location is marked as a blue triangle.

2.2. Data Revisiting

Given that the pressure gradient is the major driving factor of the HEF, we hypothesize that the HEF rates could be empirically related to the head difference between the river stage and the inland water table. We tested this hypothesis by revisiting a published data set [43]. The data was collected at half-hourly time steps at Spring 9 (Figures 1 and 2), a site inside our study domain, from August 2004 to June 2005. It included temperature time series for surface water and a location 19 cm deep in the riverbed (Figure 3A) as well as a vertical flux time series at the riverbed (Figure 3B) derived using Darcy's Law from measured Vertical Hydraulic Gradient (VHG, by piezometers) and hydraulic conductivity estimated by slug tests at this location. We also extracted the inland water level from the monitoring well mentioned above and the corresponding river stage data for the same time frame (Figure 3C). Strong linear correlations were detected between head difference and the Darcy's Law-based vertical fluxes at both half-hourly ($r = 0.971$) and daily ($r = 0.979$) time steps (Figure 4). The scatter plots also showed that the fitted linear model based on the daily time series can be used to describe a similar relationship at the half-hourly time scale with high goodness-of-fit values. This indicates that the linear relation based on daily fluxes and the corresponding head difference can be used to estimate sub-daily fluxes if the continuous sub-daily head difference data are available. Further statistical analyses indicated that at the daily time scale, the fitted sum square of HEF rates based on head difference is 2.61×10^{-10}, with a residual sum square of 1.28×10^{-11}, and R^2 of 0.958. The corresponding F-test yields a nearly zero p-value of 2.58×10^{-202}, rejecting the null hypothesis that the relationship is insignificant, which confirms the significant linear relationship. This finding inspired the method we developed for HEF rate estimations (described below).

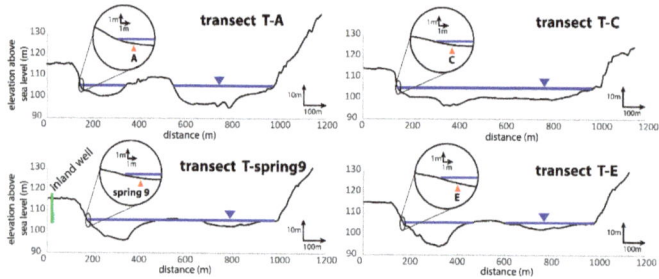

Figure 2. Transect profiles at T-A, T-C, T-spring9, and T-E at 1:10 aspect ratio (vertical : horizontal). The insets of the plots showed the locations of the measurements in zoomed in 1:1 aspect ratio plots.

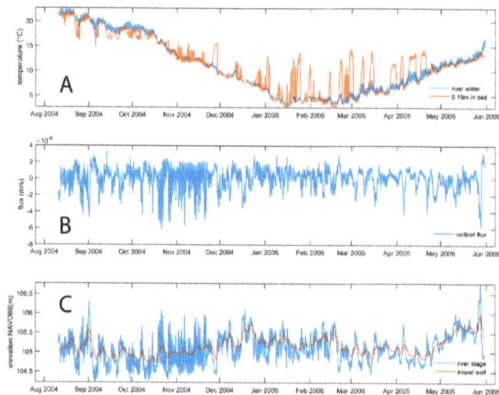

Figure 3. Temperature data collected from Spring 9 (**A**); estimated cross-bed flux rates based on piezometer data and Darcy's Law (**B**); and river stages and associated inland well water levels during the same period (**C**).

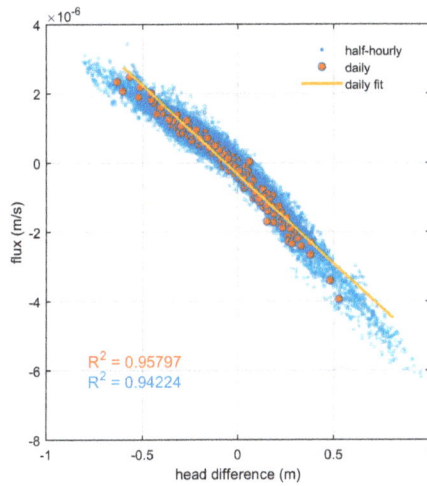

Figure 4. Vertical fluxes versus the head difference between the river stage and inland water table level at Spring 9 at daily and half-hourly scales.

2.3. Using Temperature Records to Infer Vertical HEF Rates

The method we used to quantify vertical HEF rates is based on temperature time series measured in both surface water and in the riverbed. Rapid changes in river stage lead to quick variations in magnitude and frequent directional changes in the vertical fluxes across the river bed. Because no single method was capable of estimating vertical fluxes from such a complex system, we applied two different methods for time periods with different observed subsurface thermal characteristics.

The first method is a simple analytical solution of a 1D heat transfer equation described by [32], which assumes that the system is in a steady-state condition and that the direction of HEFs is constantly upward:

$$q_z = -\frac{K_s^{(1-n)} K_f^{n}}{\rho_f C_f z} \ln \frac{T_z - T_L}{T_0 - T_L} \tag{3}$$

where K_s and K_f are thermal conductivities of fluid and solid, respectively; n is porosity; $\rho_f C_f$ is fluid volumetric heat capacity; z is the depth of sensor; and T_0, T_Z, and T_L represent the surface water temperature, the temperature at the sensor location, and groundwater temperature (assumed to be constant). This method is only applicable when there is constant upward flux.

The second method is the Local Polynomial method with a Maximum Likelihood estimator (LPML) developed by Vandersteen et al. [39] and has been extensively tested with time-series data [52]. The LPML model transforms temperature signals from the temporal domain to the frequency domain and finds the best vertical flux rate to fit equations that describe the subsurface signal frequency as a response function of the surface signal frequency. The 1D heat transport Equation (2) can be written as:

$$\frac{\partial^2 T}{\partial Z^2} + \alpha \frac{\partial T}{\partial z} + \beta T + \gamma \frac{\partial T}{\partial t} = 0 \tag{4}$$

With

$$\alpha = -\frac{q_z \rho_w c_w}{D \rho c} \tag{5}$$

$$\beta = 0 \tag{6}$$

$$\gamma = -\frac{1}{D} \tag{7}$$

where ρ and ρ_w are densities of the fluid-sediment matrix and the fluid; c and c_w are the specific heat of the fluid-sediment matrix and the fluid; D is the effective thermal diffusion coefficient that can be estimated from the bulk thermal conductivity (K) through $D = K/\rho c$; and α, β, and γ are parameters in the frequency response function (FRF) that converts the surface temperature signal to the signal in the sediment at depth z in the frequency domain. These parameters can be fitted using a maximum likelihood estimator based on observations and therefore can be used to determine the vertical flux:

$$q_z = -\frac{\alpha D \rho c}{\rho_w c_w} \tag{8}$$

The LPML model can incorporate the full or any portion of the temperature time series at daily to seasonal time steps. It can also deal with transient temperature signals by separating them into periodic, non-periodic, and additive noise portions using the local polynomial (LP) method. However, similar to other frequency-based methods such as those described by Hatch et al. [37] and Keery et al. [38], LPML works well when both surface and subsurface temperature time series share the same frequency, especially for the fundamental frequency range (e.g., diurnal cycles). This method, therefore, is more applicable when the subsurface temperature variations follow a pattern comparable, and have a fundamental frequency similar to that of the surface water temperature.

To identify periods suitable for the two methods, we employed the Dynamic Harmonic Regression (DHR) model [53] to decompose the observed surface water and subsurface time series into signals that represent the general trend and fundamental frequency. The DHR model has been successfully applied to detect the amplitude and phase of thermal signals in other studies (e.g., [38,42]). It treats observed temperature time series (y_t) as the sum of a general trend (T_t), a fundamental signal and its associated harmonics (C_t), and a white noise component (e_t):

$$y_t = T_t + C_t + e_t \tag{9}$$

$$C_t = \sum_{i=1}^{N} [a_{i,t} \cos(\omega_i t) + b_{i,t} \sin(\omega_i t)] \tag{10}$$

where $a_{i,t}$ and $b_{i,t}$ are time-varying parameters, ω_1 is the fundamental frequency, and ω_i is the harmonics of ω_1 with $\omega_i = i \times \omega_1$. The decomposition of the time series using the DHR model was conducted by applying the CAPTAIN toolbox using an auto-regression technique [54].

The decomposed general trend and amplitude of the fundamental frequency (i.e., diurnal) from surface and subsurface locations were used to first identify periods with clear upward groundwater fluxes when the analytical solution method is applicable. The surface water trend line follows a clear seasonal cycle that gradually changes between 2 °C and 24 °C within a year, while the deep groundwater temperature at this location is fixed at about 16 °C in our study reach. In summer and winter seasons, the temperature difference between surface and groundwater could reach up to 10 °C. As a result, each period with significant disparities between the two trend lines from the water and river bed temperatures indicates a groundwater discharge event and can be considered a clear upward flux period. Here we define "significant departure" as periods when the difference between surface and subsurface trend lines is more than three times the standard deviation of the fundamental (diurnal) signal. During these periods, the shallow subsurface temperature is highly affected by the upwelling of deep groundwater and the diurnal signal from the solar radiation is dampened or even disappeared (when the upward flux is very strong), which makes the frequency-based method impractical. Therefore, in these periods of strong upward fluxes, we assume that the system remains at a quasi-steady state at the daily scale and is suitable for the analytical solution (Equation (3)) for the daily flux estimations.

Given that the temperature at the bottom of the subsurface remains nearly constant and the diurnal temperature signal is forced by solar radiation heating the river water, the amplitude of the diurnal signal in the subsurface should be always less than that in the river water. However, the diurnal temperature amplitude of the subsurface decomposed by the DHR model could be overestimated at the edge of strong upwelling events because of the rapid temperature variations. Here, we simply compared the diurnal amplitudes from surface water and subsurface and discarded the data with greater subsurface amplitudes to reduce the uncertainties caused by the edging effects. The data that passed the filtering were then identified as inputs to the LPML model. Although data were screened at a daily time step, the LPML model was applied at the original 10-min interval over a three-day moving window for each day of interest. The workflow of data screening and model application are shown in Figure 5.

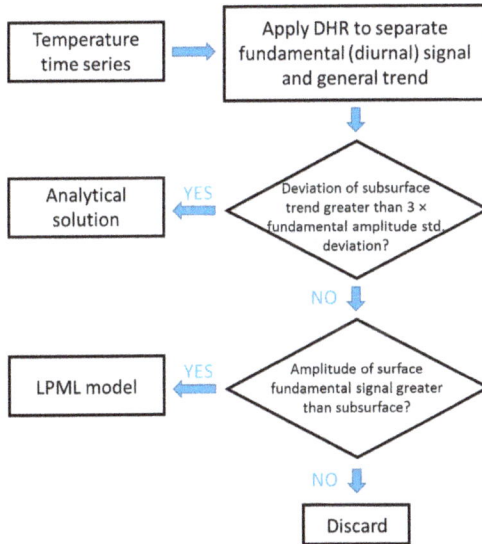

Figure 5. The framework employed to estimate hyporheic fluxes.

Data screening clearly introduces discontinuities to the inferred riverbed HEFs at the daily time scale. However, the data points would be sufficient to establish the linear relations between the daily fluxes and the corresponding head difference between the river stage and the inland water table. As we demonstrated in Section 2.2, the same relationship can also be further used to estimate sub-daily fluxes at the riverbed.

We tested the approaches on the Spring 9 temperature data set (Figure 2) and compared the estimated fluxes with the published data based on Darcy's Law. In data filtering, out of the total of 293 days, 122 days were selected to apply the analytical method, 103 days were selected for to apply the LPML model, and 68 days were discarded. Figure 6 demonstrates how the periods were selected based on the criteria described above. To incorporate the uncertainties associated with sediment physical properties, we randomly generated 100 sets of parameter values within the ranges based on sediment properties provided by Ma et al. [55] (Table 1). The estimated daily vertical fluxes from the 100 realizations were plotted against the daily head difference to fit a linear model, and the range defined by the maximum and minimum slopes generally covered the fitted line derived from the observations (Figure 7). We did a linear regression test for these data points and the p-value of the F-test was 1.22×10^{-12}, indicating the linear relationship was significant at $\alpha = 0.05$. Then we applied the 100 ensemble models to the head differences obtained from the river stage and inland water level observations and compared the envelope defined by the 5th and 95th percentiles of the ensemble with the Darcy's Law-based flux (Figure 8). The comparisons revealed that the temperature-inferred vertical flux range adequately captures the direction and magnitude of the HEFs at both daily and half-hourly time scales. The same modeling framework was then used to estimate the vertical fluxes in the shallow water along the river, as introduced in Section 2.4.

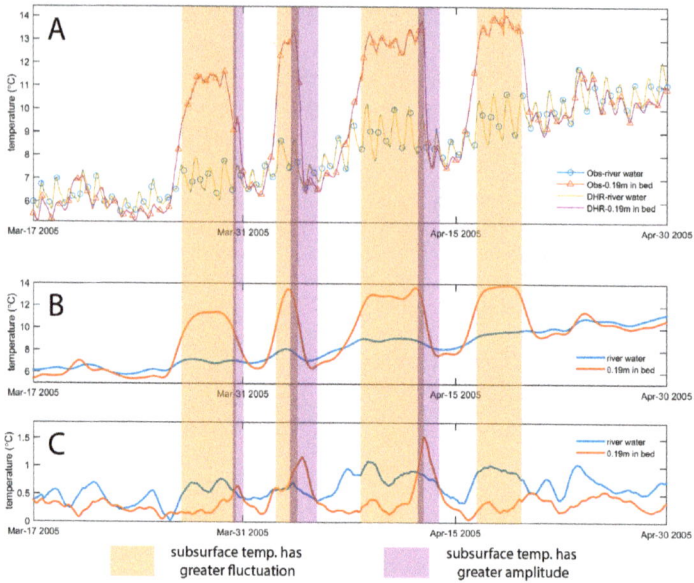

Figure 6. Demonstration of the data filtering procedure based on Spring 9 temperature data.

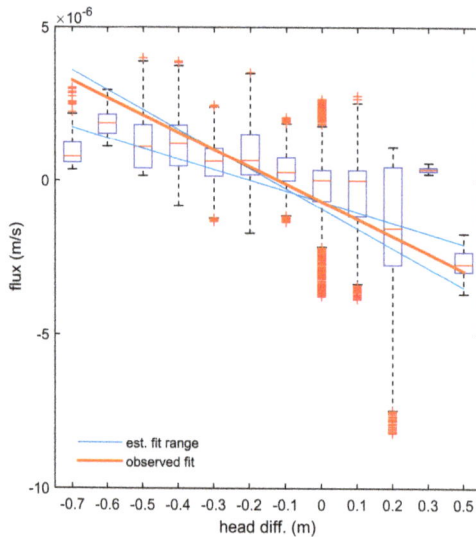

Figure 7. Relationship between model-estimated vertical flux rates and the head gradients between the river stage and inland water level based on 100 realizations of randomly selected combinations of sediment properties at Spring 9.

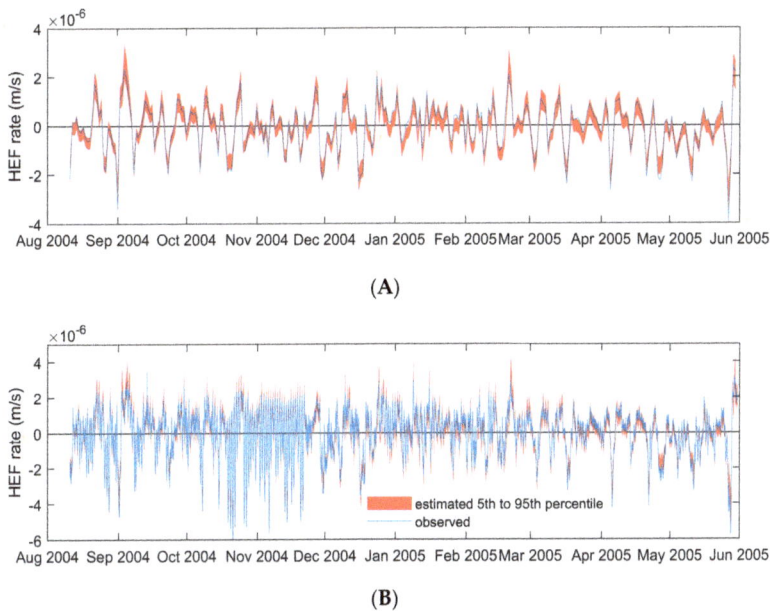

Figure 8. The range of fluxes (i.e., the 5th and 95th percentiles) estimated by the proposed model compared with the Darcy's Law-based fluxes at Spring 9 at daily (**A**) and half-hourly (**B**) time scales.

Table 1. Ranges of physical parameters used to generate parameter sets in the model.

Parameters	Range (from Ma et al. 2012)
Porosity	0.15~0.18
Solid density (kg/m^3)	2650~2760
Fluid-specific heat capacity (J/(kg K))	4186
Soild-specific heat capacity (J/(kg K))	715~920
Solid thermal conductivity (W/(m K))	1.2~2.2
Fluid thermal conductivity (W/(m K))	0.58
Thermal diffusivity (m^2/s)	5.19×10^{-7}~2.24×10^{-6}

2.4. Monitoring near Riverbed Temperature along the River

We deployed iButton® temperature sensors (Maxim Integrated Products, Model number DS1922L) at five locations along the west bank of the river (Figure 1) to record temperature time series in the river water and the riverbed. The iButton® sensor is a cylindrical, wireless device 17 mm in diameter and 6 mm thick. Its data storage capacity is 4096 values with 0.0625 °C measuring resolution and 0.5 °C accuracy. The elevations for all the monitoring locations were low enough to make sure the sensors would not be exposed to the air even in low-flow conditions. These locations were also selected based on the local hydrodynamic conditions and river bed properties and were labeled alphabetically from south to north. Sites A and B were located in the secondary channel that featured relatively low-flow velocities and small sediment sizes on the river bed, while sites D and E were located in the primary channel that featured relatively high flow velocities and large grain sizes in the alluvial sediment layer. Site C was in the transition zone between Sites B and D (Figure 2).

Temperature time series from the riverbed top and certain depth in the sediment were required for estimating the flux across the riverbed. Therefore, at each location, two sensors were secured in separate open drill holes along a 1 in. diameter, 1.5 m long solid plastic rod half planted in the river bed to record surface and subsurface temperature time series. All sensors were programmed to record every 10 min from 2 March to 30 March 2016. The coordinates and elevations of each sensor are listed in Table 2. The method described in Section 2.3 was then applied to the monitored temperature records at each location to obtain the local HEF rates at the river bed. Note that here we used the surface water sensors to approximate the temperature at the top of the river bed and the vertical HEF fluxes were estimated based on the distances between the subsurface sensors to the river bed.

Table 2. Coordinates (based on NAD 1983 StatePlane Washington South system) and depths of the iButton sensors installed at the monitoring sites.

Site	Northing (m)	Easting (m)	Surface Water Sensor (Dist. from River Bed)	Subsurface Sensor (Dist. from River Bed)
A	117,140.6	594,347.3	25 cm	25 cm
B	117,046.4	594,347.3	31 cm	19 cm
C	116,784.4	594,383.5	24 cm	26 cm
D	116,179.5	594,502.3	22 cm	28 cm
E	116,153	594,509.2	37 cm	13 cm

3. Results

The temperature data from the five iButton monitoring sites (Figure 9) showed that the river water temperature measurements were nearly identical across the sites, featuring a clear diurnal cycle over the period and a moderate raising trend with a daily mean temperature of about 5.5 °C on 2 March and 7 °C on 30 March, while the subsurface temperature differed among the sites. At Sites A and B, which are located in the secondary channel, the subsurface temperature had a similar general trend but a damped magnitude in diurnal variations compared to that in the surface water. At the sites located in the primary channel (D and E), strong variations in subsurface temperature and great departures (over 10 °C) from the surface temperature were evident, indicating strong upwelling events. Site C is located at the transition zone between the primary and the secondary channels, which has small subsurface temperature variations, with the magnitude between the primary channel and the secondary channel. Head differences at these sites were extrapolated from the reference river gauge observations at SWS-1 and inland well readings at Well 2–3 (Figure 10) based on the river stage and inland water table gradients established based on historical data. The river stage during the monitoring period ranged from 104.9 m to 105.9 m, which was quite representative because it covered about 80% of the full spectrum of the historical stage in the past 30 years that has a 5th percentile at 104.8 m and a 90th percentile at 106 m. For each site, we performed a linear regression analysis to test whether a significant linear relationship existed between the estimated HEF rate and the head difference. The p-values of the F-tests were 4.95×10^{-14}, 6.65×10^{-14}, 1.86×10^{-13}, 4.29×10^{-12}, and 5.89×10^{-13}, for the sites A–E, respectively, indicating that the linear relationship were all significant. Then, the linear model for esimating flux rates as a function of head differences between the river stage and inland water level was established at each iButton location with 100 realizations. The corresponding time series of estimated HEF rate envelopes from the 5th and 95th percentile of the ensembles are shown in Figure 11. Generally speaking, both upwelling and downwelling events were observed at all sites and extreme values are shown in response to maximum and minimum flow stages. However, the estimated flux rates show strong spatial and temporal variability and differ significantly between the sites in the primary and secondary channels. At Sites A and B the fluxes were at magnitudes of about $\pm 5 \times 10^{-7}$ m s^{-1} and $\pm 4 \times 10^{-7}$ m s^{-1}, while at sites D and E, the values were nearly one magnitude greater (i.e., up to $\pm 1.8 \times 10^{-6}$ m s^{-1}). The flux ranges at D and E were found to be comparable to fluxes at Spring 9 (about 2×10^{-6} to -4×10^{-6} m s^{-1}) derived using Darcy's Law.

Three flow conditions from the iButton monitoring period were selected for further comparison. The three conditions represented the high flow on 10 March, median flow on 28 March, and low flow on 8 March (Figure 12). We connected the median value of the predicted flux rates across the five monitoring locations under different flow conditions and found that at high and median flow conditions most of the fluxes were predicted to be negative, indicating aquifer recharging conditions; while at low flow, the aquifer was discharging at all locations. At Site B, the predicted range of flux rates were overlaping with each other across the zero flux line, showing that the uncertainty of the prediction may lead to opposite flux directions at certain locations with relatively small flux magnitudes. At Sites D and E, the predicted flux rates varied greatly (from $+1 \times 10^{-6}$ m s^{-1} to -1.5×10^{-6} m s^{-1}) with respect to different flow conditons. The results also suggested that sites located in the primary channel (i.e., D and E) had remarkably higher flux magnitudes (up to 6–9 times higer) compared to those in the secondary channel (i.e., A and B), indicating strong spatial heterogenerity induced by geomorphological features.

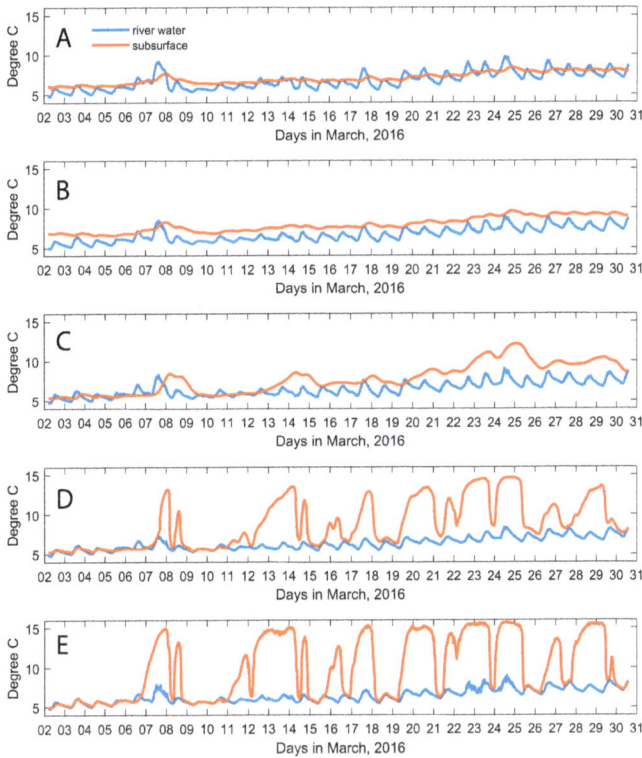

Figure 9. Time series of observed temperature at the five iButton locations.

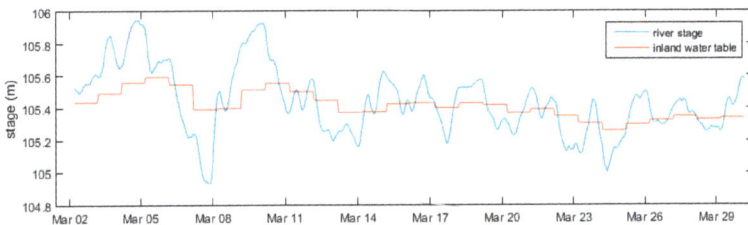

Figure 10. Time series of river stage and inland water level during the iButton monitoring period.

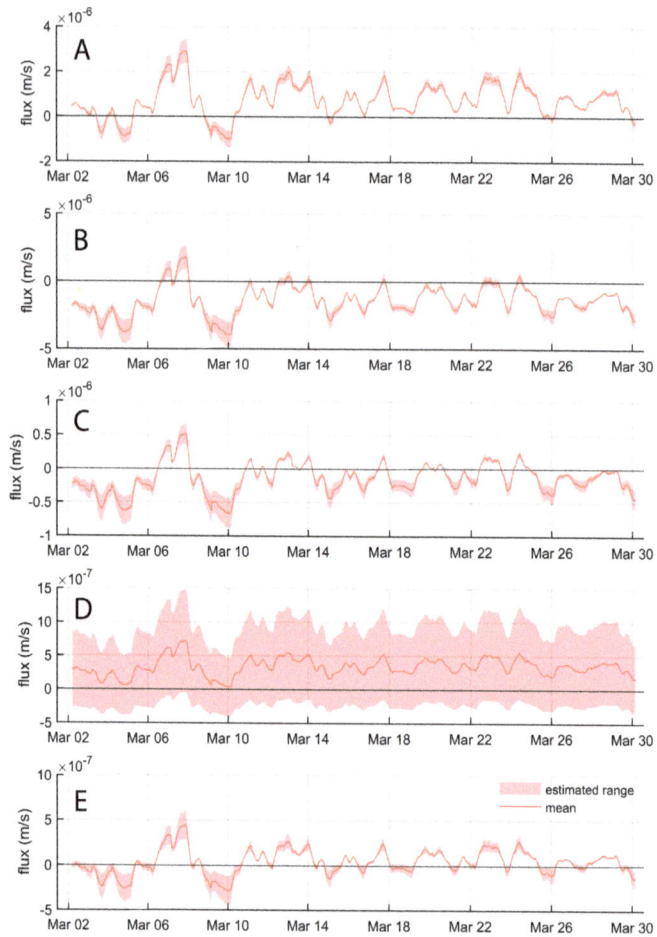

Figure 11. Time series of estimated fluxes during the monitoring period.

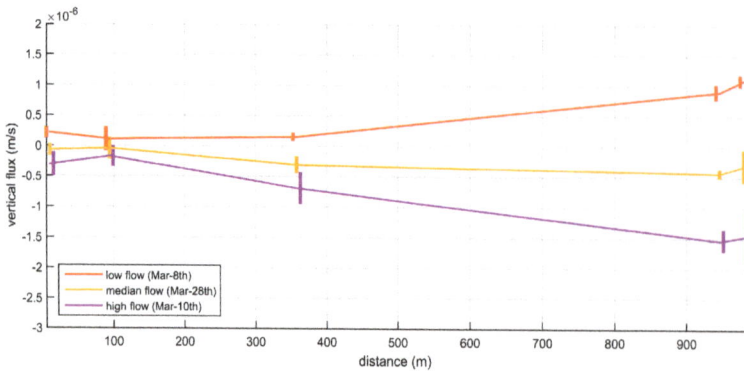

Figure 12. Estimated vertical fluxes at the five iButton locations under three representative flow conditions.

4. Discussion and Conclusions

In this study, we introduced an approach for inferring HEF rates based on temperature measurements using the stage–flux relationship. This approach allows us to obtain long-term, high spatiotemporal resolution hydrologic exchange dynamics in a large river reach based on the river flow and inland groundwater conditions with relatively easy field installations. Our predicted HEF rates vary between $+1 \times 10^{-6}$ m s^{-1} to -1.5×10^{-6} m s^{-1} under various flow conditions in the shallow water region along the river bank. This results are comparable with previous studies performed in other regulated rivers (e.g., [20,21]), in terms of the order of magnitude of the flux rates and the mechanism between the HEFs and the dam operations. The strength of this approach is that it remains robust under high-frequency river stage variations imposed by flow regulations (i.e., dam operations). Such variations rarely occur in a natural river except for during an extreme flooding event, which means it may not be generally applicable to fluvial systems with constant groundwater discharge from a confined aquifer. In small highland rivers, local geomorphological features such as riffles and pools might be a greater dominant factor than the head difference in controlling HEF rates [56,57].

Uncertainties in this approach could come from field measurements, parameterization of physical properties, model assumptions, modeling resolution, and model structure [58]. These include (1) the accuracy of the iButton sensors and the lags on temperature signal recording due to the thermal skin effects of the measuring rod might affect the inferred HEF rates [59]; (2) river bed geomorphic features (e.g., dunes) may create local perturbations on the HEF patterns and thermal processes; (3) the riverbed temperature was approximated by measurements ~20 cm above the river-bed surface due to site and equipment limitations. The water layer between the sensor and the river bed surface may create temperature signal attenuations and lead to uncertainties in flux estimates. Our future studies will evaluate and consider the attenuation effects of the water in reducing the uncertainties; (4) the ranges of the physical property values used in the LPML model parameterizations (Table 1) were from the literature, which may not fully cover the uncertainties of these parameters; (5) the approach we used to identify the time period for different models were based on a frequency analysis (DHR), which is arbitrary and might introduce errors to the analysis. In fact, the choice of applying DHR for data-screening is site-specific, which is only required in locations with strong upwelling fluxes periodically diminishing or even removing the diurnal signals from the temperature data; (6) fluxes at the horizontal and lateral directions, which are smaller in magnitude compared to the vertical fluxes, may still affect the vertical flux estimations [60]; (7) the size of the moving window and the temporal resolution used in the LPML model may introduce uncertainties to the modeling results [39]. Another limitation of this approach is that it is only valid at locations where the HEFs are dominated by the stage gradient between river and inland water table. In other words, the hydrostatic driver controls the HEF dynamics [44]. However, in some areas in the river, the HEFs are controlled by the hydrodynamic driver, which is determined by the local hydraulic conditions. For example, at the upstream tip of the island, where the hydrodynamic pressure is always high due to the speed of the river flows, the HEF rate may always be negative (i.e., downwelling).

The results of this study suggest that the magnitude of vertical fluxes in the primary channel could be about 6–9 times greater than in the secondary channel. Given that the head differences between the river stage and inland water table are similar across these locations, the major controlling factor that leads to this difference is the permeability of the riverbed sediment. We measured the flow velocity at the iButton® locations and found that the mean velocity at the primary channel sites (0.4 m/s) was about two-fold higher than at the secondary channel sites (0.2 m/s). According to the Hjulstrøm Curve [61], which demonstrates the sediment erosion, transport, and deposition with regard to the flow speed and grain size, the averaged grain size of the riverbed sediment between the primary and secondary channels might differ by about three times, which then leads to approximately nine times the difference in permeability according to a number of empirical models such as Krumbein and Monk's equation [62], or the Kozeny–Carman equation [63]. This is an overly simplified explanation of the flux differences in the two channels and requires further development and validation, but linking the HEF rate and the surface flow speed might be an informative way of estimating vertical flux in deep waters where the field measurements are not available and are hard to obtain. For example, the center of the primary channel in the study reach is as deep as 15 m under low-flow conditions, which poses a safety issue for installing iButtons® or any other instruments.

To conclude, in this study we successfully inferred the sub-daily vertical HEF rates at five shallow water sites along the shoreline based on temperature profile measurements and relationships between the riverbed hydrologic exchange rate and river flow conditions. The results reveal that the HEF rate had high spatial and temporal heterogeneities over the riverbed, and a magnitude of fluxes 6–9 times higher in the primary channel than in the secondary channel. This approach can be easily employed in other river reaches to facilitate large scale river corridor studies or to inform biogeochemical and ecological studies in highly dynamic large river reaches.

Acknowledgments: This research was supported by the U.S. Department of Energy, Office of Biological and Environmental Research (BER), as part of BER's Subsurface Biogeochemistry Research Program (SBR). This contribution originates from the SBR Scientific Focus Area at the Pacific Northwest National Laboratory, operated by Battelle Memorial Institute for the U.S. DOE under contract DE-AC05-76RLO1830. We thank Uwe Schneidewind and Christian Anibas for sharing the LPML code.

Author Contributions: T.Z. and M.H. conceived and designed the approach; all authors contributed equally to associated experiments, data analysis, and writing.

Conflicts of Interest: The authors declare no conflict of interest.

References

1. Harvey, J.; Gooseff, M. River corridor science: Hydrologic exchange and ecological consequences from bedforms to basins. *Water Resour. Res.* **2015**, *51*, 6893–6922. [CrossRef]
2. Findlay, S.; Strayer, D.; Goumbala, C.; Gould, K. Metabolism of streamwater dissolved organic carbon in the shallow hyporheic zone. *Limnol. Oceanogr.* **1993**, *38*, 1493–1499. [CrossRef]
3. Sobczak, W.V.; Findlay, S. Variation in bioavailability of dissolved organic carbon among stream hyporheic flowpaths. *Ecology* **2002**, *83*, 3194–3209. [CrossRef]
4. Anderson, M.P. Heat as a ground water tracer. *Ground Water* **2005**, *43*, 951–968. [CrossRef] [PubMed]
5. Valett, H.M.; Morrice, J.A.; Dahm, C.N.; Campana, M.E. Parent lithology, surface-groundwater exchange, and nitrate retention in headwater streams. *Limnol. Oceanogr.* **1996**, *41*, 333–345. [CrossRef]
6. Zarnetske, J.P.; Haggerty, R.; Wondzell, S.M.; Baker, M.A. Dynamics of nitrate production and removal as a function of residence time in the hyporheic zone. *J. Geophys. Res. Biogeosci.* **2011**, *116*, 1–12. [CrossRef]
7. Mollema, P.N.; Antonellini, M. Water and (bio)chemical cycling in gravel pit lakes: A review and outlook. *Earth-Sci. Rev.* **2016**, *159*, 247–270. [CrossRef]
8. Tonina, D.; Buffington, J.M. Hyporheic exchange in gravel bed rivers with pool-riffle morphology: Laboratory experiments and three-dimensional modeling. *Water Resour. Res.* **2007**, *43*. [CrossRef]
9. Morrice, J.A.; Valett, H.M.; Dahm, C.N.; Campana, M.E. Alluvial characteristics, groundwater-surface water exchange and hydrological retention in headwater streams. *Hydrol. Process.* **1997**, *11*, 253–267. [CrossRef]
10. Boulton, A.; Findlay, S.; Marmonier, P. The functional significance of the hyporheic zone in streams and rivers. *Annu. Rev. Ecol. Syst.* **1998**, *29*, 59–81. [CrossRef]
11. Winter, T.C. Relation of streams, lakes, and wetlands to groundwater flow systems. *Hydrogeol. J.* **1999**, *7*, 28–45. [CrossRef]
12. Brunke, M.; Gonser, T.; Brunke, M.; Gonser, T.; Gonser, T. The ecological significance of exchange processes between rivers and groundwater. *Freshw. Biol.* **1997**, *37*, 1–33. [CrossRef]
13. Zhou, T.; Endreny, T.A. Reshaping of the hyporheic zone beneath river restoration structures: Flume and hydrodynamic experiments. *Water Resour. Res.* **2013**, *49*, 5009–5020. [CrossRef]
14. Poff, N.L.; Allan, J.D.; Bain, M.B.; Karr, J.R.; Prestegaard, K.L.; Richter, B.D.; Sparks, R.E.; Stromberg, J.C. The Natural Flow Regime: A paradigm for river conservation and restoration N. *Bioscience* **1997**, *47*, 769–784. [CrossRef]
15. Arntzen, E.V.; Geist, D.R.; Dresel, P.E. Effects of fluctuating river flow on groundwater/surface water mixing in the hyporheic zone of a regulated, large cobble bed river. *River Res. Appl.* **2006**, *22*, 937–946. [CrossRef]

16. Vörösmarty, C.J.; Sahagian, D.L. Anthropogenic disturbance of the terrestrial water cycle. *Bioscience* **2000**, *50*, 753–765. [CrossRef]

17. Hanasaki, N.; Kanae, S.; Oki, T. A reservoir operation scheme for global river routing models. *J. Hydrol.* **2006**, *327*, 22–41. [CrossRef]

18. Zhou, T.; Nijssen, B.; Gao, H.; Lettenmaier, D.P. The contribution of reservoirs to global land surface water storage variations. *J. Hydrometeorol.* **2016**, *17*, 309–325. [CrossRef]

19. Ward, J.V.; Stanford, J.A. Ecological connectivity in alluvial river ecosystems and its disruption by flow regulation. *Regul. Rivers Res. Manag.* **1995**, *11*, 105–119. [CrossRef]

20. Cardenas, M.B.; Markowski, M.S. Geoelectrical imaging of hyporheic exchange and mixing of river water and groundwater in a large regulated river. *Environ. Sci. Technol.* **2011**, *45*, 1407–1411. [CrossRef] [PubMed]

21. Gerecht, K.E.; Cardenas, M.B.; Guswa, A.J.; Sawyer, A.H.; Nowinski, J.D.; Swanson, T.E. Dynamics of hyporheic flow and heat transport across a bed-to-bank continuum in a large regulated river. *Water Resour. Res.* **2011**, *47*, 1–12. [CrossRef]

22. Sawyer, A.; Bayani Cardenas, M.; Bomar, A.; Mackey, M. Impact of dam operations on hyporheic exchange in the riparian zone of a regulated river. *Hydrol. Process.* **2009**, *23*, 2129–2137. [CrossRef]

23. Cranswick, R.H.; Cook, P.G. Scales and magnitude of hyporheic, river-aquifer and bank storage exchange fluxes. *Hydrol. Process.* **2015**, *29*, 3084–3097. [CrossRef]

24. Kalbus, E.; Reinstorf, F.; Schirmer, M. Measuring methods for groundwater, surface water and their interactions: A review. *Hydrol. Earth Syst. Sci. Discuss.* **2006**, *3*, 1809–1850. [CrossRef]

25. Murdoch, L.C.; Kelly, S.E. Factors affecting the performance of conventional seepage meters. *Water Resour. Res.* **2003**, *39*. [CrossRef]

26. Paulsen, R.J.; Smith, C.F.; O'Rourke, D.; Wong, T.F. Development and evaluation of an ultrasonic ground water seepage meter. *Ground Water* **1997**, *39*, 904–911. [CrossRef]

27. Lee, D.R. A device for measuring seepage flux in lakes and estuaries. *Limnol. Oceanogr.* **1977**, *22*, 140–147. [CrossRef]

28. Rosenberry, D.O. A seepage meter designed for use in flowing water. *J. Hydrol.* **2008**, *359*, 118–130. [CrossRef]

29. Freeze, A.; Cherry, J. *Groundwater*; Prentice-Hall: Englewood Cliffs, NJ, USA, 1979.

30. Suzuki, S. Percolation measurements based on heat flow through soil with special reference to paddy fields. *J. Geophys. Res.* **1960**, *65*, 2883–2885. [CrossRef]

31. Stallman, R.W. Steady one-dimensional fluid flow in a semi-infinite porous medium with sinusoidal surface temperature. *J. Geophys. Res.* **1965**, *70*, 2821–2827. [CrossRef]

32. Turcotte, D.; Schubert, G. *Geodynamics*; Cambridge University Press: Cambridge, UK, 2002.

33. Healy, R.W.; Ronan, A.D. *Documentation of Computer Program VS2DH for Simulation of Energy Transport in Variably Saturated Porous Media—Modification of the U.S. Geological Survey's Computer Program VS2DT*; U.S. Department of the Interior, U.S. Geological Survey: Reston, VA, USA, 1996.

34. Silliman, S.E.; Booth, D.F. Analysis of time-series measurements of sediment temperature for identification of gaining vs. losing portions of Juday Creek, Indiana. *J. Hydrol.* **1993**, *146*, 131–148. [CrossRef]

35. Constantz, J.; Stewart, A.E.; Niswonger, R.; Sarma, L. Analysis of temperature profiles for investigating stream losses beneath ephemeral channels. *Water Resour. Res.* **2002**, *38*, 1–13. [CrossRef]

36. Anibas, C.; Fleckenstein, J.H.; Volze, N.; Buis, K.; Verhoeven, R.; Meire, P.; Batelaan, O. Transient or steady-state? Using vertical temperature profiles to quantify groundwater-surface water exchange. *Hydrol. Process.* **2009**, *23*, 2165–2177. [CrossRef]

37. Hatch, C.E.; Fisher, A.T.; Revenaugh, J.S.; Constantz, J.; Ruehl, C. Quantifying surface water-groundwater interactions using time series analysis of streambed thermal records: Method development. *Water Resour. Res.* **2006**, *42*. [CrossRef]

38. Keery, J.; Binley, A.; Crook, N.; Smith, J.W.N. Temporal and spatial variability of groundwater–surface water fluxes: Development and application of an analytical method using temperature time series. *J. Hydrol.* **2007**, *336*, 1–16. [CrossRef]

39. Vandersteen, G.; Schneidewind, U.; Anibas, C.; Schmidt, C.; Seuntjens, P.; Batelaan, O. Determining groundwater-surface water exchange from temperature-time series: Combining a local polynomial method with amaximum likelihood estimator. *Water Resour. Res.* **2015**, *51*, 922–939. [CrossRef]

40. Schneidewind, U.; Anibas, C.; Vandersteen, G.; Schmidt, C.; Joris, I.; Seuntjens, P.; Batelaan, O. Delineating Groundwater-Surface Water Exchange Flux Using Temperature- Time Series Analysis Methods. In Proceedings of the 2nd European symposium on Water Technology and Management Symposium, Leuven, Belgium, 20–21 November 2013.

41. Swanson, T.E.; Cardenas, M.B. Ex-Stream: A MATLAB program for calculating fluid flux through sediment–water interfaces based on steady and transient temperature profiles. *Comput. Geosci.* **2011**, *37*, 1664–1669. [CrossRef]

42. Gordon, R.P.; Lautz, L.K.; Briggs, M.A.; McKenzie, J.M. Automated calculation of vertical pore-water flux from field temperature time series using the VFLUX method and computer program. *J. Hydrol.* **2012**, *420–421*, 142–158. [CrossRef]

43. Fritz, B.G.; Arntzen, E.V. Effect of rapidly changing river stage on uranium flux through the hyporheic zone. *Ground Water* **2007**, *45*, 753–760. [CrossRef] [PubMed]

44. Boano, F.; Harvey, J.W.; Marion, A.; Packman, A.I.; Revelli, R.; Ridolfi, L.; Wörman, A. Hyporheic flow and transport processes: Mechanisms, models, and bioghemical implications. *Rev. Geophys.* **2014**, 1–77. [CrossRef]

45. Krause, S.; Boano, F.; Cuthbert, M.O.; Fleckenstein, J.H.; Lewandowski, J. Understanding process dynamics at aquifer-surface water interfaces: An introduction to the special section on new modeling approaches and novel experimental technologies. *Water Resour. Res.* **2014**, *50*, 1847–1855. [CrossRef]

46. Johnson, T.; Versteeg, R.; Thomle, J.; Hammond, G.; Chen, X.; Zachara, J. Four-dimensional electrical conductivity monitoring of stage-driven river water intrusion: Accounting for water table effects using a transient mesh boundary and conditional inversion constraints. *Water Resour. Res.* **2015**, *51*, 6177–6196. [CrossRef]

47. Boano, F.; Revelli, R.; Ridolfi, L. Reduction of the hyporheic zone volume due to the stream-aquifer interaction. *Geophys. Res. Lett.* **2008**, *35*, 1–5. [CrossRef]

48. Hartman, M.J. *Groundwater Quality Assessment Plan for the 1324-N/NA Site*; Pacific Northwest National Lab.: Richland, WA, USA, 1998; Volume 30.

49. Hartman, M.J.; McDonald, J.P.; Dresel, P.E.; Newcomer, D.R.; Lindberg, J.W.; Thornton, E.C. *Fiscal Year 2003 Integrated Monitoring Plan for the Hanford Groundwater Monitoring Project*; United States Department of Energy: Washington, DC, USA, 2002.

50. Sanford, W.E.; Selnick, D.L. Estimation of Evapotranspiration Across the Conterminous United States Using a Regression With Climate and Land-Cover Data1. *JAWRA J. Am. Water Resour. Assoc.* **2013**, *49*, 217–230. [CrossRef]

51. Fayer, M.J.; Keller, J.M. *Recharge Data Package for Hanford Single-Shell Tank Waste Management Areas*; Pacific Northwest National Laboratory: Richland, WA, USA, 2007.

52. Anibas, C.; Schneidewind, U.; Vandersteen, G.; Joris, I.; Seuntjens, P.; Batelaan, O. From streambed temperature measurements to spatial-temporal flux quantification: Using the LPML method to study groundwater-surface water interaction. *Hydrol. Process.* **2015**, *216*. [CrossRef]

53. Young, P.C.; Pedregal, D.; Tych, W. Dynamic harmonic regression. *J. Forecast.* **1999**, *18*, 369–394. [CrossRef]

54. Young, P.C.; Tych, W.; Taylor, C.J. *The CAPTAIN Toolbox for Matlab*; IFAC: New York, NY, USA, 2009; Volume 15.

55. Ma, R.; Zheng, C.; Zachara, J.M.; Tonkin, M. Utility of bromide and heat tracers for aquifer characterization affected by highly transient flow conditions. *Water Resour. Res.* **2012**, *48*, 1–18. [CrossRef]

56. Crispell, J.K.; Endreny, T.A. Hyporheic exchange flow around constructed in-channel structures and implications for restoration design. *Hydrol. Process.* **2009**, *1168*, 1158–1168. [CrossRef]

57. Byrne, P.; Binley, A.; Heathwaite, A.L.; Ullah, S.; Heppell, C.M.; Lansdown, K.; Zhang, H.; Trimmer, M.; Keenan, P. Control of river stage on the reactive chemistry of the hyporheic zone. *Hydrol. Process.* **2014**, *28*, 4766–4779. [CrossRef]

58. Sun, N.; Hong, B.; Hall, M. Assessment of the SWMM model uncertainties within the generalized likelihood uncertainty estimation (GLUE) framework for a high-resolution urban sewershed. *Hydrol. Process.* **2014**, *28*, 3018–3034. [CrossRef]

59. Cardenas, M.B. Thermal skin effect of pipes in streambeds and its implications on groundwater flux estimation using diurnal temperature signals. *Water Resour. Res.* **2010**, *46*, 1–12. [CrossRef]

60. Lautz, L.K. Impacts of nonideal field conditions on vertical water velocity estimates from streambed temperature time series. *Water Resour. Res.* **2010**, *46*, 1–14. [CrossRef]
61. Hjulstrom, F. Transportation of Detritus by Moving Water: Part 1. Transportation. American Advancing the World of Petroleum. Special Volumes. 1939. Available online: http://archives.datapages.com/data/specpubs/sedimen1/data/a142/a142/0001/0000/0005.htm (accessed on 14 September 2017).
62. Krumbein, W.; Monk, G. Permeability as a Function of Size parameters of Unconsolidated Sand. *Trans. AIME* **1943**, *151*. [CrossRef]
63. Carman, P. *Flow of Gases through Porous Media*; Academic press Inc.: New York, NY, USA, 1956.

water MDPI

Article

Calculation of Steady-State Evaporation for an Arbitrary Matric Potential at Bare Ground Surface

Xin Liu and Hongbin Zhan *

Department of Geology and Geophysics, Texas A&M University, College Station, TX 77843-3115, USA; lx20010531@tamu.edu
* Correspondence: zhan@geos.tamu.edu; Tel.: +1-979-574-4819; Fax: +1-979-845-6162

Received: 10 July 2017; Accepted: 19 September 2017; Published: 22 September 2017

Abstract: Evaporation from soil columns in the presence of a water table is a long lasting subject that has received great attention for many decades. Available analytical studies on the subject often involve an assumption that the potential evaporation rate is much less than the saturated hydraulic conductivity of the soil. In this study, we develop a new semi-analytical method to estimate the evaporation rate for an arbitrary matric potential head at bare soil surface without assuming that the potential evaporation rate is much less than the saturated hydraulic conductivity of the soil. The results show that the evaporation rates calculated by the new solutions fit well with the HYDRUS-1D simulation. The new solutions also can reproduce the results of potential evaporation rate calculated from previous equations under the special condition of an infinite matric potential head at bare soil surface. The developed new solutions expand our present knowledge of evaporation estimation at bare ground surface to more general field conditions.

Keywords: steady-state vertical flux; evaporation calculation; unsaturated flow; semi-analytical solution

---

1. Introduction

Evapotranspiration is one of the most important components in the study of the hydrological cycle, as it controls the exchange of mass and energy among the soil–water–vegetation system, the atmosphere and the groundwater system. Understanding water loss from soil by evapotranspiration due to the upward water flow from a water table is an important topic in many disciplines such as soil sciences [1], hydrology [2], and plant physiology [3]. The evapotranspiration comprises three processes: evapotranspiration from vegetated surfaces, evaporation from bare ground surfaces, and evaporation from open water bodies. The water exchange among soil, vegetation and atmosphere determines to a large extent the regional climate and the behavior of the hydrological cycle [4]. Land use is considered a local environmental issue and has a great influence on the hydrological circle [5,6], especially on the evapotranspiration as it alters the upper soil layer and land cover. Land cover features are part of the hydrological system, and the changes of land cover can impact evapotranspiration with the increasing expansion of urban areas, which causes permeable land reduction and increased flooding [5]. Furthermore, some previous studies show that vegetation and agriculture tend to increase evapotranspiration, and urbanization tends to decrease evaporation [7,8]. These effects of land use change on evapotranspiration need to be considered in water management study. The accurate estimation of evaporation from bare ground surface is important in determining the overall processes of evapotranspiration and will be the primary focus of this study. Direct measurement of soil evaporation is difficult and the most commonly used method involves a weighing lysimeter [9]. Although water evaporation in actual field setting is a highly complex process, a nearly steady upward flow from a water table to a bare soil surface may be established if the daily evaporative demand is reasonably uniform for a long period of time [10]. Water content in soil controls vertical water flow direction and velocity, and soil property affects water holding capacity and water flow velocity as well. While the soil moisture and matric potential head at the ground surface depend on atmospheric and soil conditions, the actual flux through the soil surface should be limited by the ability of the porous medium to transmit water from the unsaturated zone.

In terms of evaporation calculation, two scenarios are commonly seen. The first scenario deals with an evaporative front that is located at the ground surface. This is usually true when the water table is relatively shallow and the upward capillary liquid flow is sufficient to maintain a hydraulic connection between the water table and the ground surface. The second scenario deals with an evaporative front that is below the ground surface. For such a scenario, the upward capillary flow may not be strong enough to overcome the downward gravity, resulting in discontinuity of the vertical liquid profile at a certain level below the soil surface [11]. In other words, a water-depletion dry layer will be formed between the ground surface and the evaporative front, which marks the discontinuity of the liquid profile. In this case, the evaporative process is composed of an upward capillary liquid flow from the water table to the evaporative front (or vaporization plane) and a vapour diffusion through the upper dry layer. Many studies have been devoted to this subject, such as Rose et al. [12], Gowing et al. [13], Il'ichev et al. [14], Lehmann et al. [15], Shokri et al. [16], and Shokri and Salvucci [11].

The results from Gowing et al. [13] showed that the vaporization plane usually resided 4–14 cm below the ground surface for saline or non-saline soils. Assouline et al. [17] pointed out that the thickness of the dry layer above the vaporization plane varied within a relatively narrow range from 3 to 20 mm. Some studies have specifically discussed the maximum depth of the water table hydraulically connected to the ground surface via continuous capillary liquid pathways [11,18,19]. When the water table is deeper than the maximum depth hydraulically connected to the surface, the capillary flow will be disrupted. Lehmann et al. [20] and Sadeghi et al. [21] discussed a method to estimate the maximum matric potential at the bare ground surface that would maintain the hydraulic connections of the vertical unsaturated profile above the water table. Hayek [22] presented an analytical model for the steady-state vertical flow condition that could estimate the length of liquid phase and gas phase in the unsaturated zone.

Several researchers developed steady-state water flow solutions through a soil profile from a water table to bare ground surface (e.g., [12,13,23–25]). For instance, Gardner [23] developed a steady-state solution to show the relationship among the maximum evaporation rate, the capillary conductivity and the water table depth. Warrick [25] modified the form of unsaturated conductivity function as that of Brooks and Corey [26] and developed an exact solution for the maximum evaporation rate. Salvucci [24] developed approximate evaporation rate solutions on the basis of the Gardner and the Brooks–Corey functions, and such approximate solutions agreed considerably well with experimental results of three types of soils, as reported in Bras [27]. Rose et al. [12] and Gowing et al. [13] studied the evaporation flow from saline groundwater considering the salt accumulation effect. Gowing et al. [13] developed a solution when both liquid-phase and vapour-phase flows were of concern.

This study will address the first scenario with the evaporative front at the ground surface. The second scenario (with an evaporative front below the ground surface) will be addressed in a future investigation. In semi-arid or arid regions where surface vegetation is sparse or sometimes non-existent, water loss from the bare ground surface may be a grave concern in terms of water management and ecological conservation, among many other aspects. Precise determination of evaporation from bare ground surface in semi-arid or arid regions then becomes critically important, which will be the focus of this article.

The purpose of this article is to develop a new semi-analytical approach for calculating evaporation from a bare surface with an arbitrary matric potential at the surface. These newly developed solutions are more general than previously available solutions, which often involve an assumption that the matric potential at the surface is infinitely large, an assumption that is questionable in actual field applications. This study only considers the liquid flow phase of the evaporative process and does not consider the vapor flow phase, which is often much smaller than the liquid flow, as shown in many previous studies [12,15]. As bare ground is of concern here, plant transpiration is also excluded.

The paper is organized as follows. A mathematical model is built and solutions are developed in Section 2, followed by analysis of results in Section 3, including testing the new solutions against previous benchmark solutions under various special conditions, and comparing the new solutions with numerical simulations using HYDRUS-1D. Sections 4 and 5 discuss the results of comparison among new solutions, HYDRUS-1D and previous works, and summarize the major conclusions of this study.

2. Mathematical Model

2.1. Background and Problem Description

In many previous studies of the steady-state evaporation from bare surface, Gardner's unsaturated hydraulic conductivity model [23] has been used widely. When the unsaturated hydraulic conductivity–depth relationship is known, the steady-state upward water flow across the soil profile follows the Buckingham–Darcy flow law [28] as

$$z = -\int \frac{dh}{1 + E/K(h)},$$ (1)

where the z-axis is positive upward with $z = 0$ at ground surface, h is the matric potential head (negative) [L], $K(h)$ is the unsaturated hydraulic conductivity [LT^{-1}], and E is the upward water flux [LT^{-1}], which is equal to the value of evaporation rate at bare ground surface under the steady-state flow condition. A fixed water table is assumed to be below the ground surface at a distance L ($z = -L$). Only the vertical flow is of concern here, and the lateral flow is assumed to be secondary and negligible. The soil profile is assumed to be vertically homogeneous, thus soil layering and heterogeneity is not considered at this study. However, soil heterogeneity is an important feature and may be considered in a future investigation on the basis of this study.

Gardner [23] developed two widely used unsaturated hydraulic conductivity models. The first one was:

$$K(h) = K_s \exp(\alpha h),$$ (2)

where K_s is the saturated hydraulic conductivity [LT^{-1}] and α is a fitting parameter related to the pore size distribution of soil [L^{-1}]. The second one was

$$K(h) = AK_s(|h|^N + B)^{-1},$$ (3)

where A, B, and N are positive empirical factors related to soil texture and $|h|$ is a sign of absolute value [29].

Equation (1) shows that the evaporation rate often depends on the depth to the water table, the matric potential at the soil surface, and the hydraulic conductivity of soil. Gardner [23] revealed that when the matric potential at the surface was infinity, the evaporative flux would approach a maximum rate that was a function of the saturated hydraulic conductivity, the fitting parameters N, A, B in Equation (1) and the depth to the water table.

Haverkamp et al. [30] amended the parameters of Equation (3) and made the equation more concise, which is denoted as the modified Gardner model hereinafter. Jury and Horton [10] proposed a method of calculating the potential evaporation rate above a water table on the basis of the modified Gardner model [30] and an assumption that the potential evaporation rate was much less than the saturated hydraulic conductivity. Such an assumption was debatable in many field conditions. For instance, when the water table is reasonably shallow, the potential evaporation rate should increase to a value that is close to or even larger than the value of the saturated hydraulic conductivity.

Nevertheless, Jury and Horton [10] presented a one-dimensional model to describe water flow from a shallow water table upward to an evaporating surface using such an assumption.

The modified Gardner model by Haverkamp et al. [30] involves an unsaturated soil hydraulic conductivity as a function of the matric potential head h [L] (negative) as follows:

$$K(h) = \frac{K_s}{1 + (h/a)^N},$$

(4)

where a is a characteristic length [L] (negative). It is worthwhile to point out that Equation (4) is a special case of Equation (3) by setting $B = A$ and $A = |a|^N$ in Equation (3). Equation (4) is called the modified Gardner model hereinafter.

When steady-state flow is of concern, applying the Buckingham–Darcy law to vertical flow, one has:

$$E = -K(h)\left(\frac{dh}{dz} + 1\right).$$

(5)

Reorganizing Equation (5) into an integral, one has:

$$\int_{h_1}^{h_2} \frac{dh}{1 + E/K(h)} = -\int_{z_1}^{z_2} dz = z_1 - z_2,$$

(6)

where $h_1 = h(z_1)$ and $h_2 = h(z_2)$ are two matric potential heads at two different elevations z_1 and z_2, respectively. In the problem studied below, we set $z_1 = -L$ (water table) and $h_1(-L) = 0$; $z_2 = 0$ (ground surface) and $h_2(0) = h_0$, which is a constant matric potential head at ground surface. Substituting Equation (4) into Equation (6) leads to

$$\int_0^{h_0} \frac{dh}{1 + (E/K_s)\left[1 + (h/a)^N\right]} = -L.$$

(7)

Defining a new parameter $\mu = h/a$ and $\mu_0 = h_0/a$, which are positive, and substituting them into Equation (7) one has:

$$\int_0^{\mu_0} \frac{d\mu}{1 + (E/K_s) + (E/K_s)\mu^N} = -L/a.$$

(8)

Be aware that $-L/a$ on the right side of Equation (8) is positive because a is a negative constant.

Defining the following new parameters: $\varepsilon = \left(\frac{E/K_s}{1 + E/K_s}\right)^{1/N}$, $\sigma = \varepsilon\mu$ and $\sigma_0 = \varepsilon\mu_0$, one transforms Equation (8) into:

$$\int_0^{\sigma_0} \frac{d\sigma}{1 + \sigma^N} = (-L/a)\varepsilon(1 + E/K_s).$$

(9)

For the special case of calculating the potential evaporation rate, one may apply the negatively infinite matric potential head at ground surface or a positively infinite σ_0 in Equation (9). Under this condition, one can employ the following identity [31]:

$$\int_0^{\infty} \frac{d\sigma}{1 + \sigma^N} = \frac{\pi}{N\sin(\pi/N)}.$$

(10)

Substituting Equation (10) into Equation (9) for the case of $\sigma_0 \to \infty$ and recalling the definition of ε will lead to the following equation:

$$\left(\frac{E_p}{K_s}\right)^{\frac{1}{N}}\left(1 + \frac{E_p}{K_s}\right)^{1-\frac{1}{N}} = \frac{-a\pi}{NL\sin(\pi/N)},$$

(11)

where E_p in Equation (11) represents the potential evaporation rate hereinafter.

Equation (11) can be used to calculate the potential evaporation rate. The form of Equation (11) does not permit a direct analytical estimation of E_p for a general soil type, and one has to seek help from a numerical root-searching method. Under the special condition that E_p/K_s is much less than 1, one can obtain a closed-form solution for E_p based on Equation (11):

$$E_p = K_s \left(\frac{-a\pi}{NL \sin(\pi/N)} \right)^N.$$

(12)

The purpose of Equation (12) is to simplify the calculation process of Equation (11) so that one can obtain a closed-form analytical solution for E_p, as reported by Jury and Horton [10]. However, as the assumption that E_p/K_s is much less than 1 may not hold in actual field conditions, one must be cautious for using Equation (12) for estimating E_p for cases where E_p/K_s is not much less than 1. One question is that of how much error may be introduced for using Equation (12) for a certain E_p/K_s value that is not too much less than 1. To answer this question, one needs to develop an accurate solution for a more general case that E_p/K_s is permitted to vary over a wide range of allowable values, which is developed in the following section.

2.2. New Solutions of Evaporation with Arbitrary Surface Matric Potentials

We now extend the steady-state evaporation solution to a general case with an arbitrary matric potential at the bare ground surface without using the assumption that E_p/K_s is much less than 1. Our solutions are based on two popular unsaturated hydraulic conductivity models: the modified Gardner [23] model and the Brooks–Corey [26] model in describing the unsaturated zone flow processes.

2.2.1. Calculation of Evaporation Rate E with the Modified Gardner [23] Model

When the modified Gardner [23] model is of concern (see Equation (4)), for an arbitrary matric potential head at the ground surface, σ_0 is positively finite rather than infinite (as in Section 2.1) and it is given as $\sigma_0 = \left(\frac{E/K_s}{1+E/K_s} \right)^{1/N} \frac{h_0}{a}$. The integration in Equation (9) can be separated into two components: one for $0 < \sigma_0 < 1$ and one for $\sigma_0 \geq 1$.

If $\sigma_0 \geq 1$, Equation (9) can be written as:

$$\int_0^{x_0} \frac{dx}{1+x^N} = \int_0^1 \frac{d\sigma}{1+\sigma^N} d\sigma + \int_1^{x_0} x^{-N} \frac{d\sigma}{1+\sigma^{-N}} = \int_0^1 \sum_{n=0}^{\infty} (-\sigma^N)^n d\sigma + \int_1^{\sigma_0} \sigma^{-N} \sum_{n=0}^{\infty} (-\sigma^{-N})^n d\sigma$$
$$= \sum_{n=0}^{\infty} \frac{(-1)^n(-2Nn-N)}{(Nn+1)(-Nn-N+1)} + \sum \frac{(-1)^n \sigma_0^{-Nn-N+1}}{-Nn-N+1} = (-L/h_0)\sigma_0 \left(1 + \frac{E}{K_s} \right).$$

(13)

If $0 < \sigma_0 < 1$, one can similarly obtain:

$$\sum_{n=0}^{\infty} (-1)^n \frac{\sigma_0^{Nn+1}}{Nn+1} = (-L/h_0)\sigma_0 \left(1 + \frac{E}{K_s} \right).$$

(14)

Therefore, one can calculate the evaporation rate E on the basis of Equation (13) or Equation (14) with a given water table depth L, and parameters N, K_s, and a known from the Gardner [23] model. As E is embedded in the definition of σ_0, such a calculation cannot be carried out using a closed-form solution except for the special case that E/K_s is much less than 1. Rather, a numerical root-searching method such as the Newton–Raphson algorithm [32] may be used.

There are two technical issues that must be taken care of for the computation. The first issue is that since one is unclear if σ_0 is greater or less than 1 before the determination of E, thus one is also unsure whether to use Equation (13) or Equation (14) to perform the computation. To address this issue, we recommend the following steps. Firstly, one should compute E from Equation (13) using the Newton–Raphson method for root-searching [32]. Secondly, after obtaining E, one will check the σ_0

value with the obtained E value. If the σ_0 value is indeed greater than or equal to 1, then Equation (13) is valid. If the σ_0 value is less than 1, then Equation (13) is invalid and one has to use Equation (14) to calculate E.

The second issue is that one has to approximate the infinite series of summation in Equation (13) or Equation (14) by a finite series of summation with sufficient accuracy. By doing so, one can first compute E with a finite M terms approximation of Equation (13) or Equation (14), denoted as E_M. After that, one will repeat the computation of E with $(M + 1)$ terms approximation of Equation (13) or Equation (14), denoted as E_{M+1}. Then, one can check the difference of E_M and E_{M+1}. If $|(E_M - E_{M+1})/(E_M + E_{M+1})|$ is less than a pre-determined small criterion such as 10^{-6}, one can say that the infinite series of summation can be approximated with the finite M terms series of summation with sufficient accuracy. Our numerical exercises show that the M value is usually around 10–50.

2.2.2. Calculation of Evaporation Rate E with the Brooks–Corey [26] Model

The Brooks–Corey function is also widely used for water flow in unsaturated zone. It is commonly associated with Burdine's pore size distribution model [28], leading to the hydraulic conductivity function as follows:

$$K(S) = K_s S^{p+2+2/\lambda}, \tag{15}$$

$$S = \left(\frac{h_v}{h}\right)^{\lambda}, |h_v| < |h|, \tag{16}$$

$$S = 1, |h_v| \geq |h|, \tag{17}$$

where p (positive) is a soil specific parameter that accounts for the tortuosity of the flow [dimensionless], λ (positive) is the pore size distribution index [dimensionless], S is the degree of saturation [dimensionless] and h_v (negative) [L] is the air-entry value of h (negative). The p-value was assumed to be 1.0 in the original study of Brooks and Corey [26].

Defining the following new parameters: $w = p\lambda + 2\lambda + 2$, $\zeta = \left(\frac{E}{K_s}\right)^{1/w}$, $\varphi = \frac{h}{h_v}\left(\frac{E}{K_s}\right)^{1/w}$ and $\varphi_0 = \frac{h_0}{h_v}\left(\frac{E}{K_s}\right)^{1/w}$, where $\varphi_0 \geq \zeta$. As demonstrated in Appendix A, three equations can be obtained from Equation (6):

When $\zeta < 1 \leq \varphi_0$, one has:

$$\sum_{n=0}^{\infty} \frac{(-1)^n \varphi_0^{-wn-w+1}}{-wn-w+1} + \sum_{n=0}^{\infty} \frac{(-1)^n(-2wn-w)}{(wn+1)(-wn-w+1)} - \sum_{n=0}^{\infty} \frac{(-1)^n\left(\frac{\varphi_0 h_v}{h_0}\right)^{wn+1}}{wn+1} = \left(-\frac{L}{h_0}\right)\varphi_0 - \frac{\varphi_0 h_v}{h_0 + h_0\left(\frac{\varphi_0 h_v}{h_0}\right)^{w}}. \tag{18}$$

When $1 \leq \zeta \leq \varphi_0$, one has:

$$\sum_{n=0}^{\infty} \frac{(-1)^n \varphi_0^{-wn-w+1}}{-wn-w+1} - \sum_{n=0}^{\infty} \frac{(-1)^n\left(\frac{\varphi_0 h_v}{h_0}\right)^{-wn-w+1}}{-wn-w+1} = \left(-\frac{L}{h_0}\right)\varphi_0 - \frac{\varphi_0 h_v}{h_0 + h_0\left(\frac{\varphi_0 h_v}{h_0}\right)^{w}}. \tag{19}$$

When $0 < \varphi_0 < 1$, one has:

$$\sum_{n=0}^{\infty} (-1)^n \frac{\varphi_0^{wn+1}}{wn+1} - \sum_{n=0}^{\infty} \frac{(-1)^n\left(\frac{\varphi_0 h_v}{h_0}\right)^{wn+1}}{wn+1} = \left(-\frac{L}{h_0}\right)\varphi_0 - \frac{\varphi_0 h_v}{h_0 + h_0\left(\frac{\varphi_0 h_v}{h_0}\right)^{w}}. \tag{20}$$

The evaporation rate E can be determined from Equations (18)–(20), following the same procedures explained in above Section 2.2.1 for the modified Gardner model, and will not repeat here.

3. Results

3.1. Check of Applicability of Equations (11) and (12)

In the past decades, several researchers [10,24,25] have studied the effect of water table depth on evaporation from ground surface. Salvucci [24] showed that when the fitting parameter N (see Equation (4)) increased, the magnitude of relative evaporation rate should decrease. Warrick [25] presented the values of the relative evaporation (E/K_s) based on the work of Gardner [23] and Brooks–Corey [26], and Warrick [25] reduced the parameter B in the Brooks–Corey model (Equation (3)) to 0 in order to make the problem analytically amendable, thus it can be regarded as a special case of the Brooks–Corey model that may not be applicable to soils with B value not equaling to or very close to zero.

Jury and Horton [10] gave two equations for calculating the bare ground evaporation, which are Equations (11) and (12) of this study. However, applicability of these two equations for actual soils is not investigated in sufficient detail. Based on four different soil types listed in Table 1, the results of calculated relative potential evaporation rates (E_p/K_s) for water table depths ranging from 10 cm to 1000 cm by Equation (11) are listed in Table 2. A few interesting observations can be made from Table 2. Firstly, when the water table is as shallow as 10 cm, E_p/K_s values for all four soil types are greater than 1.0, thus Equation (12) cannot be used to calculate the evaporation rate as this equation requires that E_p/K_s is much less than 1. Secondly, for a water table depth of 50 cm, Equation (12) is still not applicable as E_p/K_s values are not much less than 1, particularly for the case of Pachappa fine sandy loam, which has an E_p/K_s value of 0.96. Thirdly, for a water table depth of 100 cm, Equation (12) should be applicable for Buckeye fine sand and Yolo Light Clay, but is not recommended for Chino Clay and Pachappa fine sandy loam. Fourthly, for water table depth greater than 300 cm, Equation (12) is applicable for all four soil types as the E_p/K_s values are all less than 0.016.

From the above analysis, one can see that whether E_p/K_s is much less than 1 or greater than 1 depends on the soil properties and the water table depth. For instance, when the water table depth is greater than 300 cm, the E_p/K_s values for the four soil types of Table 2 are all much less than 1. However, for the same soil types, the E_p/K_s values become greater than 1 when the water table depth is as shallow as 10 cm.

Table 1. Soil parameters used on the modified Gardner model.

Soil Site/Type	Parameter Value	References
Chino Clay	$N = 2$, $a = -23.8$ cm	[33]
Pachappa (fine sandy loam)	$N = 3$, $a = -63.83$ cm	[33]
Buckeye (fine sand)	$N = 5$, $a = -44.7$ cm	[34]
Yolo Light Clay	$N = 1.77$, $a = -15.3$ cm	[30]

To estimate the discrepancy of values produced by Equations (11) (without the assumption that E_p/K_s being much less than 1) and (12) (with the assumption that E_p/K_s being much less than 1), one may use the following formula: $\varepsilon = |E_{11} - E_{12}|/E_{11}$, where E_{11} and E_{12} represent E_p calculated from Equations (11) and (12), respectively. The results of discrepancy for five different soil types are listed in Table 3. Previous experimental data suggested the N values to be 2, 3, 4, 4, 5, and the a values to be -20.8 cm, -86.7 cm, -17 cm, -10.9 cm and -44.7 cm, respectively, for clay loam, silty loam, sandy loam, coarse sand and fine sand in Table 3, where the hydraulic properties of soils were measured by Ashraf [35,36], Rijtema [37] and van Hylckama [34].

Table 2. The E_p/K_s values for different water table depths (L) and soil types.

L (cm)	10	50	100	300	500	1000
Chino Clay, E_p/K_s	3.27	0.40	0.124	0.015	0.0056	<0.0001
Pachappa (fine sandy loam), E_p/K_s	7.07	0.96	0.280	0.016	0.004	0.00045
Buckeye (fine sand), E_p/K_s	4.00	0.29	0.023	0.0001	<0.0001	<0.0001
Yolo Light Clay, E_p/K_s	2.38	0.29	0.096	0.014	0.006	0.002

Note: The soil types of this table are the same as those in Table 1.

Figure 1 [38] shows the values of E_p/K_s for a range of N and $-a/L$ values based on Equation (11). In Figure 1, six different contours of E_p/K_s ranging from 0.05 to 0.00001 are plotted. This figure may be used to quickly estimate the range of evaporation rate based on the soil type parameters a and N for a given water table depth L. By knowing the range of E_p/K_s, one can subsequently estimate the discrepancy range of the results obtained from Equations (11) and (12) (see Table 3). Such a discrepancy range will allow us to decide if Equation (12) or Equation (11) should be used. In this study, we choose 5% discrepancy as the threshold, meaning that if the discrepancy is greater than 5%, Equation (12) is not recommended and one has to use Equation (11); if the discrepancy is less than 5%, one can use Equation (12) as a good approximation of Equation (11). For instance, when E_p/K_s are 0.05 and 0.01, the discrepancy ratios between Equations (11) and (12) for Buckeye soil (fine sand) are 17.8% and 3.9%, respectively. Then, one can conclude that Equation (12) may be applicable when E_p/K_s is 0.01, but not applicable when E_p/K_s is 0.05. However, for clay loam soil, when E_p/K_s are 0.005 and 0.01, the discrepancy ratios between Equations (11) and (12) are 4.8% and 1.0%, respectively. Therefore, Equation (12) may be applicable for both E_p/K_s of 0.005 and 0.01.

The discrepancy of 5% is a matter of authors' choice. If necessary, a different criterion may be used. However, it is better to use a criterion of no more than 10% because of the following consideration. It is notable that the discrepancy discussed here is only theoretical. This means that additional discrepancy associated with the simplification of actual field condition into the conceptual model of this study is unknown and not considered. Therefore, if the theoretical discrepancy criterion is too large, for instance, 10%, the actual discrepancy of applying such a model for field problems would be greater than 10%, which will make the proposed solution less accurate and less reliable.

Now, one may answer the question of what range of E_p/K_s values can be regarded as much less than 1, which is an assumption used in the approximation of Equation (12) [10]. The answer depends on the accuracy requirement. For instance, if one can tolerate 5% of approximation error for the final estimation of the evaporation rate, E_p/K_s values less than 0.01 are acceptable.

Table 3. The discrepancy ratio ($\varepsilon = |E_{11} - E_{12}|/E_{11}$) of results calculated from Equations (11) and (12).

E/K_s Soil	0.05	0.01	0.005	0.001	0.0001	0.00001
Buckeye (fine sand)	17.8%	3.9%	2.0%	0.4%	0.04%	0.004%
clay loam	4.8%	1.0%	0.5%	0.1%	0.01%	0.001%
silty loam	9.3%	2.0%	1.0%	0.2%	0.02%	0.002%
sandy loam	13.6%	2.9%	1.5%	0.3%	0.03%	0.003%
coarse sand	13.6%	2.9%	1.5%	0.3%	0.03%	0.003%

Note: The data of Buckeye (fine sand) and sandy loam were measured by van Hylckama [34] and Ashraf [35,36], respectively, and the data for the other soils were taken from Rijtema [37].

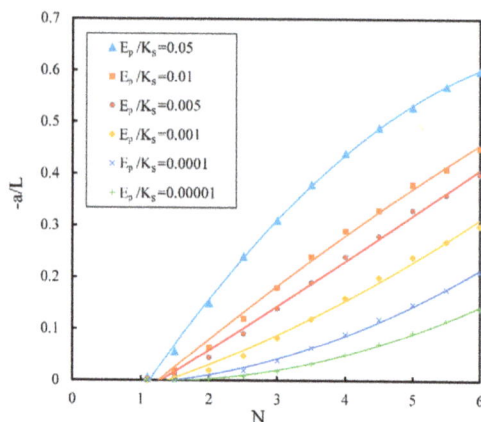

Figure 1. Different contours of E_p/K_s as a function of N and $-a/L$ computed using Equation (11) (modified from Liu [38]).

3.2. Results with Brooks–Corey and Modified Gardner Models

Sadeghi et al. [18] suggested that the Brooks–Corey soil parameters h_v equaled the Gardner fitting parameters a, and $p\lambda + 2\lambda + 2$ equaled the Gardner fitting parameter N for Chino Clay. The results of evaporation rate calculated by the modified Gardner model and the Brooks–Corey model are shown in Figure 2.

The ratios of the E values calculated by the Brooks–Corey and the modified Gardner models for a water table depth of 50 cm and the surface matric potential heads of -100 cm, -200 cm, -300 cm, -500 cm and -1000 cm are 81%, 94%, 96%, 98% and 100%, respectively. The reason to include -1000 cm of surface matric potential head is to simulate the potential evaporation rate (E_p). If changing the water table depth to 100 cm, the ratios of the E values calculated by the Brooks–Corey and the modified Gardner models are 82%, 88%, 89% and 91% for the matric potential heads of -200 cm, -300 cm, -500 cm and -1000 cm, respectively. If further changing the water table depth to 150 cm, the ratios of the E values calculated by the Brooks–Corey versus the modified Gardner models are 72%, 85%, 89%, 92% for the matric potential heads of -200 cm, -300 cm, -500 cm, -1000 cm, respectively.

A few observations can be made for the comparison of the Brooks–Corey versus the modified Gardner models. First, the calculated E values from both models are not too far apart, even for the relatively small matric potential head at the surface. The smallest ratio of the E values between the Brooks–Corey model and the modified Gardner model is 72% for the water table depth of 150 cm and a matric potential head of -200 cm. Second, such a ratio increases with the magnitude of the surface matric potential head for a given water table depth. Third, the E_p values (corresponding to the -1000 cm surface matric potential head) calculated from these two models are very close to each other. For instance, for the shallower water table depth of 50 cm, the E_p values calculated from both models are essentially the same. The greatest discrepancy of the E_p ratio for the water table depth of 150 cm is only 8%.

The Brooks–Corey parameters used above are closely related to the modified Gardner parameters. However, this is not always applicable for some soil types. Rawls et al. [39] summarized the Brooks–Corey fitting parameters for eleven types of soils. Among them, we selected four representative types with the details listed in Table 4. Substituting the Brooks–Corey parameters of Table 4 into Equations (18)–(20), we calculated the evaporation rates for four types of soils under different matric potential heads at ground surface and compared the results with the simulation results of the HYDRUS-1D program. The results for the water table depths of 100 cm were shown in Figure 3. Such calculated E values are very close to their simulated counterparts by HYDRUS-1D for all the cases.

Table 4. Soil parameter values for the Brooks–Corey model in Figure 3 [39].

Soil Type	K_s (cm/d)	H_v (cm)	λ	θ_r (m^{3}m^{-3})	θ_s (m^{3}m^{-3})
Clay loam	5.52	−25.9	0.194	0.075	0.390
Silty loam	16.32	−20.7	0.211	0.015	0.486
Sandy loam	146.6	−8.69	0.474	0.035	0.401
Coarse sand	504.0	−4.92	0.592	0.020	0.417

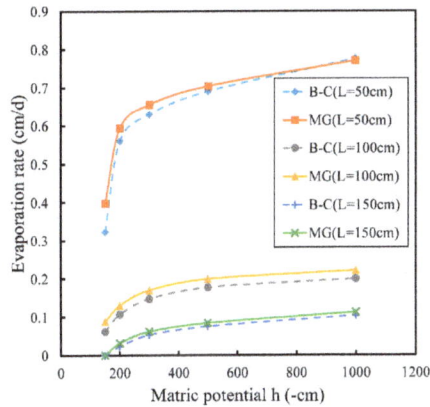

Figure 2. The evaporation rate (cm/d) calculated by the modified Gardner [23] model and the Brooks–Corey [26] model versus the surface matric potential (-cm) for the Chino Clay (see Table 1). MG and B–C represent the modified Gardner's [23] model and the Brooks–Corey [26] model in the figure, respectively.

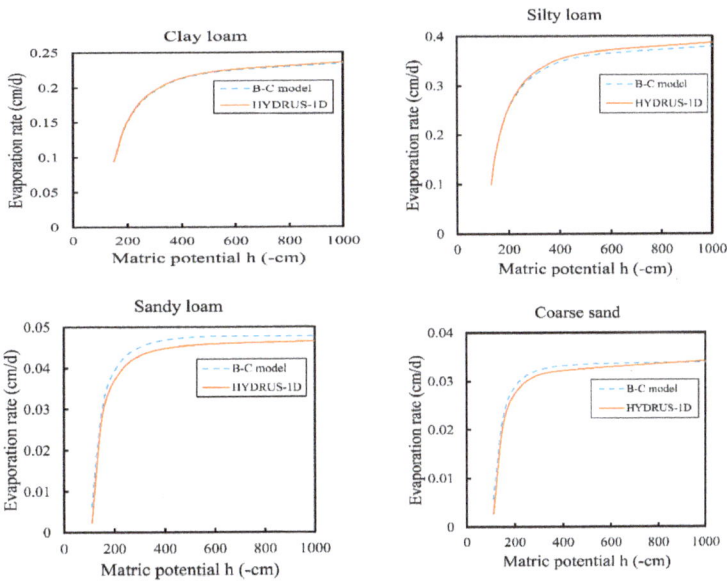

Figure 3. A comparison of the semi-analytical solutions (solid line) calculated with Equations (18)–(20) using the Brooks–Corey model and the results of HYDRUS-1D simulation (dashed-line) for four soils in Table 4.

3.3. Comparison with Previous Work of Sadeghi et al. [18]

Sadeghi et al. [18] developed a closed-form solution to calculate the steady-state evaporation rate by the Brooks–Corey function. The closed-form solution is based on an assumption of "2 + (2 + p)λ is much larger than 1". From the soil water properties estimated by Rawls et al. [39], the values of 2 + (2 + p)λ are actually not much larger than 1. From the soil properties used in Sadeghi et al. [18], there are a few soils that have large values of "2 + (2 + p)λ". Figure 4 compares results calculated by the solution of Sadeghi et al. [18] and our solutions of Equations (18)–(20) for clay loam for a water table depth of 100 cm, where E_{SSJ} is the evaporation rate calculated by the solution of Sadeghi et al. [18], and E_{LZ} is our solutions of Equations (18)–(20). The values of "2 + (2 + p)λ" for four soils in Table 4 are all between 2 and 3. Figure 4 shows that the results of this study are smaller than their counterparts computed from the closed-form solution of Sadeghi et al. [18], with the ratio of E_{SSJ}/E_{LZ} varying between 1.43 and 1.28 for the surface matric potential head changing from −150 cm to −1000 cm. This implies that the closed-form solution of Sadeghi et al. [18] may not be accurate enough to calculate the evaporation rates in these soils.

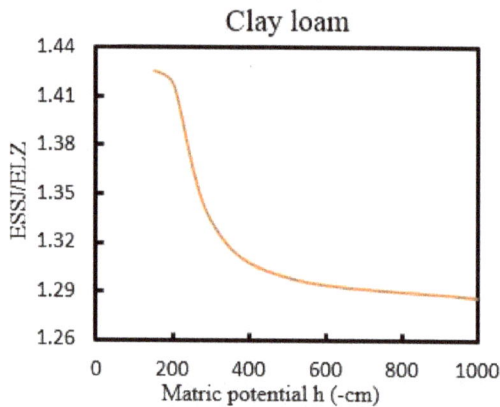

Figure 4. The ratio between Sadeghi's solution (denoted as E_{SSJ} where SSJ represents the first letters of last names of three authors of Sadeghi et al. [18]) and our solution of Equations (18)–(20) (denoted as E_{LZ}, where LZ represents the first letters of last names of this paper) for clay loam and 100 cm depth of water table.

4. Discussion

In this study, the modified Gardner model, the Brooks–Corey model and HYDRUS-1D have been used to calculate the steady-state evaporation rate for an arbitrary matric potential head at ground surface with the presence of a water table. This study is different from most previous analytical and semi-analytical studies that usually focused on estimating the potential evaporation rate at ground surface (with infinitely large matric suction at ground surface). In actual field conditions, the surface suction may be affected by the humid climate, invalidating the infinity matric suction assumption, or the actual evaporation rate may be much less than the potential evaporation rate. This study fills a gap for providing a semi-analytical method to calculate the evaporation rate under an arbitrary surface suction. The new solutions established here may also be used to estimate the difference of the potential and actual evaporation rates for a variety of conditions.

This study selects one type of soil (Chino Clay in Table 1) to demonstrate the application of the proposed method for the modified Gardner model and the Brooks–Corey model. For this soil, the fitting parameters of a and N in the modified Gardner model are directly related to the fitting parameters of p, $λ$, h_v of the Brooks–Corey model in a fashion of $N = pλ + 2λ + 2$ and $h_v = a$ [18], the E values obtained from these two models fit reasonably well with some small but consistent discrepancies (Figure 2). Such small discrepancies probably come from different functions employed for describing these two models.

The comparison between the Brooks–Corey model and the HYDRUS-1D simulation is very good for all the cases. The results show that the method developed in this study is useful for general evaporation rate estimation. We also compare the solution with the analytical solution of Sadeghi et al. [18] (see Equation (11) in Sadeghi et al. [18]), and the result is less satisfactory. This is probably because the assumption that "2 + (2 + p)λ is much larger than 1" as used in Sadeghi et al. [18] is not satisfied for the type of soils used in this study (see Table 4) [39]. This statement is confirmed after a careful inspection of Sadeghi et al. [18], which also discussed the sensitivity of the p-value (which is essentially the same term as "2 + (2 + p)λ" in this paper) for their solution. In Equation (11) of Sadeghi et al. [18], if one sets z on the left side of the equation as the water table depth, one can calculate the r term that is included inside the logarithmic functions on the right side of the equation, where r is the ratio of the evaporation rate over the saturated hydraulic conductivity of the soil. However, Equation (11) of Sadeghi et al. [18] was obtained with the assumption that the p-values were much larger than 1. Therefore, it is not surprising to see that the comparison of this study and Equation (11) of Sadeghi et al. [18] is less satisfactory because the p-values for the type of soils used in this study are not much larger than 1, violating the assumption required for using Equation (11) of Sadeghi et al. [18].

It is worthwhile to carry out well-controlled experiments in the laboratory and/or the field to directly measure the evaporation rate and to compare the experimental results with the theoretical results of this study for a variety of soil types. Such experimental works are not included in this study but should be pursued in the future.

In this study, all solutions are derived under steady-state flow condition, the assumption that the temperature at the soil profile is constant, and the vadose zone being homogeneous. This study does not consider the vapor phase diffusive flow.

Since the semi-analytical approach of this study is based on very few soil samples, its applicability to a broad range of soil samples is still unclear and should be checked in the future. Furthermore, caution should be taken to apply the solutions to deal with realistic field problems, which could be much more complex than the conceptual model of this study. For instance, soil heterogeneity, including multiple layered soil profiles that exist in some actual applications, has not been considered in this study. Preferential flow, which has been seen in some soils, has also not been taken into consideration. The influence of such complex field conditions on evaporation deserves further investigation in the future.

5. Conclusions

The following conclusions can be drawn from this study:

A new mathematical model and associated solutions have been established for computing the evaporation rate at a bare ground surface with an arbitrary surface matric potential for the modified Gardner [23] model and the Brooks–Corey [26] model. The new solutions have been tested extensively against previous solutions (under special conditions) and numerical solutions obtained with HYDRUS-1D and are shown to be robust and accurate.

The comparison of the new solutions and the previous solution of Jury and Horton [10] for computing the potential evaporation rate (see Equations (11) and (12) of this study) indicates that the often used assumption of the relative potential evaporation rate (which is the ratio of potential evaporation rate over the saturated hydraulic conductivity of the soil) being much less than 1 is not established in some conditions, and the actual evaporation rate could deviate considerably from what is calculated on the basis of such an assumption.

The results of this study are smaller than their counterparts computed from the closed-form solution of Sadeghi et al. [18] with the ratio of E_{SSJ}/E_{LZ} varying between 1.43 to 1.28 for the surface matric potential head changing from −150 cm to −1000 cm, where E_{SSJ} and E_{LZ} denote the solution of

Sadeghi et al. [18] and the solution of this study, respectively. This implies that the closed-form solution of Sadeghi et al. [18] may not be accurate enough to calculate the evaporation rates in some soils.

The mathematical model of this study provides a new and straightforward analytical approach to estimate evaporation rate based on a few soil parameters and matric potential on the bare ground surface. This approach is easier to use than the numerical solutions such as HYDRUS-1D, which requires a great number of input parameters and a proper discretization of the domain of interest. The solution developed in this study can also be used as a benchmark to test the numerical solutions that may suffer from various numerical errors. The limitation of the analytical approach is that it cannot deal with complex field conditions such as soil heterogeneity, preferential flow, flow transiency, and irregular boundary conditions, which may be handled by proper numerical models.

Acknowledgments: We sincerely thank four anonymous reviewers for their constructive comments which help us improve the quality of the manuscript.

Author Contributions: Hongbin Zhan and Xin Liu designed the research; Xin Liu performed the research; Hongbin Zhan and Xin Liu discussed the results and wrote the manuscript.

Conflicts of Interest: The authors declare no conflict of interest associated with this manuscript.

Appendix A

Derivation of Equations (18)–(20)

Instead of using the modified Gardner equation, one can use the Brooks and Corey [26] function in Equation (6) to get:

$$\int_0^{h_0} \frac{dh}{1 + E / \left(K_s(S)^{p+2+2/\lambda} \right)} = -L, \tag{A1}$$

where E is the evaporation rate. Defining the following new parameters: $w = p\lambda + 2\lambda + 2$, substituting Equations (16) and (17) into Equation (A1) leads to:

$$\int_{h_v}^{h_0} \frac{dh}{1 + \frac{Eh^w}{K_s h_v{}^w}} + \int_0^{h_v} \frac{dh}{1 + \frac{E}{K_s}} = -L. \tag{A2}$$

Equation (A2) can be simplified as follows:

$$\int_{h_v}^{h_0} \frac{dh}{1 + \frac{Eh^w}{K_s h_v{}^w}} = -L - \frac{h_v K_s}{E + K_s}. \tag{A3}$$

Defining the following new parameters: $w = p\lambda + 2\lambda + 2$, $\zeta = \left(\frac{E}{K_s}\right)^{1/w}$, $\varphi = \frac{h}{h_v}\left(\frac{E}{K_s}\right)^{1/w}$ and $\varphi_0 = \frac{h_0}{h_v}\left(\frac{E}{K_s}\right)^{1/w}$, where $\varphi_0 \geq \zeta$, Equation (A3) then becomes:

$$\int_\zeta^{\varphi_0} \frac{d\varphi}{1 + \varphi^w} = \left(-\frac{L}{h_0}\right)\varphi_0 - \frac{\varphi_0 h_v}{h_0 + h_0\left(\frac{\varphi_0 h_v}{h_0}\right)^w}. \tag{A4}$$

The integration on the left side of Equation (A4) can be dealt with using the same procedures employed in Section 2.2.1 for the modified Gardner model.

When $\zeta < 1 \leq \varphi_0$, one has:

$$\sum_{n=0}^\infty \frac{(-1)^n(-2wn-w)}{(wn+1)(-wn-w+1)} - \sum_{n=0}^\infty \frac{(-1)^n\left(\frac{\varphi_0 h_v}{h_0}\right)^{wn+1}}{wn+1} + \sum_{n=0}^\infty \frac{(-1)^n \varphi_0{}^{-wn-w+1}}{-wn-w+1} = \left(-\frac{L}{h_0}\right)\varphi_0 - \frac{\varphi_0 h_v}{h_0 + h_0\left(\frac{\varphi_0 h_v}{h_0}\right)^w}. \tag{A5}$$

When $1 \leq \xi < \varphi_0$, one has:

$$\sum_{n=0}^{\infty} \frac{(-1)^n \varphi_0^{-wn-w+1}}{-wn-w+1} - \sum_{n=0}^{\infty} \frac{(-1)^n \left(\frac{\varphi_0 h_v}{h_0}\right)^{-wn-w+1}}{-wn-w+1} = \left(-\frac{L}{h_0}\right)\varphi_0 - \frac{\varphi_0 h_v}{h_0 + h_0 \left(\frac{\varphi_0 h_v}{h_0}\right)^w}. \tag{A6}$$

When $0 < \varphi_0 < 1$, one has:

$$\sum_{n=0}^{\infty} (-1)^n \frac{\varphi_0^{wn+1}}{wn+1} - \sum_{n=0}^{\infty} \frac{(-1)^n \left(\frac{\varphi_0 h_v}{h_0}\right)^{wn+1}}{wn+1} = \left(-\frac{L}{h_0}\right)\varphi_0 - \frac{\varphi_0 h_v}{h_0 + h_0 \left(\frac{\varphi_0 h_v}{h_0}\right)^w}. \tag{A7}$$

References

1. Chanzy, A.; Bruckler, L. Significance of soil surface moisture with respect to daily bare soil evaporation. *Water Resour. Res.* **1993**, 29, 1113–1125. [CrossRef]
2. Morton, F.I. Operational estimates of areal evapotranspiration and their significance to the science and practice of hydrology. *J. Hydrol.* **1983**, 66, 1–76. [CrossRef]
3. Asbjornsen, H.; Goldsmith, G.R.; Alvarado-Barrientos, M.S.; Rebel, K.; Van Osch, F.P.; Rietkerk, M.; Chen, J.; Gotsch, S.; Tobon, C.; Geissert, D.R.; et al. Ecohydrological advances and applications in plant–water relations research: A review. *J. Plant Ecol.* **2011**, 4, 3–22. [CrossRef]
4. Feddes, R.A.; de Rooij, G.H.; van Dam, J.C. *Unsaturated-Zone Modeling: Progress, Challenges and Applications;* Kluwer Academic Publishers: Dordrecht, The Netherlands, 2004; 364p.
5. Apollonio, C.; Balacco, G.; Novelli, A.; Tarantino, E.; Piccinni, A.F. Land Use Change Impact on Flooding Areas: The Case Study of Cervaro Basin (Italy). *Sustainability* **2016**, 8, 996. [CrossRef]
6. Novelli, A.; Tarantino, E.; Caradonna, G.; Apollonio, C.; Balacco, G.; Piccinni, F. Improving the ANN Classification Accuracy of Landsat Data Through Spectral Indices and Linear Transformations (PCA and TCT) Aimed at LU/LC Monitoring of a River Basin. In Proceedings of the 16th International Conference on Computational Science and Its Applications, Beijing, China, 4–7 July 2016.
7. Dow, C.L.; DeWalle, D.R. Trends in evaporation and Bowen ratio on urbanizing watersheds in eastern United States. *Water Resour. Res.* **2000**, 36, 1835–1843. [CrossRef]
8. Kalnay, E.; Cai, M. Impact of urbanization and land-use change on climate. *Nature* **2003**, 423, 528. [CrossRef] [PubMed]
9. Plauborg, F. Evaporation from bare soil in a temperate humid climate—Measurement using micro-lysimeters and time domain reflectometry. *Agric. For. Meteorol.* **1995**, 76, 1–17. [CrossRef]
10. Jury, W.A.; Horton, R. *Soil Physics;* John Wiley and Sons: Hoboken, NY, USA, 2004.
11. Shokri, N.; Salvucci, G.D. Evaporation from porous media in the presence of a water table. *Vadose Zone J.* **2011**, 10, 1309–1318. [CrossRef]
12. Rose, D.A.; Konukcu, F.; Gowing, J.W. Effect of watertable depth on evaporation and salt accumulation from saline groundwater. *Aust. J. Soil Res.* **2005**, 43, 565–573. [CrossRef]
13. Gowing, J.W.; Konukcu, F.; Rose, D.A. Evaporative flux from a shallow watertable: The influence of a vapour–liquid phase transition. *J. Hydrol.* **2006**, 321, 77–89. [CrossRef]
14. Il'ichev, A.T.; Tsypkin, G.G.; Pritchard, D.; Richardson, C.N. Instability of the salinity profile during the evaporation of saline groundwater. *J. Fluid Mech.* **2008**, 614, 87–104. [CrossRef]
15. Lehmann, P.; Assouline, S.; Or, D. Characteristic lengths affecting evaporative drying of porous media. *Phys. Rev. E* **2008**, 77, 56309. [CrossRef] [PubMed]
16. Shokri, N.; Lehmann, P.; Or, D. Evaporation from layered porous media. *J. Geophys. Res.* **2010**, 45, W10433. [CrossRef]
17. Assouline, S.; Tyler, S.W.; Selker, J.S.; Lunati, I.; Higgins, C.W.; Parlange, M.B. Evaporation from a shallow water table: Diurnal dynamics of water and heat at the surface of drying sand. *Water Resour. Res.* **2013**, 49, 4022–4034. [CrossRef]
18. Sadeghi, M.; Shokri, N.; Jones, S.B. A novel analytical solution to steady-state evaporation from porous media. *Water Resour. Res.* **2012**, 48, W09516. [CrossRef]

19. Sadeghi, M.; Tuller, M.; Gohardoust, M.R.; Jones, S.B. Column-scale unsaturated hydraulic conductivity estimates in coarse-textured homogeneous and layered soils derived under steady-state evaporation from a water table. *J. Hydrol.* **2014**, *519*, 1238–1248. [CrossRef]

20. Lehmann, P.; Assouline, S.; Or, D. Comment on "Column-scale unsaturated hydraulic conductivity estimates in coarse-textured homogeneous and layered soils derived under steady-state evaporation from a water table" by M. Sadeghi, M. Tuller, MR Gohardoust and SB Jones. *J. Hydrol.* **2015**, *529*, 1274–1276. [CrossRef]

21. Sadeghi, M.; Tuller, M.; Gohardoust, M.R.; Jones, S.B. Reply to comments on "Column-scale unsaturated hydraulic conductivity estimates in coarse-textured homogeneous and layered soils derived under steady-state evaporation from a water table" [J. Hydrol. 2014, 519, 1238–1248]. *J. Hydrol.* **2015**, *529*, 1277–1281. [CrossRef]

22. Hayek, M. An analytical model for steady vertical flux through unsaturated soils with special hydraulic properties. *J. Hydrol.* **2015**, *527*, 1153–1160. [CrossRef]

23. Gardner, W.R. Some steady-state solutions of the unsaturated moisture flow equation with application to evaporation from a water table. *Soil Sci.* **1958**, *85*, 228–232. [CrossRef]

24. Salvucci, G.D. An approximate solution for steady vertical flux of moisture through an unsaturated homogeneous soil. *Water Resour. Res.* **1993**, *29*, 3749–3753. [CrossRef]

25. Warrick, A.W. Additional solutions for steady-state evaporation from a shallow water table. *Soil Sci.* **1988**, *146*, 63–66. [CrossRef]

26. Brooks, R.H.; Corey, A.T. *Hydraulic Properties of Porous Medium*; Colorado State University: Fort Collins, CO, USA, 1964.

27. Bras, R.L. *Hydrology: An Introduction to Hydrologic Science*; Addison-Wesley: Boston, MA, USA, 1990.

28. Burdine, N.T. Relative permeability calculations from pore size distribution data. *Int. Pet. Trans.* **1953**, *198*, 71–78. [CrossRef]

29. Warrick, A.W.; Or, D. Soil water concepts. *Dev. Agric. Eng.* **2007**, *13*, 27–59.

30. Haverkamp, R.; Vauclin, M.; Touma, J.; Wierenga, P.J.; Vachaud, G. A comparison of numerical simulation models for one-dimensional infiltration. *Soil Sci. Soc. Am. J.* **1977**, *41*, 285–294. [CrossRef]

31. Abramowitz, M.; Stegun, I.A. *Handbook of Mathematical Functions with Formula, Graphs and Mathematical Tables*; Dover Publications Inc.: New York, NY, USA, 1972; p. 322.

32. Press, W.H.; Teukolsky, S.A.; Vetterling, W.T.; Flannery, B.P. *Numerical Recipes: The Art of Scientific Computing*, 3rd ed.; Cambridge University Press: New York, NY, USA, 2007.

33. Gardner, W.R.; Fireman, M. Laboratory studies of evaporation from soil columns in the presence of a water table. *Soil Sci.* **1958**, *85*, 244–249. [CrossRef]

34. Van Hylckama, T.E.A. Evaporation from Vegetated and Fallow Soils. *Water Resour. Res.* **1966**, *2*, 99–103. [CrossRef]

35. Ashraf, M. Dynamics of Soil Water under Non-Isothermal Conditions. Ph.D. Thesis, University of Newcastle, Tyne, UK, 1997.

36. Ashraf, M. Water movement through soil in response to water-content and temperature gradients: Evaluation of the theory of Philip and de Vries (1957). *J. Eng. Appl. Sci. Pakistan* **2000**, *19*, 37–51.

37. Rijtema, P.E. *Soil Moisture Forecasting*; University of Wageningen: Wageningen, The Netherlands, 1969.

38. Liu, X. Calculation of Steady-State Evaporation for an Arbitrary Matric Potential at Ground Surface. Master's Thesis, Texas A&M University, College Station, TX, USA, 2014.

39. Rawls, W.J.; Brakensiek, D.L.; Saxtonn, K.E. Estimation of soil water properties. *Trans. Am. Soc. Agric. Eng.* **1982**, *25*, 1316–1320. [CrossRef]

water **MDPI**

Article

An Analytical Model of Fickian and Non-Fickian Dispersion in Evolving-Scale Log-Conductivity Distributions

Marilena Pannone

School of Engineering, University of Basilicata, 85100 Potenza, Italy; marilena.pannone@unibas.it;
Tel.: +39-971-205-147

Received: 14 August 2017; Accepted: 26 September 2017; Published: 30 September 2017

Abstract: The characteristics of solute transport within log-conductivity fields represented by power-law semi-variograms are investigated by an analytical Lagrangian approach that accounts for the automatic frequency cut-off induced by the initial contaminant plume size. The transport process anomaly is critically controlled by the magnitude of the Péclet number. Interestingly enough, unlike the case of fast-decaying correlation functions (i.e., exponential or Gaussian), the presence of intensive transverse diffusion acts as an antagonist mechanism in the process of Fickian regime achievement. On the other hand, for markedly advective conditions and finite initial plume size, even the ergodic longitudinal dispersion coefficient turns out to be asymptotically constant, and the corresponding expected concentration distribution can therefore be obtained by conventional mathematical methods.

Keywords: solute longitudinal dispersion; evolving-scale log-conductivity; first-order analytical approach; stochastic Lagrangian framework

1. Introduction

Modeling the considerable spatial variability exhibited by the hydraulic properties of natural porous formations such as oil reservoirs and aquifers is a key requirement for the monitoring and control of the related flow and transport processes. The classic stochastic theories (e.g., [1,2]) assume that the log-conductivity $Y(\mathbf{x}) = \ln K(\mathbf{x})$, where K is the local hydraulic conductivity and $\mathbf{x}$ is the vector of spatial coordinates, is a homogeneous random space function, normally distributed and completely characterized by constant mean and variance and by a fast-decaying correlation function. The above characteristics imply the existence of a single representative scale of heterogeneity, i.e., the so-called integral scale.

However, in the case of regional transport processes (i.e., over horizontal distances of the order of tens to hundreds aquifer thicknesses), due to the involvement of several geological units called facies, solute transport is typically influenced by several scales of structural variability. According to Neuman [3], the log-K distribution resulting from the coexistence of geological facies might be represented by a complex hierarchy of scales, progressively coming into play as the travel distance increases. In this case, the semi-variogram of Y (i.e., the variance of log-K spatial increments) tends to increase with no asymptotic threshold and likely in a discontinuous fashion. To describe such media in a mathematical framework, the authors of [3] adopted the simple-scaling model, represented by a power-law semi-variogram, $\gamma_Y(r) = ar^b$, where a is a dimensional constant and one-half of the scaling exponent $H = b/2$ is known as the Hurst coefficient [4].

The presence of significant log-conductivity trends affects flow and transport in aquifers to an extent that depends on the complexity of the geologic formation. Additionally, at the regional scale, the more appropriate models of flow and transport are two-dimensional. In this case, the hydraulic properties of the heterogeneous formation, suitably described by the log-transmissivity

$Y(\mathbf{x}) = \ln T(\mathbf{x})$, where the transmissivity T indicates the vertical average of K, are critically influenced by the depositional process. In such context, although the existence of several scales of heterogeneity seems reasonable, there is no direct experimental evidence supporting the power-law model. Neuman [3] provided an indirect justification of it, analyzing the scaling behavior exhibited by the longitudinal dispersivity of tracer plumes.

The spreading of solutes is usually investigated in terms of a longitudinal dispersion coefficient. There are two different types of dispersion coefficient. The ergodic dispersion coefficient is given by the time-rate of change of the single-particle position covariance and incorporates the uncertainty related to the plume centroid location; the effective dispersion coefficient consists in the time-rate of change of the expected longitudinal central spatial moment and only accounts for the heterogeneity at the plume scale. The difference between them was discussed by Fischer et al. [5] in the context of turbulent mixing; by Kitanidis [6], Dagan [7], and Rajaram and Gelhar [8] in the context of transport in natural single-scale porous formations; and by Glimm et al. [9] in the context of transport in natural evolving-scale porous formations. The investigation of longitudinal dispersion in a stochastic framework was later extended by Pannone to river-flow solute transport in the presence of random morphological heterogeneities based on a first-order formulation [10,11], and to strictly uniform river-flows by an exact closed-form solution based on the Taylor-Aris method of moments [12].

The implications of the ergodic or non-ergodic assumption for transport in evolving-scale formations were further discussed by Dagan [13]. In this study, among other things, the author obtained first-order analytical solutions for the effective dispersion coefficient in formations described by power-law semi-variograms. The main finding of the paper was that the ergodic assumption, implicit in most theoretical results of stochastic models, cannot be assumed as generally valid. As a consequence, rather than showing an anomalous continuous growth, effective dispersion in evolving-scale formations reaches a Fickian asymptotic limit for $b < 1$. The transport anomaly was only recovered for $b > 1$, but, in this case, the hypothesis of stationarity (on which the analytical treatment of transport was based) becomes strongly questionable.

The analytical solutions provided by Bellin et al. [14] substantially confirmed Dagan's conclusions concerning the occurrence of anomalous dispersion for $b > 1$ and asymptotically Fickian dispersion for $b < 1$. Additionally, the assumption of linearity typical of any analytical approach was relaxed and tested by solving fully nonlinear flow and transport problems. For $b < 1$, the analytical and numerical solutions provided by [14] were in good agreement. Conversely, for large b (i.e., $b = 1.75$), the numerical solutions slightly overestimated the analytical ones. The anomalous dispersion predicted by the analytical solution was confirmed by the numerical results, at least in the explored range of variability of the log-conductivity variance ($\sigma_Y^2 < 2$). Furthermore, the effective dispersion coefficient turned out to be not affected by the travel-distance cutoff, which was a consequence of the use of finite 2-D field dimensions in the numerical simulations.

Anomalous transport manifests itself in different forms, typically appearing as long tails in the spatial and/or temporal distributions of solute concentrations at given locations [15]. This tailing is classically interpreted as a result of peculiar solute spreading, associated with the existence of multiple scales of medium heterogeneity. Establishing a direct connection between continuous time random walk parameters and randomly heterogeneous hydraulic conductivity within uniform- mean flow domains, the authors of [15] showed that the features of transport cannot be explained by the structural disorder of the geologic formations only. On the contrary, dynamic/flow factors such as low-conductivity transition zones and preferential flow paths critically control the process. Based on that, it can easily be understood how the occurrence of non-Fickian behavior is a highly probable event in cases of transport through fractures [16,17].

Using first-order approximations in velocity fluctuations, the study by Suciu et al. [18] showed how anomalous super-diffusive behavior may result from the linear combination of independent random fields characterized by short-range correlation functions and increasing integral scales, e.g., [3,19]. According to the same type of linear decomposition of log-conductivity fluctuations, the present work proposes a new theoretical approach for the determination of an ergodic macro-dispersion coefficient depending on the initial plume size and the Péclet number magnitude, which is a measure of the relative importance of advective and diffusive-like transport mechanisms.

Detailed investigations of the relationship between injection modes and heterogeneity in the dispersion of solute particles in fracture networks able to store parts of solute mass were recently conducted by [20–22]. Based on the results in [20,21], the type of injection mode has a significant persistent impact on dispersion in a fracture network; more so for larger heterogeneity. The late arrivals for resident injection are several orders of magnitude larger than those related to flux injection, indicating that dispersion along the macroscopic flow direction is significantly enhanced by resident injection. Conversely, the study by [22] showed that, after accounting for a pre-asymptotic regime, the mean travel time of particles inserted using both resident and flux-weighted injection conditions scales linearly, and the tails of the corresponding breakthrough curves exhibit almost identical power laws. Demmy et al. [23] had previously studied the effect of injection modes in heterogeneous porous media. They had found that, for a general case of three-dimensional heterogeneity, uniform resident injection was associated with the nonlinear propagation of mean arrival time, whereas injection in flux was associated with linear propagation. Both injection modes yielded nonlinear arrival time variances, tending to some common asymptotic linear behavior. The moments of the flux-weighted curves were persistently lower than those characterizing the uniform resident injection case, although their propagation rates were converging toward some common value. Dentz et al. [24] developed a continuous time random walk approach for the evolution of Lagrangian velocities in steady heterogeneous flows based on a stochastic relaxation process for the streamwise particle velocities. They predicted Lagrangian particle dynamics starting from an arbitrary initial condition based on the Eulerian velocity distribution and a characteristic correlation scale. The main result of the study can be synthesized in the detection of strong Lagrangian correlation and anomalous dispersion for velocity distributions that are tailed toward low values, as well as in pronounced differences depending on the initial conditions. Note that all mentioned studies ([20–24]) investigate the relationship between injection modes and advective heterogeneity on solute dispersion without discussing the effect of local dispersion magnitude. This effect is expected to be definitely more pronounced for transport in porous media, which is the focus of the present study.

The crucial role of local dispersion (sometimes simply named 'diffusion') in solute macro-dispersion and dilution was already explored in the context of subsurface flow and transport by Pannone and Kitanidis [25] and in the context of river-flow and transport by Pannone [26,27]. Overall, these studies showed that, in the case of heterogeneous structures characterized by short-range correlations, macro-dispersion and dilution are singularly driven by the interplay of advective heterogeneities and diffusive-like mechanisms.

The results of the present investigation, which focuses on the interplay between evolving-scale heterogeneity and diffusion in longitudinal dispersion for uniform instantaneous injections of different sizes, and invariably predicts asymptotically Fickian macro-dispersion for purely advective regimes and super-diffusive transport in the presence of non-negligible local dispersion (regardless of scaling exponent value), are partly in contrast with what has previously been found by similar studies on this topic.

2. Formulation

Let us consider a first type of isotropic power-law semi-variogram, with an exponent ranging between 0 and 1:

$$\gamma_Y(r) = ar^b \quad 0 < b < 1 \tag{1}$$

It can be shown that such a semi-variogram is obtained by the superposition of an infinite hierarchy of independent stationary log-conductivity fields characterized by an exponential covariance function, an increasing integral scale, and variance proportional to it (see also [3]). The total fluctuation of the random function $Y(x) = \ln K(x) = <Y> + Y'(x)$, where the angle brackets indicate the (assumed constant) ensemble mean and the prime indicates the deviation about that mean, is given by:

$$Y'(x) = \sum_{\Lambda=0}^{\infty} Y'_{\Lambda}(x) \tag{2}$$

where Y'_{Λ} is the fluctuation associated with the Λth order of the hierarchy. The generic heterogeneous sub-unit is therefore represented by the superposition of a given finite number of stationary fields of increasing order:

$$Y^{(m)}(x) = \sum_{\Lambda=m}^{\infty} Y'_{\Lambda}(x) \tag{3}$$

For each stationary field, one has:

$$\gamma_{Y\Lambda}(r) = \sigma^2_{Y\Lambda}(I_{Y\Lambda})[1 - \exp(-r/I_{Y\Lambda})] \tag{4}$$

where the integral scale $I_{Y\Lambda}$ can be viewed as the inverse of the wave-number Λ, which represents the spatial periodicity of the associated log-conductivity heterogeneity:

$$I_{Y\Lambda} = \frac{1}{\Lambda} \tag{5}$$

Let us assume that the corresponding variance $\sigma^2_{Y\Lambda}$ is a negative power of Λ:

$$\sigma^2_{Y\Lambda} = \frac{c}{\Lambda^{1+\beta}} \quad 0 < \beta < 1 \tag{6}$$

where c is a dimensional constant. Integrating over all the possible wave-numbers:

$$\gamma_Y(r) = \int_0^{\infty} \frac{c}{\Lambda^{1+\beta}}[1 - \exp(-r\Lambda)]d\Lambda \tag{7}$$

one obtains [28]:

$$\gamma_Y(r) = -c\Gamma(-\beta)r^{\beta} \tag{8}$$

where the symbol Γ indicates a Gamma function. Equation (8) coincides with Equation (1) if:

$$b = \beta \quad a = -c\Gamma(-b) \tag{9}$$

For semi-variogram exponents ranging between 1 and 2:

$$\gamma_Y(r) = ar^b \quad 1 \le b < 2 \tag{10}$$

it can be shown that the statistically independent components of the hierarchy must be characterized by a Gaussian log-conductivity covariance function (see also [18]):

$$\gamma_{Y\Lambda}(r) = \sigma^2_{Y\Lambda}(I_{Y\Lambda})\left[1 - \exp\left(-r^2/l^2_{Y\Lambda}\right)\right] \tag{11}$$

where $l_{Y\Lambda}$ indicates the corresponding correlation length ($l_{Y\Lambda} = 2I_{Y\Lambda}/\sqrt{\pi}$) and the variance is given by:

$$\sigma^2_{Y\Lambda} = \frac{c}{\Lambda^{2+\beta}} \quad 0 \le \beta < 1 \tag{12}$$

Integrating over Λ:

$$\gamma_Y(r) = \int_0^\infty \frac{c}{\Lambda^{2+\beta}}\left[1 - \exp\left(-\pi r^2 \Lambda^2/4\right)\right]d\Lambda \tag{13}$$

one obtains [28]:

$$\gamma_Y(r) = -c\Gamma\left(-\frac{\beta+1}{2}\right)\left(\frac{\pi}{4}\right)^{\frac{\beta+1}{2}}r^{\beta+1} \tag{14}$$

Equation (14) coincides with Equation (10) if:

$$b = \beta + 1 \quad a = -c\Gamma\left(-\frac{b}{2}\right)\left(\frac{\sqrt{\pi}}{2}\right)^b \tag{15}$$

Given the mutual independence of the short-range log-conductivity fields, one can consider the whole hierarchy as decomposed into two different macroscopic components:

$$Y'(\mathbf{x}) = \tilde{Y}(\mathbf{x}) + \overline{Y}(\mathbf{x}) \tag{16}$$

where

$$\tilde{Y}(\mathbf{x}) = \sum_{\Lambda=N+1}^{\infty} Y'_\Lambda(\mathbf{x}) \tag{17}$$

identifies the log-conductivity fluctuation due to the $(N+1)^{th}$ heterogeneous sub-unit and

$$\overline{Y}(\mathbf{x}) = \sum_{\Lambda=0}^{N} Y'_\Lambda(\mathbf{x}) \tag{18}$$

is the log-conductivity fluctuation induced by the larger scales of heterogeneity.

From Darcy's law combined with the equation of continuity for incompressible fluids in incompressible solid matrices, one obtains the following steady flow equation:

$$\begin{aligned}\nabla^2 h(\mathbf{x}) + \nabla h(\mathbf{x}) \cdot \nabla Y(\mathbf{x}) \\ = \nabla^2 h'(\mathbf{x}) - \mathbf{J} \cdot \nabla Y'(\mathbf{x}) + \nabla h'(\mathbf{x}) \cdot \nabla Y'(\mathbf{x}) = 0\end{aligned} \tag{19}$$

where h indicates the hydraulic head and

$$\mathbf{J} = -\nabla < h(\mathbf{x}) > \tag{20}$$

indicates the mean head loss. In the Fourier domain, Equation (19) becomes:

$$|\mathbf{k}|^2 \hat{h}(\mathbf{k}) + \int_{\mathbf{k}'} \mathbf{k}' \cdot (\mathbf{k} - \mathbf{k}')\hat{Y}(\mathbf{k} - \mathbf{k}')\hat{h}(\mathbf{k}')d\mathbf{k}' = \frac{j}{2\pi}\mathbf{J} \cdot \mathbf{k}\hat{Y}(\mathbf{k}) \tag{21}$$

where $j = \sqrt{-1}$ and the circumflex accent indicates Fourier transforms:

$$\hat{h}(\mathbf{k}) = \int_{\mathbf{x}} h'(\mathbf{x}) \exp(-j2\pi\mathbf{k} \cdot \mathbf{x})d\mathbf{x} \tag{22}$$

$$\hat{Y}(\mathbf{k}) = \int_{\mathbf{x}} Y'(\mathbf{x}) \exp(-j2\pi\mathbf{k} \cdot \mathbf{x})d\mathbf{x} \tag{23}$$

Let us assume that constant c is so small that, even for the larger sub-units, the log-conductivity variance is always finite and not larger than 1. Physically speaking, that means that the present theory concerns geologic formations made of nested (fractal-like), though mildly heterogeneous, porous structures. In this case, flow and transport linear theory (e.g., [1]) applies and the convolution term in Equation (21) can be neglected:

$$|\mathbf{k}|^2 \hat{h}(\mathbf{k}) = \frac{j}{2\pi} \mathbf{J} \cdot \mathbf{k}\hat{Y}(\mathbf{k}) \tag{24}$$

Incidentally, as mentioned above, the 2-D numerical simulations by Bellin et al. [14] allowed it to be established that linear theory practically applies for log-transmissivity variance up to two. Similarly, from the first-order Darcy's law (e.g., [1]):

$$\hat{v}_i(\mathbf{k}) = \hat{Y}(\mathbf{k}) \left(U_i - \sum_{j=1}^{3} \frac{k_i k_j}{|\mathbf{k}|^2} U_j \right) \tag{25}$$

where $\hat{v}_i$ is the Fourier transform of the ith component of velocity fluctuation and $\mathbf{U} = (U_1, U_2, U_3)$ is the constant ensemble mean velocity. From Equations (24) and (25) one can see that, at the first-order in the log-conductivity variance, each Fourier component of the log-conductivity field corresponds to a single component of hydraulic head and velocity. Therefore, the same properties of linear superposition holding for Y (see Equation (2)) apply to h and $\mathbf{v}$ as well. Consider now an initial solute injection at a uniform concentration C_0, confined within a volume

$$V_0 = l_0^3 \tag{26}$$

such that

$$I_{YN+1} < l_0 < I_{YN} \tag{27}$$

The trajectory of the generic solute particle belonging to the dispersing plume can be expressed as:

$$\begin{aligned} \mathbf{X}(t, \mathbf{a}) &= \mathbf{a} + \int_0^t \mathbf{v}[\mathbf{X}(\tau, \mathbf{a})]d\tau + \mathbf{X}_B(t) \\ &= \mathbf{a} + \int_0^t \langle \mathbf{v}[\mathbf{X}(\tau, \mathbf{a})] \rangle d\tau + \int_0^t \mathbf{v}'[\mathbf{X}(\tau, \mathbf{a})]d\tau + \mathbf{X}_B(t) \end{aligned} \tag{28}$$

where vector $\mathbf{a}$ represents its initial position within V_0, t is the time, and $\mathbf{X}_B$ the zero-mean Brownian component. Given Equations (16) and (25), it is also:

$$\mathbf{X}(t, \mathbf{a}) = \mathbf{a} + \mathbf{U}t + \mathbf{X}'(t, \mathbf{a}) + \mathbf{X}_B(t) = \mathbf{a} + \mathbf{U}t + \tilde{\mathbf{X}}(t, \mathbf{a}) + \overline{\mathbf{X}}(t, \mathbf{a}) + \mathbf{X}_B(t) \tag{29}$$

where, at the first order:

$$\tilde{\mathbf{X}}(t, \mathbf{a}) = \sum_{\Lambda=N+1}^{\infty} \mathbf{X}'_\Lambda(t, \mathbf{a}) = \int_0^t \tilde{\mathbf{v}}[\mathbf{X}(\tau, \mathbf{a})]d\tau \cong \int_0^t \tilde{\mathbf{v}}[\mathbf{a} + \mathbf{U}\tau + \mathbf{X}_B(\tau)]d\tau \tag{30}$$

and

$$\overline{\mathbf{X}}(t, \mathbf{a}) = \sum_{\Lambda=0}^{N} \mathbf{X}'_\Lambda(t, \mathbf{a}) \cong \int_0^t \overline{\mathbf{v}}[\mathbf{a} + \mathbf{U}\tau + \mathbf{X}_B(\tau)]d\tau \tag{31}$$

Notice that, due to the linearization involved in Equations (30) and (31), the solute particles sample the velocity distribution along longitudinal deterministic trajectories ($\mathbf{a} + \mathbf{U}\tau$), disturbed only by the local-dispersive contribution represented by $\mathbf{X}_B$. Such an assumption is common to all first-order (linearized) analytical formulations of subsurface flow and transport. Its physical meaning is that one neglects the self-feeding mechanism of advective dispersion that would emerge from the solution of the exact integro-differential equation:

$$X\prime(t, \mathbf{a}) = \int\limits_0^t \mathbf{v}'[\mathbf{a} + \mathbf{U}\tau + \mathbf{X}'(\tau, \mathbf{a}) + \mathbf{X}_B(\tau)]d\tau \tag{32}$$

A possible theoretical justification of it is that, overall, the reduced spreading of the particle sampling cloud left to the only Brownian component of fluctuation balances the more persistent correlation (and, therefore, the slower spreading) that would be induced by the advective fluctuation.

The concentration spatial moments, i.e., total mass, centroid and inertia, can respectively be calculated from:

$$M = \int nCd\mathbf{x} \tag{33}$$

$$\mathbf{R}(t) = \frac{1}{M}\int nC\mathbf{x}d\mathbf{x} \tag{34}$$

and

$$S_{ij}(t) = \frac{1}{M}\int nC(x_i - R_i)(x_j - R_j)d\mathbf{x} \tag{35}$$

where $C = C(\mathbf{x}, t)$ indicates the concentration in $\mathbf{x}$ at time t. For a single-particle injection in $\mathbf{x} = \mathbf{a}$ (e.g., [1]):

$$\Delta C(\mathbf{x}, t) = \frac{\Delta M}{n}\delta[\mathbf{x} - \mathbf{X}(t, \mathbf{a})] = \frac{C_0(\mathbf{a})n_0 d\mathbf{a}}{n}\delta[\mathbf{x} - \mathbf{X}(t, \mathbf{a})] \tag{36}$$

where δ indicates Dirac's distribution, ΔM is the associated mass, C_0 is the associated initial concentration, n is the generic volume porosity, and n_0 is the volume porosity at injection site. Integrating over the whole initial volume V_0 for $n \cong n_0$ gives:

$$C(\mathbf{x}, t) = \int\limits_{V_0} C_0(\mathbf{a})\delta[\mathbf{x} - \mathbf{X}(t, \mathbf{a})]d\mathbf{a} \tag{37}$$

The substitution of Equation (37) into Equations (34) and (35) for C_0 = const and $M = n_0C_0V_0$ yields:

$$\mathbf{R}(t) = \frac{1}{V_0}\int\limits_{V_0} \mathbf{X}(t, \mathbf{a})d\mathbf{a} = \bar{\mathbf{a}} + \mathbf{U}t + \frac{1}{V_0}\int\limits_{V_0} [\mathbf{X}'(t, \mathbf{a}) + \mathbf{X}_B(t)]d\mathbf{a} \tag{38}$$

and

$$S_{ij}(t) = \frac{1}{V_0}\int\limits_{V_0} [X_i(t, \mathbf{a}) - R_i(t)][X_j(t, \mathbf{a}) - R_j(t)]d\mathbf{a}$$

$$= S_{ij}(0) + \frac{1}{V_0}\int\limits_{V_0} \left[X_i'(t, \mathbf{a}) - \frac{1}{V_0}\int\limits_{V_0} X_i'(t, \mathbf{a}')d\mathbf{a}'\right]\left[X_j'(t, \mathbf{a}) - \frac{1}{V_0}\int\limits_{V_0} X_j'(t, \mathbf{a}'')d\mathbf{a}''\right]d\mathbf{a} \tag{39}$$

$$+ X_{Bi}(t)X_{Bj}(t)$$

where

$$S_{ij}(0) = \frac{1}{V_0}\int\limits_{V_0} (a_i - \bar{a}_i)(a_j - \bar{a}_j)d\mathbf{a} \tag{40}$$

is the generic initial inertia moment and

$$\bar{a}_r = \frac{1}{V_0}\int\limits_{V_0} a_r d\mathbf{a} \tag{41}$$

is the rth component of initial centroid vector position. Ensemble averaging will be performed on Equation (39) recalling that, by definition, the ensemble mean of a random fluctuation is zero and that Equation (16) and flow linear treatment allow it to be assumed that:

$$\langle \widetilde{\mathbf{X}}(t, \mathbf{a}) \overline{\mathbf{X}}(t, \mathbf{a}) \rangle = 0 \tag{42}$$

Additionally, we know that, for Brownian displacements:

$$\langle X_{Bi}(t) X_{Bj}(t) \rangle = 2D\delta_{ij}t \tag{43}$$

where D is the coefficient of the (assumed isotropic) local dispersion and δ_{ij} is Kronecker's Delta. Based on Equation (31), for negligible local dispersion and provided that the integral scales of all fields of the $\overline{Y}$-hierarchy are larger than the initial plume size, the corresponding components of particle positions can be viewed as almost fully correlated (i.e., as if the particles were concentrated in a single point) at any time:

$$\langle \overline{X}_i(t, \mathbf{a}') \overline{X}_j(t, \mathbf{a}'') \rangle \cong \langle \overline{X}_i(t, \mathbf{a}') \overline{X}_j(t, \mathbf{a}') \rangle \tag{44}$$

Thus:

$$\begin{aligned}
\langle S_{ij}(t) \rangle &= S_{ij}(0) + \frac{1}{V_0} \int_{V_0} \langle \overline{X}_i(t, \mathbf{a}) \overline{X}_j(t, \mathbf{a}) \rangle d\mathbf{a} - \frac{1}{V_0^2} \int_{V_0} \int_{V_0} \langle \overline{X}_i(t, \mathbf{a}') \overline{X}_j(t, \mathbf{a}'') \rangle d\mathbf{a}' d\mathbf{a}'' \\
&+ \frac{1}{V_0} \int_{V_0} \langle \widetilde{X}_i(t, \mathbf{a}) \widetilde{X}_j(t, \mathbf{a}) \rangle d\mathbf{a} - \frac{1}{V_0^2} \int_{V_0} \int_{V_0} \langle \widetilde{X}_i(t, \mathbf{a}') \widetilde{X}_j(t, \mathbf{a}'') \rangle d\mathbf{a}' d\mathbf{a}'' \\
&= S_{ij}(0) + \widetilde{X}_{ij}(t) - \widetilde{R}_{ij}(t)
\end{aligned} \tag{45}$$

where

$$\widetilde{X}_{ij}(t) = \frac{1}{V_0} \int_{V_0} \langle \widetilde{X}_i(t, \mathbf{a}) \widetilde{X}_j(t, \mathbf{a}) \rangle d\mathbf{a} \tag{46}$$

indicates the single-trajectory covariance due to the $\widetilde{Y}$-hierarchy and

$$\widetilde{R}_{ij}(t) = \frac{1}{V_0^2} \int_{V_0} \int_{V_0} \langle \widetilde{X}_i(t, \mathbf{a}') \widetilde{X}_j(t, \mathbf{a}'') \rangle d\mathbf{a}' d\mathbf{a}'' \tag{47}$$

indicates the related centroid covariance. The $\widetilde{Y}$-hierarchy is equivalent to a geological sub-unit characterized by a finite correlation length (or integral scale), which is smaller than the initial plume size. Therefore, at large times, and unlike the case of the $\overline{Y}$-hierarchy, the corresponding components of different particle positions will tend to become asymptotically uncorrelated. As a result:

$$\widetilde{R}_{ij}(t) \to \frac{1}{V_0^2} \int_{V_0} \int_{V_0} \langle \widetilde{X}_i(t, \mathbf{a}') \rangle \langle \widetilde{X}_j(t, \mathbf{a}'') \rangle d\mathbf{a}' d\mathbf{a}'' = 0 \tag{48}$$

In this case:

$$\langle S_{ij}(t) \rangle = S_{ij}(0) + \widetilde{X}_{ij}(t) \tag{49}$$

and the effective large-time macro-dispersion coefficient coincides with the ergodic macro-dispersion coefficient:

$$D_{Mij}(t) = \frac{1}{2} \frac{d\langle S_{ij} \rangle}{dt} = \frac{1}{2} \frac{d\widetilde{X}_{ij}}{dt} \tag{50}$$

The novelty of Equations (49) and (50) as compared to the results by Kitanidis [6], Dagan [7], and Rajaram and Gelhar [8] in the context of transport in natural, single-scale, porous formations consists in the definition of the ergodic dispersion coefficient. Indeed, in the case of the evolving-scale heterogeneity represented by power-law semi-variograms and based on a first-order approach, there is an automatic cutoff in terms of the scales of heterogeneity actually involved in the dispersion process.

Such cutoff, the effects of which are synthesized by Equation (44) for negligible local dispersion, is determined by the finite initial plume size.

Due to the linearity implied by Equations (2), (24), and (25), the derivation of D_{Mij} for log-conductivity fields characterized by evolving scales of heterogeneity and power-law semi-variograms can be pursued by computing the corresponding coefficient for any stationary field of the continuous hierarchy (exponential covariance for $0 < b < 1$ and Gaussian covariance for $1 \leq b < 2$) and by integrating the result over the truncated frequency domain. See Appendix A for the derivation of $D_{Mij}(\infty, \Lambda)$ in the case of 3-D stationary exponential and Gaussian log-K covariance. To obtain the global asymptotic macro-dispersion coefficient, given by the linear combination of the macro-dispersion coefficients characterizing the heterogeneous single-scale fields of the hierarchy, one has to compute:

$$\tilde{D}_{Mij}(\infty) = \lim_{t \to \infty} \left(\frac{1}{2} \frac{d\tilde{X}_{ij}}{dt} \right) = \lim_{t \to \infty} \left(\frac{1}{2} \sum_{\Lambda=N+1}^{\infty} \frac{dX_{ij\Lambda}}{dt} \right) \tag{51}$$

where

$$\lim_{t \to \infty} \left(\frac{1}{2} \frac{dX_{ij\Lambda}}{dt} \right) = D_{Mij}(\infty, I_{Y\Lambda}) \tag{52}$$

and

$$\lim_{t \to \infty} \left(\frac{1}{2} \sum_{\Lambda=N+1}^{\infty} \frac{dX_{ij\Lambda}}{dt} \right) = \int_0^{l_0} D_{Mij}(\infty, I_{Y\Lambda}) dI_{Y\Lambda} = \int_{\Lambda_0}^{\infty} D_{Mij}(\infty, \Lambda) d\Lambda \tag{53}$$

with $\Lambda_0 = 1/l_0$. From Equation (A7), respectively for exponential and Gaussian hierarchy:

$$D_{M11}(\infty, \Lambda) = \frac{cU}{\Lambda^{2+\beta}} \tag{54}$$

and

$$D_{M11}(\infty, \Lambda) = \frac{cU}{\Lambda^{3+\beta}} \tag{55}$$

Thus:

$$\tilde{D}_{M11}(\infty) = \int_{\Lambda_0}^{\infty} \frac{cU}{\Lambda^{2+\beta}} d\Lambda = -\frac{aU}{\Gamma(-b)} \frac{1}{(1+b)\Lambda_0^{1+b}}$$
$$= -\frac{aU}{\Gamma(-b)} \frac{l_0^{1+b}}{(1+b)} \quad 0 < b < 1 \tag{56}$$

and

$$\tilde{D}_{M11}(\infty) = \int_{\Lambda_0}^{\infty} \frac{cU}{\Lambda^{3+\beta}} d\Lambda = -\frac{aU}{\Gamma(-\frac{b}{2})\left(\frac{\sqrt{\pi}}{2}\right)^b} \frac{1}{(1+b)\Lambda_0^{1+b}}$$
$$= -\frac{aU}{\Gamma(-\frac{b}{2})\left(\frac{\sqrt{\pi}}{2}\right)^b} \frac{l_0^{1+b}}{(1+b)} \quad 1 \leq b < 2 \tag{57}$$

Therefore, for any value of exponent b, transport turns out to be asymptotically ergodic and Fickian.

Note that the truncated semi-variograms for the exponential and the Gaussian hierarchy are respectively given by:

$$\gamma_{Yt}(r) = \int_{\Lambda_0}^{\infty} \frac{c}{\Lambda^{1+b}} [1 - \exp(-r\Lambda)] d\Lambda \quad 0 < b < 1 \tag{58}$$

and

$$\gamma_{Yt}(r) = \int_{\Lambda_0}^{\infty} \frac{c}{\Lambda^{1+b}} \left[1 - \exp\left(-\frac{\pi r^2 \Lambda^2}{4} \right) \right] d\Lambda \quad 1 \leq b < 2 \tag{59}$$

with the following large-distance approximations:

$$\gamma_{Yt}(r) \cong -\frac{al_0^b}{b\Gamma(-b)}\left[1 - \frac{bl_0^{1-b}\exp(-r/l_0)}{r^{1+b}}\right] \quad 0 < b < 1 \tag{60}$$

$$\gamma_{Yt}(r) \cong -\frac{al_0^b}{b\Gamma\left(-\frac{b}{2}\right)\left(\frac{\sqrt{\pi}}{2}\right)^b}\left[1 - \left(\frac{2^{1+b}}{\pi^{1+b/2}}\right)\frac{bl_0^{2-b}\exp\left(-r^2\pi/4l_0^2\right)}{r^{2+b}}\right] \quad 1 \le b < 2 \tag{61}$$

where

$$\lim_{r\to\infty}\gamma_{Yt}(r) = \sigma_{Y0}^2 = -\frac{al_0^b}{b\Gamma(-b)} \quad 0 < b < 1 \tag{62}$$

$$\lim_{r\to\infty}\gamma_{Yt}(r) = \sigma_{Y0}^2 = -\frac{al_0^b}{b\Gamma\left(-\frac{b}{2}\right)\left(\frac{\sqrt{\pi}}{2}\right)^b} \quad 1 \le b < 2 \tag{63}$$

indicate the truncated-field variances. Thus,

$$\tilde{D}_{M11}(\infty) = \frac{\sigma_{Y0}^2 bUl_0}{(1+b)} \quad 0 < b < 2 \tag{64}$$

In dimensionless terms:

$$D_{M11} = \frac{\tilde{D}_{M11}(\infty)}{\sigma_{Y0}^2 Ul_0} = \frac{b}{(1+b)} \quad 0 < b < 2 \tag{65}$$

In the case of non-negligible local dispersion, the subdivision expressed by Equations (30) and (31) is a mobile one. Indeed, even if the $\overline{Y}$-hierarchy is characterized by integral scales larger than l_0, particles' displacement includes a local-dispersive component that makes the original distances increase as $\sim\sqrt{2Dt}$. Thus, the threshold sub-unit corresponding to the subdivision into $\tilde{X}$- and $\overline{X}$-displacement hierarchy changes in time, and the boundary wave-number is now $\Lambda_0' = \left(l_0 + \chi\sqrt{2Dt}\right)^{-1}$, with χ indicating a suitable constant related to the assumed width of the Brownian-Gaussian bell. Equations (56) and (57) transform into:

$$\tilde{D}_{M11}(t) = \int_{\Lambda_0'}^{\infty}\frac{cU}{\Lambda^{2+\beta}}d\Lambda + D = -\frac{aU}{\Gamma(-b)}\frac{\left(l_0 + \chi\sqrt{2Dt}\right)^{1+b}}{(1+b)} + D \quad 0 < b < 1 \tag{66}$$

and

$$\tilde{D}_{M11}(t) = \int_{\Lambda_0'}^{\infty}\frac{cU}{\Lambda^{3+\beta}}d\Lambda + D = -\frac{aU}{\Gamma\left(-\frac{b}{2}\right)\left(\frac{\sqrt{\pi}}{2}\right)^b}\frac{\left(l_0 + \chi\sqrt{2Dt}\right)^{1+b}}{(1+b)} + D \quad 1 \le b < 2 \tag{67}$$

or

$$\tilde{D}_{M11}(t) = \frac{\sigma_{Y0}^2 bU}{(1+b)}\frac{\left(l_0 + \chi\sqrt{2Dt}\right)^{1+b}}{l_0^b} + D \quad 0 < b < 2 \tag{68}$$

In dimensionless terms:

$$D_{M11}(\tau) = \frac{\tilde{D}_{M11}(t)}{\sigma_{Y0}^2 Ul_0} = \frac{b}{(1+b)}\left(1 + \chi\sqrt{\frac{2\tau}{Pe_0}}\right)^{1+b} + \frac{1}{\sigma_{Y0}^2 Pe_0} \quad 0 < b < 2 \tag{69}$$

with

$$Pe_0 = \frac{Ul_0}{D} \tag{70}$$

and

$$\tau = \frac{tU}{l_0} \tag{71}$$

Thus, in these conditions, transport tends to be asymptotically non-ergodic, due to the mobile threshold wave number $\Lambda_0' = \left(l_0 + \chi\sqrt{2Dt}\right)^{-1}$ that makes the travel time needed to achieve condition (48) larger and larger, and non-Fickian or super-diffusive (D_{M11} increases in time).

Equation (69) can be integrated to obtain the dimensionless particle covariance:

$$X_{11}(\tau) = \frac{\tilde{X}_{11}(t)}{\sigma_{Y0}^2 l_0^2} = 2\int_0^\tau D_{M11}d\tau = \frac{2bPe_0}{\chi(1+b)(2+b)}\left(1+\chi\sqrt{\frac{2\tau}{Pe_0}}\right)^{2+b}\sqrt{\frac{2\tau}{Pe_0}} +$$
$$\frac{2bPe_0}{\chi^2(1+b)(2+b)(3+b)}\left(1+\chi\sqrt{\frac{2\tau}{Pe_0}}\right)^{3+b} + \frac{2bPe_0}{\chi^2(1+b)(2+b)(3+b)} \quad 0 < b < 2 \tag{72}$$

Finally, it should be emphasized that, due to the invariance of the longitudinal large-time macro-dispersion coefficient for both types of short-range correlations (exponential and Gaussian) with respect to the dimensionality of the flow domain, the longitudinal macro-dispersion coefficient obtained for isotropic 3-D evolving-scale formations (Equations (56) and (57), Equations (66)–(67)) applies also to 2-D cases, with Y referring now to log-transmissivity.

3. Results

Based on the statistical equivalence between the evolving-scale log-conductivity fields represented by power-law semi-variograms and the superposition of independent log-conductivity fields characterized by short-range correlations (exponential or Gaussian) and increasing integral scale, the present work allowed it to be established that:

1. Assuming the validity of the linear mathematical treatment for subsurface flow and transport, it is always possible to subdivide the solute particle displacement in two big components (Equations (30) and (31)), respectively influenced by the fields of the log-conductivity hierarchy characterized by integral scales smaller and larger than the initial plume size.

2. In the presence of markedly advective regimes and negligible local dispersion (or diffusion), the second component of the displacement hierarchy referred to different particles is characterized by almost perfect correlation at any time. For that reason, it is possible to rewrite a well-known relation involving ensemble mean inertia moment, particle position covariance, and centroid covariance (e.g., [6,7]) in a formally identical but substantially different way:

$$\langle S_{ij}(t) \rangle = S_{ij}(0) + \tilde{X}_{ij}(t) - \tilde{R}_{ij}(t) \tag{73}$$

where the statistical moments on the right-hand side are now dependent on the truncated log-conductivity hierarchy only. Thus, unlike what was inferred by Dagan [13], there is no need to invoke the non-ergodicity of the process and to *a priori* define the longitudinal macro-dispersion coefficient as one-half of the time-derivative of $<S_{ij}>$ in order to cut-off the long tail of the log-conductivity spectrum, obtaining an asymptotically constant value. As a matter of fact, the centroid covariance $\tilde{R}_{ij}$ is affected by a restricted range of heterogeneity scales and tends to zero at large times. As a consequence, the time derivative of $<S_{ij}>$ tends to coincide with the time-derivative of $\tilde{X}_{ij}$, which envisions an asymptotically ergodic transport process. Additionally, the assumed linearity of the problem and the integration of the asymptotic macro-dispersion coefficient obtained for a generic single-scale log-conductivity field over the truncated hierarchy domain lead to an invariably constant asymptotic macro-dispersion coefficient and, therefore, to Fickian transport conditions. It should be in any case emphasized that the never-decaying dependence of this coefficient on the initial plume size l_0 in Equation (64) which means that the system is characterized by persistent memory.

3. The nature of solute transport in evolving-scale heterogeneous formations is critically controlled by the magnitude of the initial Péclet number, intended as the ratio of the product between the ensemble

mean velocity and the initial plume size to the local dispersion coefficient. Generally speaking, the Péclet number is a measure of the relative importance of advective and diffusive-like transport mechanisms. In this specific case, the initial Péclet number can be interpreted as the ratio of the largest period of log-conductivity intercepted by the initial plume to the solid matrix diffusivity. When the Péclet number is not exceedingly large and the diffusive-like transport mechanisms play a non-negligible role in the dispersion process, the truncation of the hierarchy cannot be univocally defined. As a matter of fact, the boundary between the first and the second component of the displacement hierarchy (Equations (30) and (31)) tends to change in time due to the advection-free effect of diffusion that increases the distance between different particles, even when they undergo highly correlated (and, therefore, identical) advective displacements.

4. For a finite Péclet number, the dimensionless longitudinal macro-dispersion coefficient and the trajectory second-order statistical moment (particle covariance) are respectively given by Equations (69) and (72).

Equations (69) and (72) are graphically represented in Figures 1 and 2, respectively, for $\chi = 3$ and a travel distance equal to $30l_0$. As one can see, the effect produced by high Péclet (Pe_0) numbers is opposite to the effect produced by high scaling exponents (b). Specifically, the higher the Péclet number, the closer the transport process to asymptotic Fickian conditions, represented by a constant longitudinal macro-dispersion coefficient (although, for $\tau \to 0$, the dispersion coefficient is higher for higher Péclet numbers due to the larger number of heterogeneity scales initially sampled). Conversely, the higher the scaling exponent, the faster the macro-dispersion coefficient increases. Notice that, in Figure 2, the red dotted line represents the Fickian, linear behavior corresponding to a single-scale log-conductivity field (exponential or Gaussian log-K covariance) characterized by the truncated-hierarchy variance and by an integral scale equal to the initial plume size. Thus, unlike the case of fast-decaying correlation functions, the coexistence of evolving-scale advective heterogeneity and intensive diffusive mixing acts as an antagonist mechanism in the process of solute dilution and Fickian regime achievement. Such a behavior, which is in definite contrast with what was previously found for stationary porous media, was detected here for the first time.

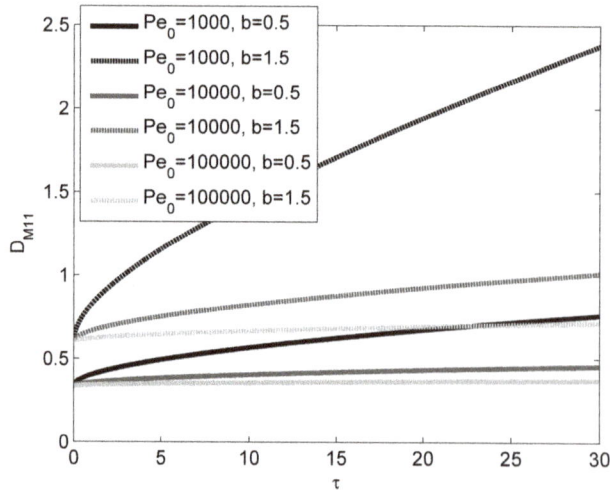

Figure 1. Longitudinal macro-dispersion coefficient for variable Péclet numbers and scaling exponents.

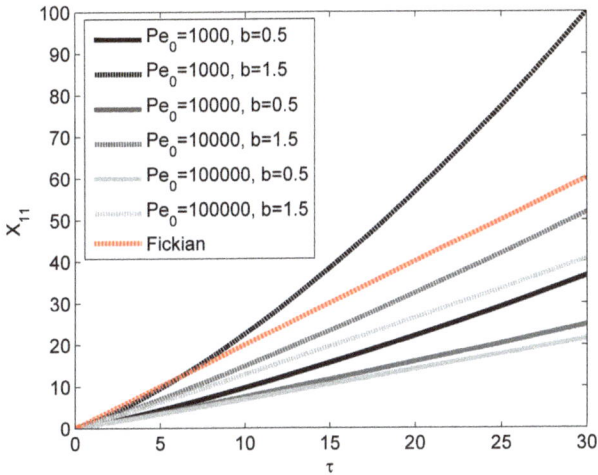

Figure 2. Longitudinal particle trajectory covariance for variable Péclet numbers and scaling exponents. The red-dotted line refers to Fickian linear behavior corresponding to a single-scale log-conductivity field (exponential or Gaussian log-K covariance) characterized by the truncated-hierarchy variance and by an integral scale equal to the initial plume size.

4. Discussion and Conclusions

The present study has shown that the dispersion of solutes in evolving-scale heterogeneous porous formations represented by power-law semi-variograms can be ergodic and Fickian or non-ergodic and super-diffusive, based on the magnitude of the Péclet number, where the Péclet number is intended as the ratio of the product of the ensemble mean velocity and the initial plume size to the local dispersion coefficient, and the scaling exponent value.

Specifically, the larger the Péclet number, the closer the transport process to asymptotically ergodic-Fickian conditions. Conversely, the higher the scaling exponent, the closer the transport process to a non-ergodic super-diffusive regime. In the limit for Péclet tending to infinity (negligible local dispersion), transport is always asymptotically Fickian (regardless of the scaling exponent), with the longitudinal macro-dispersion coefficient that increases with the initial plume size and scaling exponent.

The most important result is that, whereas in weakly heterogeneous formations characterized by short-range correlation functions the local dispersive/diffusive transport mechanisms enhance solute mixing and dilution by a faster achievement of Fickian conditions, the opposite happens in the case of heterogeneous porous formations characterized by persistent correlations represented by power-law semi-variograms.

This result is different from what has been found by previous studies on the same topic. Dagan [13] predicted invariably non-ergodic conditions (the uncertainty related to the centroid location, which is directly dependent on the incomplete dilution, never decayed) and adopted the effective macro-dispersion coefficient as representative of longitudinal dispersion. Dispersion turned out to be asymptotically Fickian for $0 < b < 1$ and anomalous for $1 \leq b < 2$. On the other hand, Suciu et al. [18] did not deal with ergodicity issues, defined the macro-dispersion coefficient as one-half of the time-derivative of the particle trajectory covariance X_{11}, and detected invariably super-diffusive regimes.

The different conclusions reached by the present study come from the acknowledgement of the effective-heterogeneity frequencies selection imposed by the finite initial plume size in such very peculiar hydraulic conductivity fields. Practically speaking, the present study recognizes that, in these cases, it is the initial dilution degree that plays a crucial role in terms of further possible dilution.

Indeed, a very concentrated solute spot in high-Péclet number transport processes would remain a very concentrated solute spot at any time: in the presence of long-range log-K correlations, a small solute body does not intercept a band of heterogeneity frequencies sufficient to induce solute dispersion about the center of mass. Conversely, the presence of a non-negligible local dispersive contribution would determine a continuous though slow involvement of larger heterogeneity scales, with the band of effective frequencies that becomes wider and wider. In these conditions, the evolving cut-off would tend to become ineffective in that all scales of heterogeneity would gradually affect both the macro-dispersion coefficient and the uncertainty related to the centroid location. In both cases, however, transport would be non-ergodic:

$$\frac{d\langle S_{11}\rangle}{dt} \neq \frac{d\tilde{X}_{11}}{dt} \tag{74}$$

and anomalous. In the first case, Pe $\rightarrow \infty$ and there is no frequency selection due to the infinitesimal plume size:

$$\frac{d\langle S_{11}\rangle}{dt} = \frac{dX_{11}}{dt} - \frac{dR_{11}}{dt} = 0 \tag{75}$$

In the second case, Pe is finite and the frequencies cut off is mobile:

$$\frac{d\langle S_{11}\rangle}{dt} = \frac{d\tilde{X}_{11}}{dt} - \frac{d\tilde{R}_{11}}{dt} + 2D \neq 0 \tag{76}$$

Only with a finite initial plume size (already diluted solute) and negligible local dispersion, one would recover asymptotically ergodic conditions:

$$\frac{d\langle S_{11}\rangle}{dt} \rightarrow \frac{d\tilde{X}_{11}}{dt} \tag{77}$$

and a Fickian regime:

$$\frac{d\tilde{X}_{11}}{dt} \rightarrow 2D_{m11} = \text{const} \tag{78}$$

Since local dispersion is related to laboratory-scale formation structure, the local analysis of long-range correlation porous media is a crucial issue in predicting the fate of fluids or pollutants subject to release and migration through them. Additionally, it must be emphasized that the linear treatment of flow and transport may hide some relevant second-order effects such as those coming from the interplay of the heterogeneity scales allowed by the complete solution of the flow equation (see Equation (21)) and from the Lagrangian solution of transport based on the complete integro-differential Equation (32), instead of on its linearized version:

$$\mathbf{X}'(t, \mathbf{a}) \cong \int_0^t \mathbf{v}'[\mathbf{a} + \mathbf{U}\tau + \mathbf{X}_B(\tau)]d\tau \tag{79}$$

Finally, provided that any numerical approach would be to some extent biased by the dimensions of the computational domain, it would be desirable that field experiments in highly heterogeneous and complex porous formations validate the theoretical conclusions.

Conflicts of Interest: The author declares no conflicts of interest.

Appendix A

From Pannone and Kitanidis [29], we know that the general expression of the asymptotic macro-dispersion coefficient in the case of short-range correlations and the mean velocity U directed along x_1 is:

$$D_{Mij}(\infty) = \int_{-\infty}^{\infty} \int_{-\infty}^{\infty} \int_{-\infty}^{\infty} \frac{\left(4\pi^2 D \sum_{r=1}^{3} \lambda_r^2 / I_Y^2\right) S_{uij}(\lambda_1, \lambda_2, \lambda_3)}{(2\pi\lambda_1 U / I_Y)^2 + \left(4\pi^2 D \sum_{r=1}^{3} \lambda_r^2 / I_Y^2\right)^2} \frac{d\lambda_1 d\lambda_2 d\lambda_3}{I_Y^3} + D \tag{A1}$$

where $\lambda_r = k_r I_Y$ and S_{uij} is the velocity spectrum. If the axes of the Cartesian reference frame (x_1, x_2, x_3) coincide with the principal axes of the dispersing plume, the velocity spectrum and log-conductivity spectrum S_Y are related by the following expressions (e.g., [1]):

$$S_{u11}(\lambda_1, \lambda_2, \lambda_3) = U^2 S_Y(\lambda_1, \lambda_2, \lambda_3) \left(1 - \frac{\lambda_1^2}{\lambda_1^2 + \lambda_2^2 + \lambda_3^2}\right)^2 \tag{A2}$$

$$S_{u22}(\lambda_1, \lambda_2, \lambda_3) = U^2 S_Y(\lambda_1, \lambda_2, \lambda_3) \left(\frac{\lambda_1 \lambda_2}{\lambda_1^2 + \lambda_2^2 + \lambda_3^2}\right)^2 \tag{A3}$$

$$S_{u33}(\lambda_1, \lambda_2, \lambda_3) = U^2 S_Y(\lambda_1, \lambda_2, \lambda_3) \left(\frac{\lambda_1 \lambda_3}{\lambda_1^2 + \lambda_2^2 + \lambda_3^2}\right)^2 \tag{A4}$$

The spectrum of the isotropic log-conductivity can, in turn, be derived from the definition of Fourier transform. For the exponential covariance:

$$S_Y(k) = \int_r C_Y(r) \exp(-j2\pi \mathbf{k} \cdot \mathbf{r}) d\mathbf{r} = \frac{2}{k} \int_0^{\infty} \sigma_Y^2 \exp\left(-\frac{r}{I_Y}\right) \sin(2\pi k r) r dr = \frac{8\pi I_Y^3 \sigma_Y^2}{(1 + 4\pi^2 k^2 I_Y^2)^2} \tag{A5}$$

while, for the Gaussian counterpart:

$$S_Y(k) = \frac{2}{k} \int_0^{\infty} \sigma_Y^2 \exp\left(-\frac{r^2}{I_Y^2}\right) \sin(2\pi k r) r dr = \pi^{3/2} \sigma_Y^2 I_Y^3 \exp(-\pi^2 I_Y^2 k^2)$$
$$= 8 I_Y^3 \sigma_Y^2 \exp(-4\pi I_Y^2 k^2) \tag{A6}$$

where $k^2 = k_1^2 + k_2^2 + k_3^2$. The advective part of D_{M11} for both exponential and Gaussian covariance is therefore:

$$D_{M11A}(\infty, I_{Y\Lambda}) = \sigma_{Y\Lambda}^2 U I_{Y\Lambda} \tag{A7}$$

Equation (A7) is obtained from Equation (A1) after the substitution $v_1 = \lambda_1 Pe / 2\pi$, where $Pe = U I_Y / D \gg 1$, from [28]:

$$\int_{-\infty}^{\infty} \frac{1}{v_1^2 + (\lambda_2^2 + \lambda_3^2)^2} dv_1 = \frac{\pi}{(\lambda_2^2 + \lambda_3^2)^2} \tag{A8}$$

by switching to polar coordinates for the subsequent integration over λ_2 and λ_3:

$$\begin{cases} \lambda = \sqrt{\lambda_2^2 + \lambda_3^2} \\ \lambda_2 = \lambda \cos\phi \\ \lambda_3 = \lambda \sin\phi \\ d\lambda_2 d\lambda_3 = \lambda d\lambda d\phi \end{cases} \tag{A9}$$

References

1. Dagan, G. *Flow and Transport in Porous Formations*; Springer: Berlin, Germany, 1989; p. 465.
2. Gelhar, L.W. *Stochastic Subsurface Hydrology*; Prentice Hall: Upper Saddle River, NJ, USA, 1993; p. 390.
3. Neuman, S.P. Universal scaling of hydraulic conductivities and dispersivities in geologic media. *Water Resour. Res.* **1990**, *26*, 1749–1758. [CrossRef]
4. Feder, J. *Fractals*; Plenum Press: New York, NY, USA, 1988; p. 283.
5. Fischer, H.B.; List, J.E.; Koh, C.R.; Imberger, J.; Brooks, N.H. *Mixing in Inland and Coastal Waters*; Academic Press: San Diego, CA, USA, 1979; p. 483.
6. Kitanidis, P.K. Prediction by the method of moments of transport in heterogeneous formation. *J. Hydrol.* **1988**, *102*, 453–473. [CrossRef]
7. Dagan, G. Transport in heterogeneous porous formations: Spatial moments, ergodicity, and effective dispersion. *Water Resour. Res.* **1990**, *26*, 1281–1290. [CrossRef]
8. Rajaram, H.; Gelhar, L.W. Plume scale-dependent dispersion in heterogeneous aquifers: 2. Eulerian analysis and three-dimensional aquifers. *Water Resour. Res.* **1993**, *29*, 3261–3276. [CrossRef]
9. Glimm, J.; Lindquist, W.B.; Pereira, F.P.; Zhang, Q. A theory of macrodispersion for the scale-up problem. *Transp. Porous Med.* **1993**, *13*, 97–122. [CrossRef]
10. Pannone, M. Transient Hydrodynamic Dispersion in Rough Open Channels: Theoretical Analysis of Bed-Form Effects. *J. Hydraul. Eng.* **2010**, *136*, 155–164. [CrossRef]
11. Pannone, M. Longitudinal Dispersion in River Flows Characterized by Random Large-Scale Bed Irregularities: First-Order Analytical Solution. *J. Hydraul. Eng.* **2012**, *138*, 400–411. [CrossRef]
12. Pannone, M. On the exact analytical solution for the spatial moments of the cross-sectional average concentration in open channel flows. *Water Resour. Res.* **2012**, *48*, W08511. [CrossRef]
13. Dagan, G. The significance of heterogeneity of evolving scales and of anomalous diffusion to transport in porous formations. *Water Resour. Res.* **1994**, *30*, 3327–3336. [CrossRef]
14. Bellin, A.; Pannone, M.; Fiori, A.; Rinaldo, A. On transport in porous formations characterized by heterogeneity of evolving scales. *Water Resour. Res.* **1996**, *32*, 3485–3496. [CrossRef]
15. Edery, Y.; Guadagnini, A.; Scher, H.; Berkowitz, B. Origins of anomalous transport in heterogeneous media: Structural and dynamical controls. *Water Resour. Res.* **2014**, *50*, 1490–1505. [CrossRef]
16. Kang, P.K.; Brown, S.; Juanes, R. Emergence of anomalous transport in stressed rough fractures. *Earth Planet. Sci. Lett.* **2016**, *454*, 46–54. [CrossRef]
17. Wang, L.; Cardenas, M.B. Transition from non-Fickian to Fickain longitudinal transport through 3-D rouogh fractures: Scale-(in)sensitivity and roughness dependence. *J. Contam. Hydrol.* **2017**, *198*, 1–10. [CrossRef] [PubMed]
18. Suciu, N.; Attinger, S.; Radu, F.A.; Vamos, C.; Vanderborght, J.; Vereecken, H.; Knauber, P. Solute transport in aquifers with evolving scale heterogeneity. *Versita* **2015**, *23*, 167–186. [CrossRef]
19. Gelhar, L.W. Stochastic subsurface hydrology from theory to applications. *Water Resour. Res.* **1986**, *22*, 135S–145S. [CrossRef]
20. Frampton, A.; Cvetkovic, V. Significance of injection modes and heterogeneity on spatial and temporal dispersion of advecting particles in two-dimensional discrete fracture networks. *Adv. Water Resour.* **2009**, *32*, 649–658. [CrossRef]
21. Kang, P.K.; Dentz, M.; Le Borgne, T.; Lee, S.; Juanes, R. Spatial Markov model for dispersion with variable injection modes. *Adv. Water Resour.* **2017**, *106*, 80–94. [CrossRef]
22. Hyman, J.D.; Painter, S.L.; Viswanathan, H.; Makedonska, N.; Karra, S. Influence of injection mode on transport properties kilometer-scale three-dimensional discrete fracture networks. *Water Resour. Res.* **2015**, *51*, 7289–7308. [CrossRef]
23. Demmy, G.; Berglund, S.; Graham, W. Injection mode implications for solute transport in porous media: Analysis in a stochastic Lagrangian framework. *Water Resour. Res.* **1999**, *35*, 1965–1973. [CrossRef]
24. Dentz, M.; Kang, P.K.; Comolli, A.; Le Borgne, T.; Lester, D.R. Continuous time random walks for the evolution of lagrangian velocities. *Phys. Rev. Fluid.* **2016**, *1*, 074004. [CrossRef]
25. Pannone, M.; Kitanidis, P.K. Large-time behavior of concentration variance and dilution in heterogeneous formations. *Water Resour. Res.* **1999**, *35*, 623–634. [CrossRef]

26. Pannone, M. Effect of nonlocal transverse mixing on river flows dispersion: A numerical study. *Water Resour. Res.* **2010**, *46*, W08534. [CrossRef]

27. Pannone, M. Predictability of tracer dilution in large open channel flows: Analytical solution for the coefficient of variation of the depth-averaged concentration. *Water Resour. Res.* **2014**, *50*, 2617–2635. [CrossRef]

28. Gradshteyn, I.S.; Ryzhik, I.M. *Table of Integrals, Series, and Products*, 5th ed.; Academic Press: San Diego, CA, USA, 1994; p. 1204.

29. Pannone, M.; Kitanidis, P.K. On the asymptotic behavior of dilution parameters for Gaussian and Hole-Gaussian log-conductivity covariance functions. *Transp. Porous Med.* **2004**, *56*, 257–281. [CrossRef]

water

MDPI

Article

An Integrated Approach Based on Numerical Modelling and Geophysical Survey to Map Groundwater Salinity in Fractured Coastal Aquifers

Costantino Masciopinto *, Isabella Serena Liso, Maria Clementina Caputo and Lorenzo De Carlo

National Research Council, Water Research Institute, via Francesco De Blasio 5, 70132 Bari, Italy;
serena.liso@ba.irsa.cnr.it (I.S.L.); maria.caputo@ba.irsa.cnr.it (M.C.C.); lorenzo.decarlo@ba.irsa.cnr.it (L.D.C.)
* Correspondence: costantino.masciopinto@ba.irsa.cnr.it; Tel.: +39-080-5820-537

Received: 27 July 2017; Accepted: 7 November 2017; Published: 10 November 2017

Abstract: Aquifer over-exploitation may increase coastal seawater intrusion by reducing freshwater availability. Fractured subsurface formations commonly host important freshwater reservoirs along sea coasts. These water resources are particularly vulnerable to the contamination due to seawater infiltration occurring through rapid pathways via fractures. Modeling of density driven fluid flow in fractured aquifers is complex, as their hydrodynamics are controlled by interactions between preferential flow pathways, 3D interconnected fractures and rock-matrix porosity distribution. Moreover, physical heterogeneities produce highly localized water infiltrations that make the modeling of saltwater transport in such aquifers very challenging. The new approach described in this work provides a reliable hydrogeological model suitable to reproduce local advancements of the freshwater/saltwater wedge in coastal aquifers. The proposed model use flow simulation results to estimate water salinities in groundwater at a specific depth (1 m) below water table by means of positions of the Ghyben-Herzberg saltwater/freshwater sharp interface along the coast. Measurements of salinity in 25 boreholes (i.e., salinity profiles) have been used for the model calibration. The results provide the groundwater salinity map in freshwater/saltwater transition coastal zones of the Bari (Southern Italy) fractured aquifer. Non-invasive geophysical measurements in groundwater, particularly into vertical 2D vertical cross-sections, were carried out by using the electrical resistivity tomography (ERT) in order to validate the model results. The presented integrated approach is very easy to apply and gives very realistic salinity maps in heterogeneous aquifers, without simulating density driven water flow in fractures.

Keywords: fractured aquifers; seawater intrusion; flow modeling; salinity map; groundwater ERT

1. Introduction

Seawater encroachments may lead to a consistent reduction of freshwater volume availability. Mathematical models are very useful to simulate seawater intrusion in coastal aquifers, as for instance in order to locate the freshwater/saltwater sharp interface position along coastal areas. There are specific numerical codes produced by academic institutions, such as the United States Geological Survey (USGS, Reston, VA, USA) or by commercial software houses, such as Aquanty, Inc. (Waterloo, ON, Canada), that can provide largely utilized models such as FEFLOW [1], SUTRA [2], SEAWAT-2000 [3] or HydroGeoSphere [4]. These are specific codes to study transient density driven flow of seawater inland advancements in coastal aquifers, even by 3D visualization. Anyway, the application of these codes in a fractured aquifer may have severe limitations when heterogeneities and the preferential water flow pathways in fractures are not properly taken into account in the governing equations.

This is because the representative elementary volume (REV) of the fractured groundwater, to which we must refer all model parameters (i.e., constant hydraulic conductivity and transmissivity, porosity, storativity, etc.) and variables of the flow and transport equations, it may have a very large size which renders unsuitable the application of the conventional models above mentioned. Moreover, the REV of a fractured aquifer might not exist when the constancy of parameters is not achieved in the entire size of the computational domain.

The only method to overpass this obstacle is to apply the governing flow and transport equation to a REV of the flowing fluid in each single fracture. This can be possible when the geometry of the fluid flow pathway is known a priori. Thus, specific methods are required to reproduce and address into the model all preferential flow geometries that occur in different fractures of the coastal studied aquifer. Different conceptual model can be defined in fractured media, even by taking into account of tortuosity of the flow pathways [5]. Major stochastic media idealization for modeling the flow in fractured rocks lead to: the fracture zone continuum model [6], where the fractured aquifer is considered as an equivalent heterogeneous porous medium and the "discrete" fracture networks [7–9]. Most used in coastal areas is the flow in a set of parallel and identical fractures (i.e., layered model) [10]. In this work, the stochastic method reproduced into the model, the geometry of real preferential water flow pathways of the Bari fractured aquifer. The selected stochastic method was able to transfer all real medium heterogeneities into the computation procedure. Valid stochastic methods can provide appropriate numeric model solutions not only in fractured aquifers, but also in a generic heterogeneous aquifer. The stochastic method applied in this work investigated the spatial variability of the sizes of fracture apertures of the Bari groundwater.

Groundwater flow at the regional scale (>1000 m), is usually mediated into the vertical thickness (z) by considering a prevalent mean horizontal flow (x, y), as the saturated (vertical, z) thickness is usually less than 50 m [11]. In the present work, instead, the groundwater flow modeling was addressed in a 3D set made by N_f parallel fractures, which have the same mean aperture value $2b_m$ and an impermeable rock matrix. To support the stochastic method in this work, the data derived from pumping tests on wells were necessary to implement the real heterogeneities of the filtration medium into each single fracture of the model.

The flow simulation results defined the freshwater/saltwater (50%) sharp interface position in the Bari aquifer by using the Ghyben-Herzberg theory and highlighted the part of the costal aquifer where the seawater encroachment was present. In order to validate the spatial distribution of water salinity close to seawater encroachment in groundwater, the numerical results were compared with geo-electrical measurements carried out in two separated groundwater zones. These field investigations utilize non-invasive geophysical techniques for monitoring coastal aquifer salinity dynamics. The geophysical survey, particularly, the electrical resistivity tomography (ERT), is a powerful tool to evaluate the heterogeneity of subsoil by estimating groundwater salinity at specific depths below the ground. The ERT can be very useful when few boreholes (i.e., data) are available for direct measurements by probes of groundwater salinity. Furthermore, monitored salinity profiles in boreholes are affected by water salt mixing into the water columns and by vertical saltwater stratification due to density. Thus, the real water salinity in a fracture at a specific depth can be accurate only by installing packers into the well. These devices can isolate specific water depth intervals into the borehole by providing appropriate salinity concentrations with depth. However the packers are not easy to apply because they cannot be applied in boreholes with large diameter. Moreover packer installation can be efficiently made only in unscreened wells. Subsequently, errors on direct measurements of water salinity in boreholes might increase the uncertainty of model predictions.

In the literature there are many papers [12–21] concerning the application of the ERT technique to detect the fresh/saltwater sharp interface in groundwater by visualizing the inland zone of saltwater encroachment. Usually ERT is applied to obtain a qualitative result, which is a function of the electrical resistivity contrast between the freshwater and the saltwater contained in the investigated groundwater volume. However, quantitative estimations of water salinity concentrations by using resistivity measurements have been provided by Wagner et al. [22] and Singha et al. [23]. These researchers defined a specific petro-physical relationship based on the Archie's law [24] in order to derive water salinities.

In present study a new site specific relationship resistivity/salinity has been defined in order to infer salinity data by ERT survey in two coastal sections (*y*, *z*) of the Bari groundwater. These salinity data have been then successful compared with results (i.e., salinity map) given by model at the depth of 1 m below water table. The good agreement of the trends of ground water salinity suggests that ERT is a powerful tool to provide suitable data of groundwater salinity in coastal areas by supporting flow and salt transport model validations.

2. Materials and Methods

The field tests, carried out at the Bari site (Figure 1), have been conducted at the top of a karstic fractured limestone formation that hosts the Murgia aquifer. The water table ranges from 5 to 40 m below the ground.

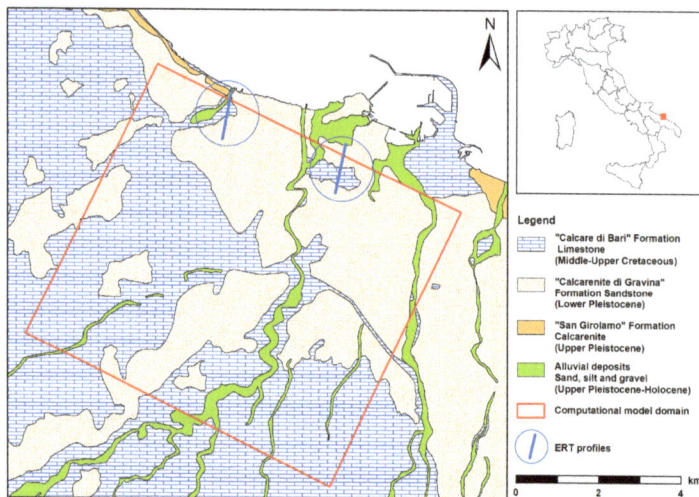

Figure 1. Bari site geological sketch and computational domain (red square) for groundwater flow and Ghyben-Herzberg simulations, and the positions of the electrical resistivity tomography (ERT) profiles of subsoil.

A detailed geological description of the Bari coastal aquifer is available in [25]. The study area lies on the eastern edge of the Murge, that represents the central part of the foreland of the Southern Apennine mountains [26], characterized by a thick Mesozoic sedimentary sequence, overlain by relatively thin and discontinuous Quaternary deposits. Locally, the "Calcare di Bari" represents the outcropped Mesozoic sequence formation (Figure 1) characterized by numerous karstic cavities of different shapes and sizes, partially or completely filled by "terra rossa" deposits. "Calcarenite of Gravina" (Lower Pleistocene) represents the Quaternary formation, consisting of litho-bioclastic sandstone. Colluvial and eluvial deposits (Upper Pleistocene-Holocene) cover stream beds (i.e., Lame), while narrow bands, as outcrops of well-cemented porous sandstone (Upper Pleistocene) appear along the coast. Limestone bedrock hosts a wide and thick aquifer. High limestone permeability is the result of the intense fracturing of rock and of the karst dissolving action. The irregular spatial distribution of the fractures and karstic channels renders the Bari aquifer very anisotropic. The groundwater flows toward the sea, under a low pressure, in different subparallel fractured layers separated by compact (i.e., not fractured) blocks of rock. In particular, along a generic water flow pathway, the fracture apertures with small sizes control the gradient line of predominant horizontal freshwater flow, i.e., small sized apertures are the bottlenecks of the freshwater flow. In the study area, hydraulic transmissivity, T [L^2/t], and conductivity, K [L/t], of the "Calcare di Bari" formation, have been determined by inverting the steady radial flow solution to a well (i.e., the Thiem's equation) [27]. Results given by thirty-six (Table 1) pumping-well tests provide the experimental variogram of fracture apertures of the coastal aquifer. The model estimated the local fracture apertures by inverting mean aquifer conductivity formula, i.e., $K = \gamma_w/\mu \times nb^2/3$, where n [-] is the effective porosity of the saturated freshwater thickness. In fact the hydraulic conductivity in a single (smoothed) fracture with aperture $2b$, was obtained by comparing the velocity defined by Hagen-Poiseuille equation, which is usually adopted [28] to determine the plane flow velocity in a fracture, with the velocity provided by Darcy equation.

Table 1. Estimations of the mean fracture apertures at the Bari coastal aquifer, by inverting the solution of the steady radial water flow to a well during pumping.

X (m)	Y (m)	K (m/s)	Well ID	2b (mm)
		Pumping Test		
654,627.88	4,548,060.78	4.05×10^{-6}	PT1	248.79
654,668.04	4,548,542.57	2.40×10^{-5}	PT2	336.14
655,009.30	4,548,492.39	2.21×10^{-5}	PT3	331.71
654,979.19	4,548,151.12	4.90×10^{-5}	PT4	378.25
656,378.54	4,549,873.53	6.64×10^{-4}	PT7	629.59
648,038.69	4,552,384.46	1.86×10^{-5}	PT8	322.25
648,794.68	4,551,906.47	1.14×10^{-5}	PT9	297.27
648,351.69	4,551,897.47	2.32×10^{-5}	PT10	334.38
647,309.71	4,551,475.47	2.76×10^{-5}	PT11	344.01
647,914.70	4,551,580.47	1.92×10^{-5}	PT12	324.11
648,341.69	4,551,249.48	5.77×10^{-5}	PT13	388.76
648,686.68	4,551,472.47	2.66×10^{-5}	PT14	342.00
653,466.00	4,552,424.00	2.71×10^{-3}	IS1	958.84
655,868.00	4,554,156.00	4.25×10^{-4}	IS2	566.99
655,515.00	4,552,586.00	9.74×10^{-5}	IS4	425.37
655,272.00	4,552,018.00	6.40×10^{-6}	IS5	269.43
654,726.00	4,551,838.00	1.19×10^{-5}	IS7	299.38
654,599.00	4,551,852.00	7.25×10^{-5}	IS8	404.13
651,985.00	4,553,569.00	3.80×10^{-3}	IS9	1089.42
651,172.00	4,552,620.00	9.25×10^{-6}	IS10	286.87
651,558.00	4,550,230.00	2.37×10^{-4}	IS11	501.40
650,678.00	4,554,592.00	3.45×10^{-5}	IS13	356.91
651,256.00	4,554,586.00	1.47×10^{-5}	IS14	310.07
647,153.00	4,553,736.00	6.26×10^{-6}	IS19	268.41
645,754.00	4,553,926.00	7.03×10^{-6}	IS21	273.81
653,419.00	4,549,497.00	3.72×10^{-5}	IS22	361.33
654,415.00	4,550,306.00	1.27×10^{-5}	IS23	302.37
654,812.00	4,550,479.00	1.26×10^{-5}	IS24	302.08
654,559.00	4,551,970.00	1.19×10^{-5}	IS25	299.38
656,315.00	4,552,223.00	5.48×10^{-5}	IS26	385.45
656,919.00	4,550,832.00	1.47×10^{-5}	IS28	310.07
652,850.93	4,553,352.70	4.17×10^{-3}	L4	1130.53
653,252.50	4,555,151.70	2.17×10^{-3}	PSUD	887.25
652,430.90	4,554,429.80	6.43×10^{-3}	L3-S	1360.49
647,930.70	4,551,813.20	1.33×10^{-4}	L5-S	449.84
654,679.50	4,555,109.10	2.29×10^{-3}	L12-S	903.60
Mean value		6.58×10^{-4}		471.69

It should be noted that during the field investigation only 25 wells of 36 listed in Table 1 were accessible to carry out water depth and salinity measurements.

2.1. Fractures Description and Flow Solutions: Experimental Tests

The fracture aperture size at grid position (x, y) was generated by the following stationary random field, ε

$$\varepsilon(x,y) = Y(x,y) - \overline{Y} \tag{1}$$

where is $Y = \log 2b$ and $\overline{Y}$ is its mean. The semi-variogram model [5] of the expected value of the variance is:

$$\gamma(\xi_{xy}) = \frac{1}{2}E\left[\{\varepsilon(x,y) - \varepsilon(x + \xi_x, y + \xi_y)\}^2\right] \tag{2}$$

which can be derived using the autocovariance function

$$R(\xi_{xy}) = \sigma^2_{xy}\exp\left[\left(\frac{\xi^2_x}{\xi^2_x} + \frac{\xi^2_y}{\xi^2_y}\right)^{1/2}\right] \tag{3}$$

where the unknown semi-variogram model parameters σ^2_{xy} (sill + nugget), ξ_x and ξ_y and (i.e., correlation lengths) can be calculated using SURFER (Golden Software Inc., Golden, CO, USA) on the basis of the spatial distribution of mean apertures determined from the results of pumping tests. For the Bari coastal aquifer, the best-fit of the experimental semi-variogram was made using the exponential model (Figure 2) with $\sigma^2_{xy} = 0.268$, $\xi_x = 1000$ m and $\xi_y = 2000$ m, using data derived from thirty-six tests (Table 1). However, at the field scale, it should also be considered the uncertainty (~15%) due to the prediction of the spatial covariance of fracture apertures, which is dependent upon the available number of field measurements (i.e., well pumping-tests). This uncertainty was due to non-ergodicity of the scholastic variable [29].

The flow rate in each channel in x direction with cross section $2b \times \Delta y$ (or $2b \times \Delta x$, in y direction) can be estimated by revising the Darcy-Welsbach equation ([28] p. 126).

$$(\phi_i - \phi_j) = Q_{ij}{}^2\left[\frac{f}{2g\Delta y}\frac{\Delta x}{\Delta y}\left(\frac{1}{(2b_i)^3} + \frac{1}{(2b_j)^3}\right)\right] \tag{4}$$

where the friction factor f [-] can be derived from the Reynolds number [22] even for non-laminar or turbulent fluxes; g is gravity acceleration and Q_{ij} [L^3/t] $= U \times 2b \times \Delta x \Delta y$, is the local discharge between grid nodes i and j into the single fracture, where Δx and Δy are the discretization grid steps. The finite difference method can be used to solve the continuity equation (i.e., $\Sigma Q = 0$) applied to each grid node of the discretized domain. The resulting set of equations was solved by the iterative successive-over-relaxation (SOR) method by using as boundary conditions the piezometric heads into the depressed areas (i.e., pumping wells) and along the border of the studied area.

Ghyben-Herzberg Freshwater/Saltwater Sharp Interface

The flow simulation results enabled the estimation of the 50% freshwater/saltwater interface positions with respect to the coastline by applying the Ghyben-Herzberg equation. Indeed, to predict the interface toe position L [L], with respect to the coastline, the resulting total groundwater freshwater outflow given by Equation (4) was managed to calculate the length of intrusion for every position along the coast [30]

$$L - L_d = K\frac{B^2 - H^2_s}{2\delta_\gamma \times Q_0} - L_d = n\frac{b^2_i}{3}\frac{\gamma_w}{\mu}\frac{(\delta_\gamma \times \phi_0)^2 - H^2_s}{2\delta_\gamma \times Q^i_0} - L_d \tag{5}$$

where L_d [L] is the distance of the contour head ϕ_0 (for instance of 1 m) from the coastline given by the flow simulation results; H_s [L] is the freshwater head at the outflow section (usually set to 0); $Q^i{}_0$ [L^3/t/L] is the groundwater (i.e., freshwater) discharge along the coast predicted by the model at grid node *i*; *B* [L] is the aquifer saturated thickness where is $\phi = \phi_0$; and $\delta_\gamma = \gamma_w/(\gamma_s - \gamma_w)$ is the ratio of the water specific weights.

In each single fracture of the Bari aquifer, the distance *d* of the generic grid node (*x*, *y*) from the sharp interface was converted into a salinity concentration by using the empirical formula [31]

$$C_{salt} = C_{s0} + A_s \left[\exp\left(-\frac{d}{D_s} \right) \right] \tag{6}$$

where the best fit constants $C_{s0} = 1.54$ g·L^{-1}, $A_s = 12.02$ g·L^{-1} and $D_s = 592.65$ m were estimated by fitting the groundwater salt concentrations measured in twenty-five boreholes of the coastal aquifer, at the depth of 1.0 m below the water table.

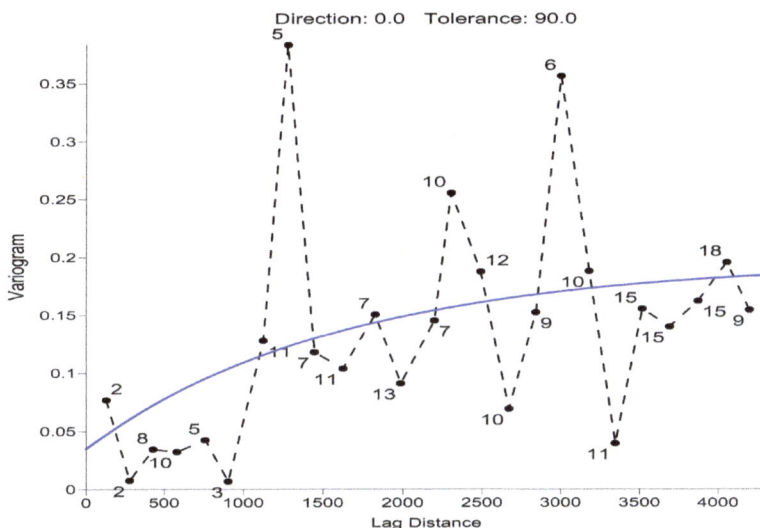

Figure 2. Experimental aperture variogram and model variogram parameters at Bari (Southern Italy) coastal fractured aquifer; Model: Exponential (Scale = 0.16; Length = 3000 m; Anisotropy: ratio = 2, angle = 64.5 degrees; Nugget Effect: error = 0.03518, micro = 0); Experimental: max lag distance = 4300 m, number of lag = 25, lag width = 172 m, vertical scale = 0.384.

2.2. ERT Survey

ERT is a non-invasive and cost-effective geophysical technique commonly used for spatial characterization of the subsoil over extended areas.

Soil electrical resistivity (i.e., the inverse of electrical conductivity) is an intrinsic parameter of the soil, which can quantify the resistance of a given porous medium to the flow of the electric current. Many factors affect the electrical resistivity of the medium, such as rock (or soil) porosity, clay content, and salinity and temperature of water. Among these, water content and water salinity are the most important parameters because the electric current flows into the rock by means of dissolved electrolyte ions (i.e., dissolved salts) of the water in the pore spaces of the soil or in fractures. In the field, electrical resistivity data are usually measured by using an array of four electrodes: two electrodes are used to insert the electric current into the ground, and other two electrodes are used to measure the difference of electrical potential into the investigated rock volume. The measurement, which is the apparent resistivity of a homogeneous bulk volume of the soil (or the rock) to which the measured electrical resistance is equivalent [32], is then calculated by multiplying the resistance of the medium by a geometrical factor, which depends on the arrangement (i.e., geometry and distance) of the electrodes of the array. The subsequent inversion process, based on an iterative numerical method, allows us to minimize the misfit between the theoretical resistivity and the measured one. This last numerical method is usually required to estimate a more realistic resistivity distribution of the investigated soil.

In the study area, the ERT survey has been collected through a sequence of ten (y, z) ERT profiles, grouped into two main subsurface cross sections. Each section was about 1 km of length in the direction perpendicular to the coastline. These sections were located in two separated coastal areas. The first area (Line A) was positioned where model results highlighted the seawater encroachment in groundwater. The second section (Line B), instead, was located where the model simulation has shown a freshwater outflow into the sea. The first section (Line A) grouped five ERT profiles, from E1 to E5, whereas the section Line B, grouped other five ERT profiles, from E6 to E10 (Figure 3). ERT was carried out using the Syscal Switch Pro 48 (IRIS Instruments, Orleans, France) resistivity-meter. A Wenner–Schlumberger configuration array provided a good arrangement by considering the depth of investigation (<30 m) below the ground and lateral resolution (<3 m). The length of each ERT profile was changed according to the required depth of the investigation below the ground, which was dependent upon the expected groundwater depth. Therefore, E1, E6 and E7 profiles were 135–160 m of length, while remaining profiles were length about 50 m. For each ERT profile, 1500 data points were collected, including both direct and reciprocal measurements required for the data quality control. This is because following the reciprocity principle when the previous current electrodes are switched to the potential electrodes the same resistivity values should be expected. The operating parameters during the acquisitions (Table 2) were settled to optimize the field electrical resistivity measurements.

Figure 3. ERT profile locations. Blue points are boreholes (six) used for salinity measurements to calibrate Equation (7) by using resistivity data provided by ERT profiles.

Table 2. Transmission (i.e., operating) parameters used in ERT surveys.

Injection Pulse Duration	250 ms
Minimum and maximum number of cycles for each measurement	3–6
Standard deviation of the measurements in a cycle	5%

The parameter optimization allowed bad data points removal when threshold values (i.e., the noise) were overcome, as is shown in Table 3. The low number of removed data (5%) confirmed the good quality of the ERT sequence. The inversion was made by using code RES2DINV (Geotomosoft Solutions, Gelugor, Malaysia).

Table 3. Error bounds during ERT data filtering.

Parameter	Lower Bound	Upper Bound
Injection current (mA)	5	1000
Potential measurement (mV)	5	5000
Deviation standard of the measurements in a cycle (%)	0	5
Percentage difference between direct and reciprocal data (%)	0	5

Usually, in order to estimate groundwater salinity from resistivity collected data a relationship based on the Archie's law could be applied. In the proposed study, Archie's law, did not yield satisfactory results, due to the wide heterogeneity of the fractured aquifer investigated, which present preferential water flow pathways. In such complex hydrogeological formations, it might be challenging to calibrate the Archie's law due to variability of the rock quality, i.e., porosity, cementation index, etc., and of the water salinity in the bulk volume investigated by ERT.

For these reasons, a new site-specific relationship of the resistivity-salinity concentration (ρ-C_{salt}) was proposed. For this investigation, additional six ERT profiles were performed close to the boreholes where the salinity of groundwater at 1 m of depth was directly measured using an electrical conductivity probe (MS5 OTT, Inc., Kempten, Germany). This means that from the twenty-five boreholes we selected six at different water salinities to carry out six additional ERT. We selected only six ERT/wells for a technical reason (i.e., low groundwater depth <3 m). In fact, to obtain a reliable empirical resistivity-salinity relationship, the ERT images must have a high resolution [32] and for this the depth of investigation must not exceed 5 m below the ground. This leads to a short inter-electrode spacing and to the high resolution ERT images.

A rock electrical resistivity value in the upper part of the aquifer at a depth of 1 m below the water table close was derived from each ERT carried out close the borehole. The measurements were correlated with the salt concentration measured at the same depth in the water of boreholes. Thus, groundwater salt concentration (g/L) as a function of the soil resistivity [33] was estimated by using

$$C_{salt} = a\rho^b \tag{7}$$

where a (= 47.02) and b (= −0.7) are two dimensionless best fit (R^2 = 0.93) constants and ρ (Ωm) is the monitored electrical resistivity of the groundwater. Despite of the few (six) boreholes considered, the spread of measured values of water salinity, which ranged from 1 to 5 g/L, allowed a high (Figure 4) correlation coefficient. The graph shows a lack of information for salt concentrations higher than 5 g/L, due to the absence of boreholes.

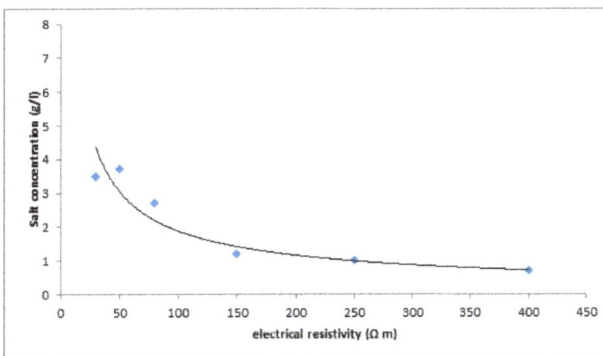

Figure 4. The relationship between electrical resistivity (ρ) and groundwater salt concentration (C_{salt}) at the Bari coastal aquifer.

3. Results and Discussion

Flow model results (Figure 5) provide freshwater heads (m) and freshwater velocity and discharge in each grid channel of discretized aquifer domain of 7200 × 6300 m^2.

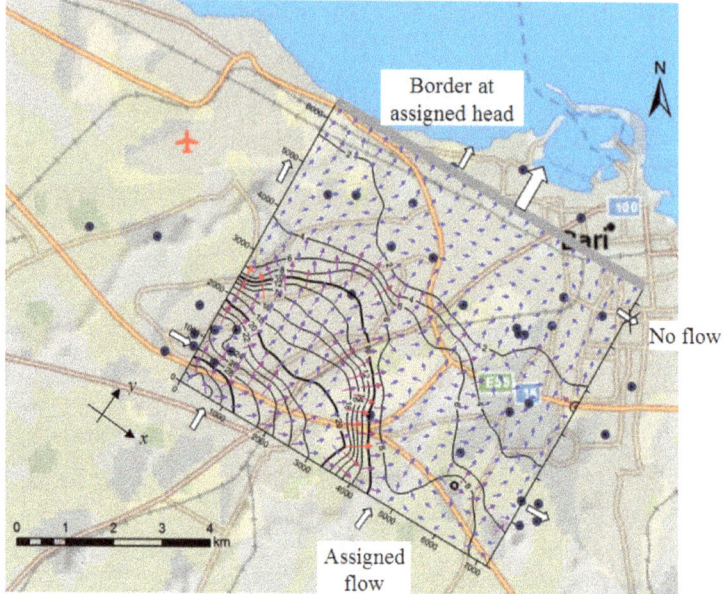

Figure 5. Model output: Freshwater heads (contour lines) in meters above sea level and water velocities (vectors) ranging from 10 (in blue) to 80 m/d (in red) in the studied area. White arrows and the gray border indicate the boundary conditions imposed during flow simulations; solid circles are wells.

Each horizontal fracture, at the regional scale, belonging to the 3D parallel set was discretized using a grid step size of $\Delta x = \Delta y = 150$ m (i.e., 49 × 43 grid nodes). The saturated aquifer thickness was 30 m, on average, and by considering an average aperture of 0.47 mm for each single fracture of the set, a total of 80 fractures were estimated for an effective porosity of 0.35% [23]. The conceptual fractured model used in this work was derived from the layered model [24] and is made up of several horizontal fractures bounded by impermeable rocks [25]. This is because vertical or sub vertical fractures that usually occur in these limestone formations, which are formed by the movements of tectonic plates, are very infrequent, so groundwater does not generally flow, as in the vertical lattice of fractures. This means that, although vertical connections between horizontal (and parallel) fractures of the coastal aquifer exist, horizontal preferential pathways dominate the water flow and each vertical fracture acts like a piezometer of an aqueduct comprising parallel pipelines: it ensures that the freshwater flows in each sub horizontal (and parallel) fracture (or pipe) with the same head gradient line. The freshwater discharge in each grid channel allows the estimation of the sharp interface position in the coastal area. For this calculation, an Excel (Microsoft) sheet was implemented with Equation (5) in order to estimate the $L - L_d$ distance for each Δx along the coast. Then, the coast distance y_d from the border domain at the assigned head (see Figure 5) was also considered to estimate all L_d distances with respect to the coastline. Finally, the application of Equation (6) to all grid nodes of the domain led to a salinity map of the Bari aquifer (Figure 6), at a depth of 1 m below the water table.

Figure 7 shows the results of the inverted ERT profiles at Line A and Line B. A common colour scale has been settled in order to point out the differences in electrical resistivity between Line A

(Figure 7a) and Line B (Figure 7b) profiles. Due to the great size of the studied area with respect to the length of the geophysical profiles, the ERT images are shown unscaled.

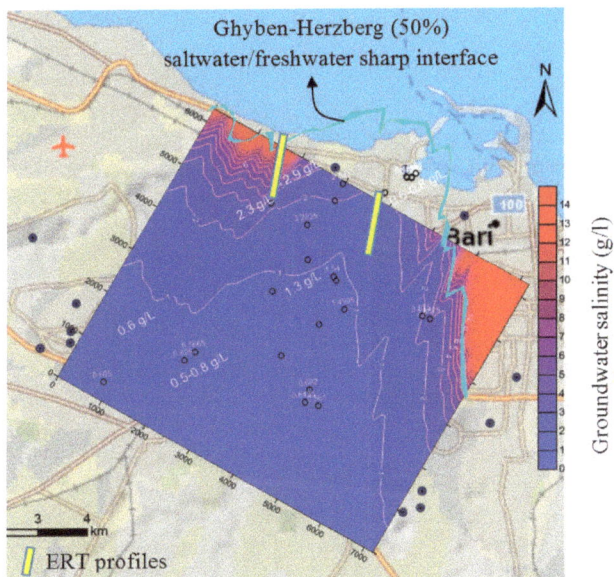

Figure 6. Groundwater salinity map (at a depth of 1.0 m below the water table) given by Ghyben-Herzberg and flow simulation results at the Bari coastal aquifer. Open circles are measured salinities in boreholes.

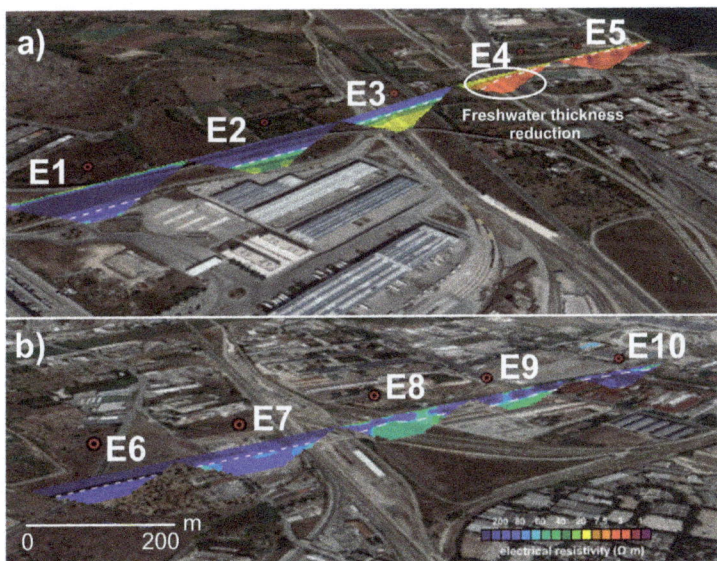

Figure 7. Inverted ERT profiles at Line A (**a**) (from E1 to E5) and at Line B (**b**) (from E6 to E10): ERT model error: <5%. E4 shows freshwater thickness reduction due to seawater intrusion.

Vertical and horizontal exaggerations in Figure 7 highlighted the change of electrical resistivity of monitored groundwater along each profile section. White dashed line shows the position of the water table at a depth of 1.5–15 m below the ground given by the flow model and the field measurements. From upstream to downstream, Line A highlights a significant decrease of aquifer resistivity, from values higher than 200–300 Ωm to values less than 5 Ωm. In particular, low resistivity values in E4–E5 (on the Line A) (see Figure 7a) show a decrease of freshwater thickness associated to the inland seawater intrusion. On the contrary, only a small range of resistivity from the high values of 200–300 Ωm associated with freshwater to the values of 40–60 Ωm, was recorded along the Line B. This is because fractures along Line B transport high freshwater water flows due to the large size of the fracture apertures, by avoiding seawater intrusion. These geophysical results agree with the model outputs, confirming ERT is a valuable technique to detect the seawater intrusion in groundwater. In order to provide a quantitative estimation of the salt concentration in groundwater a new relationship ρ-C_{salt} was implemented to convert the collected electrical resistivity into salt concentrations in groundwater. In order to compare the groundwater salinity derived from ERT profiles with modeling results, the two trends of salinities estimated by ERT into the Bari coastal aquifer have been plotted in Figure 8, together with results derived from model flow simulations (by including Ghyben-Herzberg estimations) at the depth of 1 m below the water table. This result successful validated the modeling output and at same time shows the efficacy of ERT to prove, experimentally, the seawater encroachment along the coast. Moreover direct measurements in boreholes by probes can be affected by the mixing due to water inflow coming from fractures at different depths of the water column, whereas the salinity estimations derived from ERT measurements can better represent real salt concentration into the fractures at a specific depth. This can be a valid support for modeling validations.

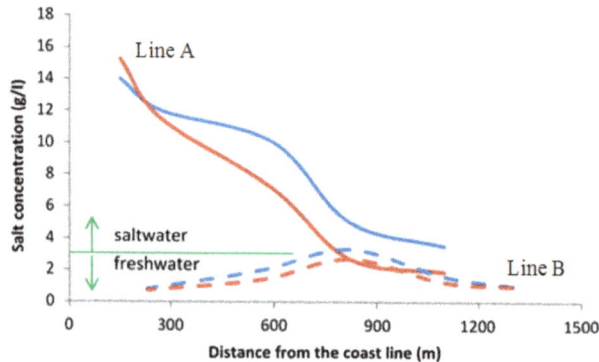

Figure 8. Comparison between salinity trends estimated by using ERT (red lines) and by using the Ghyben Herzberg model (blue lines). Solid lines represent the Line A groundwater section; dotted lines represent the Line B groundwater section.

4. Conclusions

The proposed case study deals with an innovative approach to model flow and salt transport phenomena in fractured coastal aquifers affected by seawater intrusion. The adopted procedure is based on a stochastic method able to transfer all real filtration medium heterogeneities into the numerical flow model. In particular, the stochastic method applied in this work investigated the spatial variability related to the size of the fracture apertures of the Bari groundwater. The model implemented the Ghyben-Herzberg to estimate groundwater salinity by means of sharp freshwater/saltwater position along the coast. The result provided the map of groundwater salinity at the depth of 1 m below water table, visualising the sea encroachment in groundwater and the freshwater flow in different areas along the coast. The main advantage of the proposed approach lies in the capability of the model to simulate flow and salt transport processes in complex hydrogeological systems, where the wide heterogeneous nature of the fractured rock and the small size of the fracture aperture make difficult the application of the conventional flow equation, such as Darcy's law to a groundwater representative volume. However, at the field scale, it should also be considered the uncertainty (about 15%) due to prediction of spatial covariance of fracture apertures, which is dependent upon the available number of field measurements (i.e., well pumping-tests). For this, the ERT technique has proven to be a useful tool for the validation and uncertainty reduction of the flow and transport numerical model in fractured coastal aquifers affected by seawater intrusion. Moreover, ERT derived salt concentration data may overcome the issue related to the mixing of the fracture fluxes that occur inside the borehole at different depths of the water columns. Hence, ERT may give more reliable salt concentration values in comparison with measurements carried out in boreholes without using packers. However in order to improve numerical model solutions and the accuracy of petro-physical relationship ρ-C_{salt}, a higher number (>40) of appropriate measurements in wells is required.

Author Contributions: Costantino Masciopinto developed the mathematical model, collected field pumping tests and salinity measurements, and carried out model simulations. Lorenzo De Carlo and Maria Clementina Caputo conceived and designed the experimental geophysical tests; and Lorenzo De Carlo and Isabella Serena Liso performed the geophysical survey and analyzed the data. Each author has contributed to the writing of the manuscript.

Conflicts of Interest: The authors declare no conflict of interest.

References

1. Diersch, H.J. *FEFLOW: Finite Element Modeling of Flow, Mass and Heat Transport in Porous and Fractured Media*; Springer: Heidelberg, Germany, 2009; ISBN 978-3-642-38739-5.
2. Voss, C.I. *A Finite-Element Simulation Model for Saturated-Unsaturated, Fluid-Density-Dependent Groundwater Flow with Energy Transport or Chemically-Reactive Single-Species Solute Transport: U.S. Geological Survey Water-Resources Investigations Report*; US Geological Survey: Reston, VA, USA, 2009; Volume 84, p. 4369. Available online: https://pubs.usgs.gov/wri/1984/4369/report.pdf (accessed on 8 November 2017).
3. Simpson, M.J. Software Spotlight/SEAWAT-2000: Variable-Density Flow Processes and Integrated MT3DMS Transport Processes. *Ground Water* **2004**, *42*, 642–645. [CrossRef]
4. Werner, A.D.; Bakker, M.; Post, V.E.A.; Vandenbohede, A.; Lu, C.; Ataie-Ashtiani, B.; Simmons, C.; Barry, D.A. Seawater intrusion processes, investigation and management: Recent advances and future challenges. *Adv. Water Resour.* **2013**, *51*, 3–26. [CrossRef]
5. Masciopinto, C.; Palmiotta, D. Flow and Transport in Fractured Aquifers: New Conceptual Models Based on Field Measurements. *Transp. Porous Media* **2012**, *96*, 117–133. [CrossRef]
6. Tsang, Y.W.; Tsang, C.F.; Hale, F.V.; Dverstorp, B. Tracer transport in a stochastic continuum model of fractured media. *Water Resour. Res.* **1996**, *32*, 3077–3092. [CrossRef]
7. Schwartz, F.; Smith, L.; Crowe, A. A stochastic analysis of macroscopic dispersion in fractured media. *Water Resour. Res.* **1983**, *19*, 1253–1265. [CrossRef]
8. Chiles, J.P.; De Marsily, G. Stochastic Models of Fracture System and Their Use in Flow and Transport Modeling. In *Flow and Contaminant Transport in Fractured Rock*; Bear, J., Tsang, C.F., De Marsily, G., Eds.; Academic Press: San Diego, CA, USA, 1993; pp. 169–236.
9. Nordqvist, W.A.; Tsang, Y.W.; Tsang, C.F.; Dverstorp, B.; Andersson, J. Effects of high variance of fracture transmissivity on transport and sorption at different scales in a discrete model for fractured rocks. *J. Contam. Hydrol.* **1996**, *22*, 39–66. [CrossRef]
10. Kazemi, H.; Gilman, J.R. Multiphase flow in fractured petroleum reservoirs. In *Flow and Contaminant Transport in Fractured Rock*; Bear, J., Tsang, C.-F., De Marsily, G., Eds.; Academic Press, Inc.: London, UK, 1993; Volumes 6–19, pp. 270–272.
11. Gelhar, L.W. *Stochastic Subsurface Hydrology*; Prentice-Hall, Inc.: Englewood Cliffs, NJ, USA, 1993; pp. 230–234, ISBN 978-0138467678.
12. Carter, A.R.J. Investigation into the Use of Resistivity Profiling in the Detection of the Fresh/Saline Water Interface within Coastal Settings. Bachelor's Thesis, University of Lancaster, Lancaster, UK, 2002; p. 71.

13. Cassiani, G.; Bruno, V.; Villa, A.; Fusi, N.; Binley, A.M. A saline trace test monitored via time-lapse surface electrical resistivity tomography. *J. Appl. Geophys.* **2006**, *59*, 244–259. [CrossRef]

14. Choudhury, K.; Saha, D.K. Integrated geophysical and chemical study of saline water intrusion. *Ground Water* **2004**, *42*, 671–677. [CrossRef] [PubMed]

15. Cimino, A.; Cosentino, C.; Oieni, A.; Tranchina, L. A geophysical and geochemical approach for seawater intrusion assessment in the Acquedolci coastal aquifer (Northern Sicily). *Environ. Geol.* **2008**, *55*, 1473–1482. [CrossRef]

16. De Franco, R.; Biella, G.; Tosi, L.; Teatini, P.; Lozej, A.; Chiozzotto, B.; Giada, M.; Rizzetto, F.; Claude, C.; Mayer, A.; et al. Monitoring the saltwater intrusion by time lapse electrical resistivity tomography: The Chioggia test site (Venice Lagoon, Italy). *J. Appl. Geophys.* **2009**, *69*, 117–130. [CrossRef]

17. Gnanasundar, D.; Elango, L. Groundwater quality assessment of a coastal aquifer using geoelectrical techniques. *Int. J. Environ. Hydrol.* **1999**, *7*, 21–33.

18. Nowroozi, A.A.; Horrocks, S.B.; Henderson, P. Saltwater intrusion into the freshwater aquifer in the eastern shore of Virginia: A reconnaissance electrical resistivity survey. *J. Appl. Geophys.* **1999**, *42*, 1–22. [CrossRef]

19. Sathish, S.; Elango, L.; Rajesh, R.; Sarma, V.S. Assessment of seawater mixing in a coastal aquifer by high resolution electrical resistivity tomography. *Int. J. Environ. Sci. Tech.* **2009**, *8*, 483–492. [CrossRef]

20. Shammas, M.I.; Jacks, G. Seawater intrusion in the Salalah plain aquifer. *Oman. Environ. Geol.* **2007**, *53*, 575–587. [CrossRef]

21. Urish, D.W.; Frohlich, R.K. Surface electrical resistivity in coastal groundwater exploration. *Geoexploration* **1990**, *26*, 267–289. [CrossRef]

22. Wagner, F.M.; Möller, M.; Schmidt-Hattenberger, C.; Kempka, T.; Maurer, H. Monitoring freshwater salinization in analog transport models by time-lapse electrical resistivity tomography. *J. Appl. Geophys.* **2013**, *89*, 84–95. [CrossRef]

23. Singha, K.; Gorelick, S.M. Saline tracer visualized with three-dimensional electrical resistivity tomography: Field-scale spatial moment analysis. *Water Resour. Res.* **2005**, *41*. [CrossRef]

24. Archie, G.E. The electrical resistivity log as an aid in determining some reservoir characteristics. *Trans. Am. Inst. Min. Metall. Pet. Eng.* **1942**, *146*, 54–62. [CrossRef]

25. Masciale, R.; De Carlo, L.; Caputo, M.C. Impact of a very low enthalpy plant on a costal aquifer: A case study in Southern Italy. *Environ. Earth Sci.* **2015**, *74*, 2093–2144. [CrossRef]

26. Pieri, P.; Sabato, L.; Spalluto, L.; Tropeano, M. Note illustrative della carta geologica dell'area urbana di Bari in scala 1:25.000 (notes of the geological map of the urban area of Bari at scale 1:25.000). *Rend. Online Soc. Geol. Ital.* **2011**, *14*, 26–36. [CrossRef]

27. Masciopinto, C.; La Mantia, R.; Chrysikopoulos, C.V. Fate and transport of pathogens in a fractured aquifer in the Salento area, Italy. *Water Resour. Res.* **2008**, *44*, 1–18. [CrossRef]

28. Bear, J. *Dynamics of Fluids in Porous Media*; American Elsevier Publishing Company, Inc.: New York, NY, USA, 1972; pp. 25–170.

29. Fiori, A.; Bellin, A. Non-ergodic transport of kinetically sorbing solutes. *J. Contam.* **1999**, *40*, 201–219. [CrossRef]

30. Masciopinto, C.; Liso, I.S. Assessment of the impact of sea-level rise due to climate change on coastal groundwater discharge. *Sci. Total Environ.* **2016**, *569*, 672–680. [CrossRef] [PubMed]

31. Masciopinto, C.; Palmiotta, D. A New Method to Infer Advancement of Saline Front in Coastal Groundwater Systems by 3D: The Case of Bari (Southern Italy) Fractured Aquifer. *Computation* **2016**, *4*, 9. [CrossRef]

32. Binley, A.; Kemna, A. DC Resistivity and Induced Polarization Methods. In *Hydrogeophysics*; Rubin, Y., Hubbard, S.S., Eds.; Springer: Dordrecht, The Netherlands, 2005; pp. 129–156, ISBN 101-4020-3101-7.

33. Keller, G.V.; Frischknecht, F.C. Electrical Methods in Geophysical Prospecting, Pergamon, London. 1996. Available online: https://www.eoas.ubc.ca/ubcgif/iag/foundations/properties/resistivity.htm (accessed on 11 September 2017).

Article

Groundwater Flow Determination Using an Interval Parameter Perturbation Method

Guiming Dong [1], Juan Tian [2], Hongbin Zhan [3,* and Rengyang Liu [1]

[1] School of Resource and Earth Science, China University of Mining and Technology, Xuzhou 221116, China; guiming14432@126.com (G.D.); lry@cumt.edu.cn (R.L.)
[2] Department of Environmental Science, Jiangsu Normal University, Xuzhou 221116, China; tianjuan980106@126.com
[3] Department of Geology and Geophysics, Texas A&M University, College Station, TX 77843, USA
* Correspondence: zhan@geos.tamu.edu; Tel.: +1-778-433-115

Received: 27 September 2017; Accepted: 13 December 2017; Published: 15 December 2017

Abstract: Groundwater flow simulation often inevitably involves uncertainty, which has been quantified by a host of methods including stochastic methods and statistical methods. Stochastic methods and statistical methods face great difficulties in applications. One of such difficulties is that the statistical characteristics of random variables (such as mean, variance, covariance, etc.) must be firstly obtained before the stochastic methods can be applied. The dilemma is that one is often unclear about such statistical characteristics, given the limited available data. To overcome the problems met by stochastic methods, this study provides an innovative approach in which the hydrogeological parameters and sources and sinks of groundwater flow are represented by bounded but uncertain intervals of variables called interval of uncertainty variables (IUVs) and this approach is namely the interval uncertain method (IUM). IUM requires only the maximum and minimum values of the variable. By utilizing the natural interval expansion, an interval-based parametric groundwater flow equation is established, and the solution of that equation can be found. Using a hypothetical steady-state flow case as an example, one can see that when the rate of change is less than 0.2, the relative error of this method is generally limited to less than 5%; when the rate of change is less than 0.3, the relative error of this method can be kept within 10%. This research shows that the proposed method has smaller relative errors and higher computational efficiency than the Monte Carlo methods. It is possible to use this method to analyze the uncertainties of groundwater flow when it is difficult to obtain the statistical characteristics of the hydrogeological systems. The proposed method is applicable in linear groundwater flow system. Its validity in nonlinear flow systems such as variably saturated flow or unconfined flow with considerable variation of water table will be checked in the future.

Keywords: groundwater flow model; numerical simulation; uncertainty; IUV; IUM; perturbation method; Monte Carlo; GFModel

---

1. Introduction

Over the past 40 years, numerical models of groundwater flow and solute transport have been widely used in studies of water resources management, migration of pollutants, sea water intrusion and many other applications [1–4]. However, because of the complexity of hydrogeological conditions and the limited budget and time, scientists often find it difficult to collect sufficient data to completely describe the hydrogeological systems in great details. Lack of hydrogeological data will inevitably affect the reliability of the model simulations and result in model uncertainty. Therefore, how to describe the uncertainty of model inputs and outputs and to improve the reliability of model predictions has become an important element of the numerical simulations used in studies of groundwater flow and solute transport [5–9].

Stochastic method based on a certain statistical structure is one of the most important methods currently used to study the uncertainty of numerical models of groundwater flow [2–9]. Stochastic methods include a host of mathematical methods such as the traditional random methods [7,8] and the Bayesian statistical methods [9]. The traditional random methods mainly include the Monte Carlo (MC) method, the moment equation method and Taylor expansion method or the perturbation expansion method. The Bayesian method is a widely used stochastic method in recent decades.

The MC method is first widely used to obtain the statistical characteristics of groundwater head or solute concentrations by random sampling after the probability density function (PDF) of the hydrogeological elements is known, where the hydrogeological elements refer to the aquifer parameters such as hydraulic conductivity or transmissivity, storativity, porosity, dispersivity, etc., boundary and initial conditions of flow and transport, and sinks and sources of flow and transport [10–13]. The moment equation method can be used to calculate the mean and variance of groundwater head or solute concentrations, provided that the mean and variance of known hydrogeological elements are given [14–18]. The moment equation method is used to obtain the first and second moments of the hydraulic head (for flow problems) or the solute concentration (for transport problems) by the first and second moments of the stochastic governing equation. Many investigators have carried out the uncertainty analysis of numerical simulations based on the Taylor expansion method or the perturbation expansion method, which also requires the mean and variance of the hydrogeological elements [19,20]. The Taylor expansion method or the perturbation expansion method expands head or concentration around its mean value and tries to make a connection between the mean and variance of head and concentration with the mean and variance of the hydrogeological elements [21,22]. The Taylor expansion or perturbation expansion method often involves the inversion of a large coefficient matrix, which could be computationally expensive. In addition, several investigators have studied the random theory using other methods [23–29].

Bayesian theory has also been widely used in the uncertainty analysis of numerical simulations of groundwater flow and solute transport. This theory involves three aspects including sampling methods, likelihood functions and convergence criteria [1]. A crucial step of the Bayesian theory is the sampling algorithm. There are several commonly used sampling algorithms, such as the Metropolis-Hastings algorithm, the Gibbs algorithm and the adaptive Metropolis algorithm [30–40]. Bayesian theory can obtain the posterior probability density function of hydrogeological elements, but the premise is that the distribution of hydrogeological parameters is known a priori.

Based on the random statistical characteristics of the hydrogeological elements, the stochastic method obtains the statistical characteristics of the output (mainly, the head and concentration). Unfortunately, it is sometimes difficult to know the stochastic statistical characteristics of the hydrogeological elements a priori in actual applications.

However, although precise statistical characteristics of the hydrogeological elements may be difficult to obtain, it may be easier to determine the ranges of possible values for the hydrogeological elements, probably with the help hydrogeological surveys, which is conventionally conducted in nearly all the hydrogeological investigations. If this is true, now the question is: Can we quantitatively determine the ranges of system outputs, given the bounded but uncertain inputs of the hydrogeological elements? The method used to tackle this question is named the interval uncertainty method (IUM), and it is practically appealing in terms of managing groundwater resources and conducting risk assessment of contaminated aquifers. A minor point to note is that IUM has been successfully carried out in other disciplines such as structural engineering, interval optimization method, and irrigation water distribution, but has never been applied in subsurface hydrology [41–50].

This study is the first attempt to use IUM for dealing with a linear groundwater flow system in a confined aquifer. Comparison of the presented (analytic) IUM and numerical analysis of a few hypothetical examples shows that the computational efficiency and precision of this method are very well. The validity of IUM in nonlinear subsurface flow systems such as flow in variably saturated porous media or flow in an unconfined aquifer with considerable variation of water table with time is out of the scope of this study, and will be investigated in the future.

2. Problem Description

IUM involves the interval response expression whose mathematical background is briefly explained as follows.

2.1. Interval Response Expression

We consider that the groundwater system response ω is a function of the system parameters $\alpha = (\alpha_1, \alpha_2, \ldots, \alpha_m)^T$, as stated below.

$$\omega = \omega(\alpha) = \omega(\alpha_1, \alpha_2, \ldots, \alpha_m), \tag{1}$$

where ω is a vector or scalar, m is the number of parameters, and T in the superscript is the transpose sign.

We consider that there are parameters with uncertainties in the parameter space. If we only know their uncertainty range, they can be expressed using the following interval form:

$$\alpha^I = [\underline{\alpha}, \overline{\alpha}] = \{\alpha : \underline{\alpha} \leq \alpha \leq \overline{\alpha}, \underline{\alpha}, \overline{\alpha} \in R^L\} \tag{2}$$

in which $\overline{\alpha}$ and $\underline{\alpha}$ are, respectively, the vectors of the upper and lower bounds of the vector of uncertain parameters α, R^L is a real vector space. When the bounded uncertain parameters or parameters vector α change within the range represented by Equation (2), the changing range for the system response ω can be expressed in the following form:

$$\Gamma = \{\omega : \omega = \omega(\alpha), \underline{\alpha} \leq \alpha \leq \overline{\alpha}\} \tag{3}$$

When we know the upper and lower bounds of the bounded uncertain parameters or the unascertained parameter vector α, the next task is to acquire the upper and lower bounds of the groundwater system response ω, i.e., $\overline{\omega}$ and $\underline{\omega}$.

2.2. Interval Parameter Type Groundwater Head Equation

The general form of groundwater flow governing equation used in numerical simulation may be expressed as

$$KH = F \tag{4}$$

where K is the coefficient matrix, H is the groundwater head vector at different nodes, and F is a column vector. K and F are usually obtained by matrix and column vector superposition in a finite element or finite difference framework, i.e., K and F can be decomposed into summations as follows:

$$K = K_1 + K_2 + \cdots + K_m = \sum_{i=1}^m K_i \tag{5}$$

$$F = F_1 + F_2 + \cdots + F_m = \sum_{i=1}^m F_i \tag{6}$$

where m is the number of rows of K and F which are functions of hydrogeological elements, i.e.,

$$K = K(\alpha), \quad F = F(\alpha) \tag{7}$$

Therefore, Equation (7) can be transformed into

$$K = K(\alpha) = \sum_{i=1}^m \varphi_i(\alpha) K_i', \quad F = F(\alpha) = \sum_{i=1}^m \phi_i(\alpha) F_i' \tag{8}$$

where $K_i = \varphi_i(\alpha)K_i'$, $F_i = \phi_i(\alpha)F_i'$, $\varphi_i(\alpha)$ and $\phi_i(\alpha)$ are functions of parameter α, and K_i' and F_i' are matrix and column vector, respectively. Functions $\varphi_i(\alpha)$ and $\phi_i(\alpha)$ ($i = 1, 2, \ldots, m$) may behave differently, i.e., some may be nonlinear, some may be linear, and some may be zero. If α denotes the hydraulic conductivity, the specific yield, the specific storage or parameters associated with the flow boundary, $\varphi_i(\alpha)$ and $\phi_i(\alpha)$ are linear functions, and Equation (8) can be simplified as

$$K = K(\alpha) = \sum_{i=1}^{m} \alpha_i K_i', \quad F = F(\alpha) = \sum_{i=1}^{m} \alpha_i F_i' \tag{9}$$

If α has an uncertainty within a specified range, α can be defined as α^I, where α^I is the interval parameter.

$$\alpha^I = [\underline{\alpha}, \overline{\alpha}] = \alpha_0 + \Delta\alpha^I, \Delta\alpha^I = [-\frac{\overline{\alpha} - \underline{\alpha}}{2}, \frac{\overline{\alpha} - \underline{\alpha}}{2}] \tag{10}$$

where α_0 is the average value of parameter α^I, $\overline{\alpha} - \underline{\alpha}$ is the variation of α^I, and $\frac{\overline{\alpha} - \underline{\alpha}}{2\alpha_0}$ is the rate of change of parameter α^I. According to Equation (8), we can get the following expression by utilizing the interval expansion from interval mathematics [51].

$$K\left(\alpha_i^I\right) = \sum_{i=1}^{m} \alpha_i^I K_i', \quad F\left(\alpha_i^I\right) = \sum_{i=1}^{m} \alpha_i^I F_i' \tag{11}$$

Combining Equation (11) with Equation (10), we can also obtain the following expression.

$$K\left(\alpha^I\right) = \sum_{i=1}^{m} \left(\alpha_i^0 + \Delta\alpha_i^I\right) K_i', \quad F\left(\alpha^I\right) = \sum_{i=1}^{m} \left(\alpha_i^0 + \Delta\alpha_i^I\right) F_i' \tag{12}$$

Here we can simplify Equation (12) as follows:

$$K_0 = \sum_{i=1}^{m} \alpha_i^0 K_i', \quad F_0 = \sum_{i=1}^{m} \alpha_i^0 F_i' \tag{13}$$

$$\Delta K = K\left(\Delta\alpha^I\right) = \sum_{i=1}^{m} \Delta\alpha_i^I K_i', \quad \Delta F = F\left(\Delta\alpha^I\right) = \sum_{i=1}^{m} \Delta\alpha_i^I F_i' \tag{14}$$

Finally, we can obtain the interval of groundwater head equation

$$\left(K_0 + K(\Delta\alpha^I)\right) H = F_0 + F\left(\Delta\alpha^I\right) \tag{15}$$

which is contained in the interval parameter $\Delta\alpha^I$. Equation (15) is an interval algebraic equation group concerning H, whose solution is also an interval variable. Solving Equation (15), the corresponding changes of head H interval can be obtained when α^I or $\Delta\alpha^I$ changes within a certain range.

3. Interval Response Solution

3.1. Groundwater Head Interval Response Expression

When determining the interval solution of H in Equation (15), we will first obtain the average value H_0 of H, and then calculate the variation ΔH, which is a key issue to deal with. The following discussion provides the solution to Equation (15).

The average H_0 of the groundwater head satisfies the following equation:

$$K_0 H_0 = F_0 \tag{16}$$

where K_0 is the coefficient matrix when the hydrogeological parameters are set at their averages, F_0 is the column vector when sources and sinks take their averages as well.

Meanwhile, using $K_0 + \Delta K$, $H_0 + \Delta H$ and $F_0 + \Delta F$, the following equation can be written:

$$(K_0 + \Delta K)(H_0 + \Delta H) = F_0 + \Delta F \tag{17}$$

where ΔK and ΔF are respectively the variations of K and F. Considering Equations (16) and (17), and ignoring the second order and higher order terms of ΔK, ΔH, and ΔF, we can obtain an approximate expression of the first-order disturbance about the head [51].

$$\Delta H^{(1)} = K_0^{-1}(\Delta F - \Delta K H_0) \tag{18}$$

where $\Delta H^{(1)}$ is the first-order approximation of ΔH. That is, replacing ΔH with $\Delta H^{(1)}$ in Equation (18), where H is

$$H = H_0 + \Delta H^{(1)} \tag{19}$$

The second-order approximation of ΔH and the corresponding H is

$$\Delta H^{(2)} = K_0^{-1}\Delta K\left(K_0^{-1}\Delta F - K_0^{-1}\Delta K H_0\right), \quad H = H_0 + \Delta H^{(1)} + \Delta H^{(2)} \tag{20}$$

According to the interval expansion from interval mathematics, substituting Equation (14) into Equation (18), the first order perturbation of ΔH is approximately

$$\Delta H^{(1)} = \sum_{i=1}^{m} \Delta \alpha_i^I \left(K_0^{-1}(F_i' - K_i' H_0)\right) \tag{21}$$

and the component form of Equation (21) can be written as

$$\psi = (\psi_{j,i}), \quad \psi_i = \left(K_0^{-1}(F_i' - K_i' H_0)\right), \quad i = 1, 2, \dots, m \, ; \quad j = 1, 2, \dots, k' \tag{22}$$

where m and k' are respectively the number of columns and rows of ψ in Equation (22). That is, they indicate the number of nodes to be solved. When Equation (22) is substituted into Equation (21), the following can be obtained:

$$\Delta H^{(1)} = \psi \Delta \alpha^I \tag{23}$$

According to the interval algorithm [51], the following can be obtained:

$$\Delta H^{(1)} = \psi[-\Delta \alpha, \Delta \alpha] = [-|\psi|\Delta \alpha, |\psi|\Delta \alpha] \tag{24}$$

Therefore, one can obtain the first-order perturbation of H, which is approximately expressed as

$$H = H_0 + [-|\psi|\Delta \alpha, |\psi|\Delta \alpha] \tag{25}$$

Equation (25) is obtained when $\varphi_i(\alpha)$ and $\phi_i(\alpha)$ both are linear expressions. It can be seen from the solution of the interval head equation that one only needs to find the inverse of the coefficient matrix rather than the head derivative of the changing elements to calculate the interval value of H. In addition, when the interval value of H is obtained by this method, we find the extreme of H (i.e., the maximum and minimum values) in a certain interval, which is obviously different from the general perturbation method that only finds the variation of H within a certain interval rather than its extreme value. When applied in groundwater, the general perturbation method is a method of calculating the variation of groundwater head, which is based on the hydrogeological elements rate of change is known.

3.2. Algorithmic Details

The groundwater head interval response calculation is applied in the whole process of the number simulation of groundwater flow. Because the coefficient matrixes of the subdivision element sections are first obtained based on the subdivision elements coefficient matrixes superposition, the total coefficient matrix in the number simulation program of groundwater flow is obtained. The algorithmic description of the interval parameter perturbation method is listed in Algorithm 1.

Algorithm 1. Interval Parameter Perturbation Method Algorithm.

After the head H_0 for a certain calculation period was obtained
1. Solving K_i', F_i' and $K_i'H_0$ for the section of each subdivision element;
2. Solving K_0^{-1}, and then $K_0^{-1}(F_i' - K_i'H_0)$ can be obtained;
3. Traversing each subdivision element, ψ can be obtained;
4. Calculating $|\psi|\Delta\alpha$, $\Delta H^{(1)}$ could be obtained;
Calculating above-described steps 1–4, the current time head can be obtained $\Delta H^{(1)}$

In the whole calculation, when the hydrogeological elements change within a certain interval, determining the head interval response is essentially a matter of interval optimization. In the following sections, numerical examples will be used to analyze the effectiveness of the proposed interval method.

4. Numerical Examples

To verify the effectiveness of the method in solving the interval-based parametric groundwater head equation, we establish a synthetic two-dimensional confined aquifer flow model in this section. The confined aquifer has an extent of 400 m × 400 m in horizontal dimensions, and a thickness of 10 m. The total numbers of subdivision triangular elements are 620. Taking the aquifer base as the reference surface, the top elevation of the aquifer is 10 m. The initial water head of the aquifer is uniformly set at 100 m, the initial hydraulic gradient is zero, and the left boundary of the aquifer is defined as a constant-head boundary with a value of 100 m. The remaining three boundaries of the aquifer are set to be general-head boundaries (GHB) in which the head, the hydraulic conductivity and the reciprocal of the distance are 100 m, 1 m/day and 0.1 m^{-1}, respectively. The aquifer is heterogeneous and isotropic. According to different hydraulic conductivity values chosen at different regions, the aquifer is divided into four sections.

A graphic representation of the aquifer associated with a sample discretization used in the numerical simulation is shown in Figure 1. The values of the hydraulic conductivities of different zones are shown in Table 1. From Figure 1, one can see that the pumping well (well No. 1 in Figure 1), which has coordinates of (200 m, 200 m), is located at the center of the aquifer. Meanwhile, each section contains an observation well. Considering steady-state flow, we will calculate and analyze three scenarios in the following sections. For Scenario One, the conductivities of the four sections are considered as the interval variable. On the basis of Scenario One, an injection well at coordinates of (100 m, 150 m) is added in Scenario Two. On the basis of Scenario Two, additional interval of boundary conditions is considered in Scenario Three. To verify the effectiveness of this proposed method, the degree of problem complexity increases from Scenario One to Scenario Three.

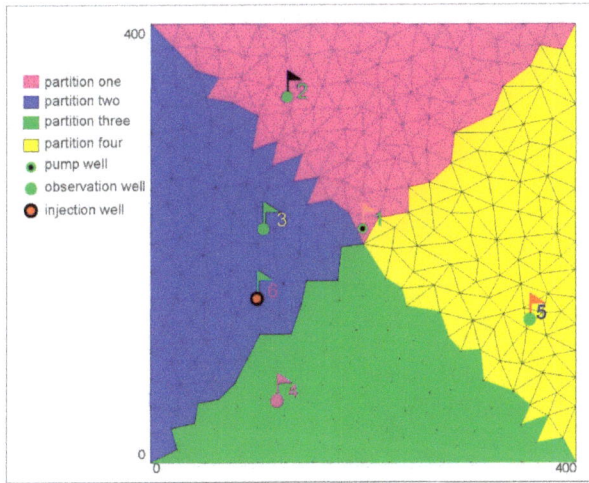

Figure 1. Division graph of ideal groundwater flow model and map showing the locations of the pumping well and the observation wells. (Pumping well: 1; observation wells: 2–5; injection well: 6; length unit: m.)

Where the numerical calculations are performed, we adopt the Groundwater Flow Model (GFModel) program which is based on an arbitrary polygon finite-difference method, developed by the first author (Guiming Dong) of China University of Mining and Technology [52]. The domain of interest is discretized into polygons. Further refinement of the discretization mesh does not provide noticeable improvement of the simulation results. Adopting the method of superposition of the osmotic matrix of the triangular element to form the coefficient matrix, the GFModel is a newly developed program, which can deal with the common boundary conditions, sources and sinks, and time-dependent hydrogeological parameters and can also carry out three-dimensional groundwater flow MC calculation, the general perturbation calculation, and interval parameter perturbation calculation.

Table 1. Hydraulic conductivity in each zone of the model.

Partition	Hydraulic Conductivity (Average Value) (m/day)
One	1
Two	10
Three	5
Four	3

4.1. Scenario One

We set the flow of the pumping wells to be 1500 m^3/day. Increasing at 0.05 intervals, the rate of change in the hydraulic conductivity changes from 0.05 to 0.4 in four zones seen in Figure 1, where the average value multiplied by the rate of change is equal to half of the variation of the parameter as referring to Equation (10). For example, the average hydraulic conductivity of the first partition is 1 m/day, when the rate of change is 0.2, then the interval of the hydraulic conductivity is [0.8, 1.2], and the variation of the parameter is 0.4 m/day. Moreover, we perform a comparative analysis of the MC method and the method proposed in this paper. The number of realizations used in the MC method is found to be 6.25 × 10^6 after some numerical exercises. The maximum, minimum and variation of the groundwater head obtained by the MC method are taken as theoretical values. Using a computer with 4 GB of memory and CPU frequency of 2.5 GHz, the calculation time of MC is 2.6 days, and the calculation time of the IUM is 4.2 s. The results are shown in Figure 2, Tables 2 and 3.

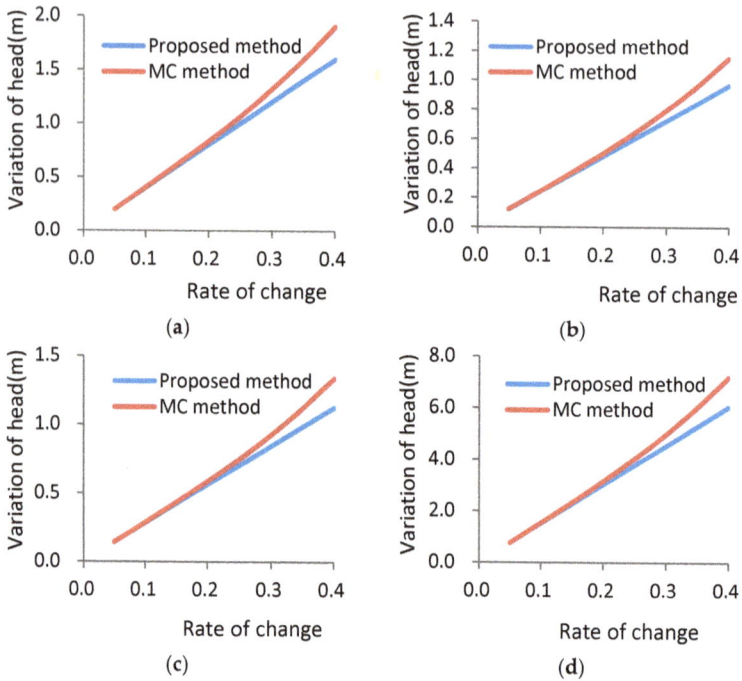

Figure 2. Comparison of two methods in Scenario One ((**a**) observation well 2; (**b**) observation well 3; (**c**) observation well 4; (**d**) observation well 5).

The smallest head and the largest head in each observation well decrease and increase respectively as the hydraulic conductivity rate of change increases, and the variations in head and the relative error increase correspondingly. The results of the proposed method show that when the rate of change is less than 0.2, the relative error of the head variation is generally limited to within 5%; while when the rate of change is less than 0.3, the relative error of the head variation is generally limited to within 10%, as seen in Table 1.

The theoretical head variation obtained from the MC method in the observation wells displays somewhat nonlinear relationship with the rate of change in the hydraulic conductivity. Instead, the relationship is best described as piecewise-linear. As can be seen from Figure 2, the theoretical head variation follows a nice linear relationship with the rate of change when the latter is less than 0.2. However, the theoretical variation starts to deviate from the linear trend when the rate of change is greater than 0.2.

For a given rate of change, when the hydraulic conductivities of four sections simultaneously take their minimum values, the head for the observation well is the same as the minimum theoretical head. Similarly, when the hydraulic conductivities of four sections simultaneously take their maximum values, the head for the observation well is also the same as the maximum theoretical head. The variations in the head calculated by the general perturbation method are the same as those calculated using the method proposed in this study. In the meantime, the corresponding values of ψ all must be positive in Equation (22), because only in this case, the general perturbation method calculated results will be the same with the computational result of Equation (25), which suggests that the general perturbation method has a close relationship with the proposed method.

Table 2. Calculation results table for Scenario One (unit: m).

Rate of Change	MC Calculation Results			IUM Calculation Results			Relative Error (%)
	Lowest Head	Highest Head	Variations in Head	Lowest Head	Highest Head	Variations in Head	
				Observation Well No. 2			
0.05	97.8684	98.0687	0.2003	97.8760	98.0758	0.1998	0.28
0.1	97.7515	98.1552	0.4037	97.7762	98.1757	0.3995	1.03
0.15	97.621	98.2342	0.6133	97.6763	98.2756	0.5993	2.28
0.2	97.4741	98.3067	0.8326	97.5764	98.3755	0.7991	4.02
0.25	97.3076	98.3733	1.0657	97.4765	98.4753	0.9988	6.27
0.3	97.1174	98.4348	1.3174	97.3766	98.5752	1.1986	9.02
0.35	96.8979	98.4917	1.5938	97.2767	98.6751	1.3984	12.27
0.4	96.6419	98.5447	1.9028	97.1769	98.7750	1.5981	16.01

Rate of Change	MC Calculation Results			IUM Calculation Results			Relative Error (%)
	Lowest Head	Highest Head	Variations in Head	Lowest Head	Highest Head	Variations in Head	
				Observation Well No. 3			
0.05	98.6456	98.7665	0.1209	98.6506	98.7711	0.1205	0.28
0.1	98.5751	98.8187	0.2436	98.5903	98.8314	0.2411	1.04
0.15	98.4963	98.8664	0.3701	98.5300	98.8916	0.3616	2.28
0.2	98.4078	98.9102	0.5024	98.4697	98.9519	0.4822	4.02
0.25	98.3074	98.9504	0.643	98.4095	99.0122	0.6027	6.26
0.3	98.1927	98.9876	0.7948	98.3492	99.0725	0.7233	9.01
0.35	98.0605	99.022	0.9615	98.2889	99.1327	0.8438	12.24
0.4	97.9062	99.054	1.1478	98.2287	99.1930	0.9643	15.98

Rate of Change	MC Calculation Results			IUM Calculation Results			Relative Error (%)
	Lowest Head	Highest Head	Variations in Head	Lowest Head	Highest Head	Variations in Head	
				Observation Well No. 4			
0.05	98.4301	98.5709	0.1408	98.4359	98.5763	0.1404	0.28
0.1	98.348	98.6318	0.2838	98.3657	98.6466	0.2809	1.04
0.15	98.2564	98.6874	0.4311	98.2955	98.7168	0.4213	2.27
0.2	98.1534	98.7385	0.5851	98.2253	98.7870	0.5617	4
0.25	98.0367	98.7854	0.7488	98.1551	98.8572	0.7021	6.23
0.3	97.9034	98.8288	0.9254	98.0848	98.9274	0.8426	8.95
0.35	97.7498	98.869	1.1193	98.0146	98.9976	0.983	12.18
0.4	97.5707	98.9064	1.3358	97.9444	99.0678	1.1234	15.9

Rate of Change	MC Calculation Results			IUM Calculation Results			Relative Error(%)
	Lowest Head	Highest Head	Variations in Head	Lowest Head	Highest Head	Variations in Head	
				Observation Well No. 5			
0.05	92.005	92.7623	0.7572	92.0305	92.7857	0.7552	0.27
0.1	91.5633	93.0893	1.526	91.6529	93.1633	1.5104	1.02
0.15	91.0697	93.3879	2.3182	91.2753	93.5409	2.2656	2.27
0.2	90.5145	93.6617	3.1472	90.8977	93.9185	3.0208	4.01
0.25	89.8852	93.9135	4.0283	90.5201	94.2961	3.776	6.26
0.3	89.1661	94.146	4.9799	90.1425	94.6737	4.5312	9.01
0.35	88.3364	94.3612	6.0249	89.7649	95.0513	5.2864	12.26
0.4	87.3684	94.5612	7.1928	89.3873	95.4289	6.0416	16

Table 3. Root mean square error (RMSE) calculation results of variations in head for Scenario One.

Rate of Change	0.05	0.1	0.15	0.2	0.25	0.3	0.35	0.4
RMSE	0.2236	0.3162	0.3873	0.4472	0.5000	0.5477	0.5916	0.6325

4.2. Scenario Two

Based on Scenario One, we install a new injection well with a flow rate of 1500 m^3/day at the coordinates (100 m, 150 m). The hydraulic conductivity rate of change is also adjusted from 0.05 to 0.4.

The number of realizations used in the MC method model is also 6.25×10^6. The calculation time of MC and the IUM is same as that on the scenario One. The results are shown in Figure 3, Tables 4 and 5.

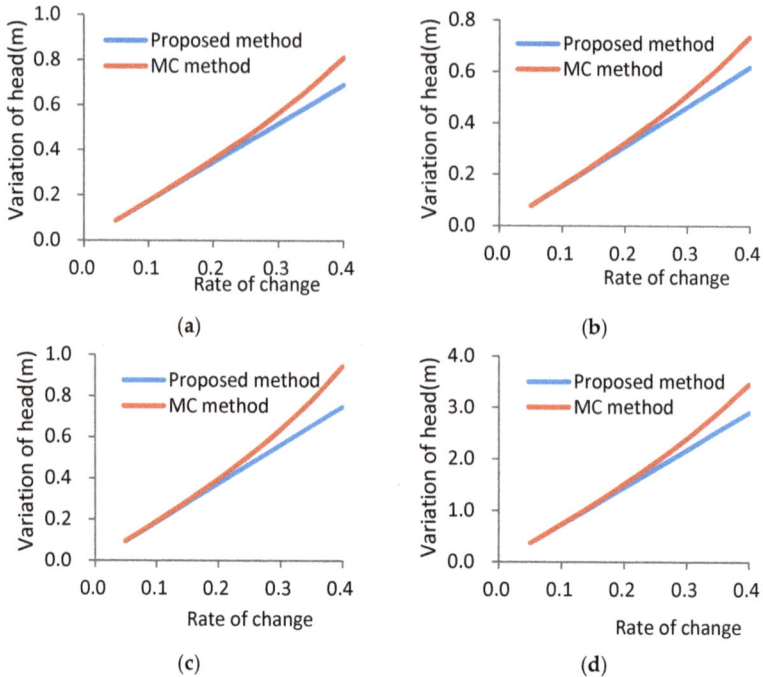

Figure 3. Comparison of the two methods in Scenario Two ((**a**) observation well 2; (**b**) observation well 3; (**c**) observation well 4; (**d**) observation well 5).

The results of Scenario Two are consistent with their counterparts in Scenario One, with some slight discrepancies in terms of specific values shown in Table 3 and Figure 3. For the same rate of change in the hydraulic conductivity, the relative errors of observation well 3 and observation well 5 are basically the same in both scenarios; while the relative error of observation well 4 in Scenario Two is slightly larger than its corresponding value in Scenario One. Moreover, the relative error of the observation well 2 under Scenario Two is slightly less than its corresponding value under Scenario One. It is obvious that although Scenario Two is more complicated hydrogeologically (as it involves an additional injection well) than that of Scenario One, the accuracy of the proposed method does not show any noticeable decline. In addition, the results of the proposed method show that when the rate of change is less than 0.2, the relative error of the head variation is generally less than 5%; while when the rate of change is less than 0.3, the relative error of the head variation is generally less than 10%.

For a given rate of change of the hydraulic conductivity, further analysis shows that when the hydraulic conductivities of four sections simultaneously take their minimum values, the head for the observation well does not correspond to the minimum theoretical head calculated from the MC method. Similar conclusion can be drawn when the hydraulic conductivities of four sections simultaneously take their maximum values. This finding is certainly quite different from that of Scenario One. Taking the observation well 4 as an example, with a rate of change of 0.2, when the hydraulic conductivities of four sections are set at their minimum and maximum values simultaneously, the heads of observation well 4 are 100.62 and 100.41 m, respectively. These values are clearly different from the minimum (100.35 m) and the maximum (100.74 m) heads calculated using the MC method.

Based on above analysis, one can conclude that the result of the general perturbation method will differ from the method proposed in this paper. That implies that the corresponding value of ψ in Equation (22) is not always positive. Using observation well 4 as an example, when the hydraulic conductivity rate of change is all 0.2 in four sections, the head variation of the observation well is 0.20 m using the general perturbation method, which is obviously different from the head variation of 0.37 m calculated using the method of this study. It also shows that the method proposed in this study can calculate the extreme values of the head at any given node, whereas the general perturbation method only calculates the head variation of the node under a certain variation, which is obviously different.

Table 4. Calculation results table for Scenario Two (unit: m).

	Observation Well No. 2						
Rate of Change	MC Calculation Results			IUM Calculation Results			Relative Error (%)
	Lowest Head	Highest Head	Variations in Head	Lowest Head	Highest Head	Variations in Head	
0.05	99.1171	99.2035	0.0864	99.1157	99.2018	0.0861	0.27
0.1	99.067	99.2409	0.1739	99.0726	99.2449	0.1723	0.95
0.15	99.0114	99.2754	0.264	99.0296	99.2880	0.2584	2.12
0.2	98.9493	99.3071	0.3578	98.9865	99.3310	0.3445	3.72
0.25	98.8793	99.3364	0.457	98.9435	99.3741	0.4306	5.78
0.3	98.7997	99.3634	0.5637	98.9004	99.4172	0.5168	8.33
0.35	98.7082	99.3886	0.6804	98.8573	99.4602	0.6029	11.39
0.4	98.6018	99.412	0.8102	98.8143	99.5033	0.689	14.95

	Observation Well No. 3						
Rate of Change	MC Calculation Results			IUM Calculation Results			Relative Error (%)
	Lowest Head	Highest Head	Variations in Head	Lowest Head	Highest Head	Variations in Head	
0.05	99.1447	99.2219	0.0772	99.1440	99.2210	0.077	0.3
0.1	99.0996	99.2552	0.1556	99.1055	99.2595	0.154	1.05
0.15	99.0493	99.2857	0.2364	99.0671	99.2980	0.2309	2.3
0.2	98.9927	99.3136	0.3209	99.0286	99.3365	0.3079	4.05
0.25	98.9286	99.3393	0.4107	98.9901	99.3750	0.3849	6.29
0.3	98.8553	99.3631	0.5078	98.9516	99.4135	0.4619	9.04
0.35	98.7708	99.385	0.6143	98.9131	99.4519	0.5388	12.28
0.4	98.6722	99.4055	0.7333	98.8746	99.4904	0.6158	16.02

	Observation Well No. 4						
Rate of Change	MC Calculation Results			IUM Calculation Results			Relative Error (%)
	Lowest Head	Highest Head	Variations in Head	Lowest Head	Highest Head	Variations in Head	
0.05	100.4527	100.5463	0.0936	100.4468	100.5401	0.0933	0.32
0.1	100.4138	100.6029	0.1891	100.4001	100.5867	0.1866	1.32
0.15	100.379	100.6676	0.2885	100.3535	100.6334	0.2799	3.01
0.2	100.3479	100.7422	0.3943	100.3069	100.6800	0.3731	5.36
0.25	100.3199	100.8288	0.5088	100.2602	100.7266	0.4664	8.34
0.3	100.2947	100.9302	0.6355	100.2136	100.7733	0.5597	11.93
0.35	100.2718	101.0502	0.7783	100.1669	100.8199	0.653	16.1
0.4	100.2511	101.1935	0.9424	100.1203	100.8666	0.7463	20.81

	Observation Well No. 5						
Rate of Change	MC Calculation Results			IUM Calculation Results			Relative Error (%)
	Lowest Head	Highest Head	Variations in Head	Lowest Head	Highest Head	Variations in Head	
0.05	96.1732	96.5366	0.3634	96.1744	96.5368	0.3624	0.28
0.1	95.9612	96.6935	0.7322	95.9933	96.7180	0.7247	1.03
0.15	95.7243	96.8368	1.1125	95.8121	96.8992	1.0871	2.29
0.2	95.4578	96.9682	1.5104	95.6309	97.0803	1.4494	4.04
0.25	95.1557	97.089	1.9333	95.4497	97.2615	1.8118	6.29
0.3	94.8105	97.2006	2.3901	95.2686	97.4427	2.1741	9.04
0.35	94.4123	97.3038	2.8916	95.0874	97.6239	2.5365	12.28
0.4	93.9476	97.3998	3.4522	94.9062	97.8050	2.8988	16.03

Table 5. RMSE calculation results of variations in head for Scenario Two.

Rate of Change	0.05	0.1	0.15	0.2	0.25	0.3	0.35	0.4
RMSE	0.2236	0.3162	0.3873	0.4472	0.5000	0.5477	0.5916	0.6325

4.3. Scenario Three

The difference between Scenario Three and Scenario Two is that the number of hydrogeological elements with interval variations, and the hydrogeological elements with interval variations, include the hydraulic conductivities of four sections, the head at the head boundary, the hydraulic conductivity, and the head at GHB. The reciprocal of the distance at GHB is constant. The hydraulic conductivity rate of change also changes from 0.05 to 0.4 in four sections. The numbers of realizations used in the MC method model are also 6.25×10^6 times. The calculation time of MC is 4.1 days, while the calculation time of the IUM is only 4.2 s. The results are shown in Figure 4, Tables 6 and 7.

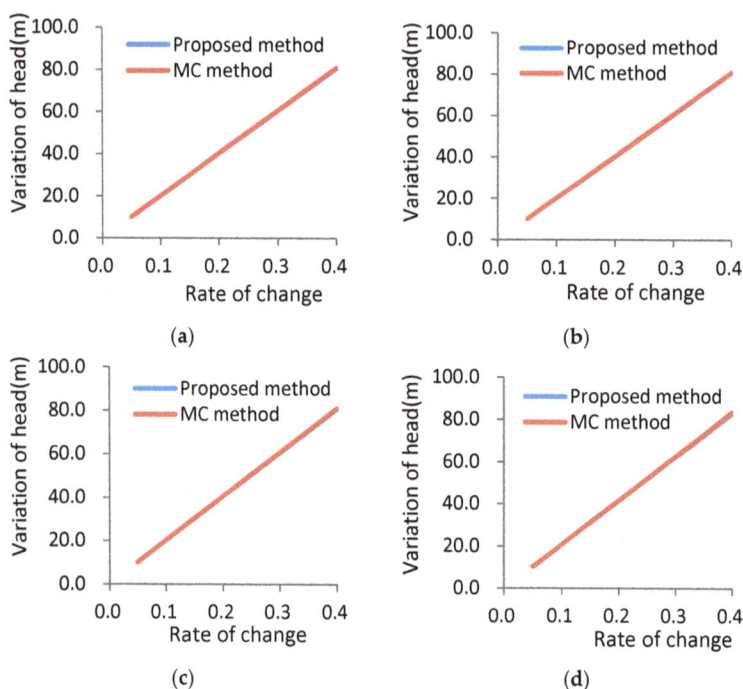

Figure 4. Comparison of two methods in Scenario Three ((**a**) observation well 2; (**b**) observation well 3; (**c**) observation well 4; (**d**) observation well 5).

Table 6. Calculation results table for Scenario Three (unit: m).

Rate of Change	Observation Well No. 2						
	MC Calculation Results			IUM Calculation Results			Relative Error (%)
	Lowest Head	Highest Head	Variations in Head	Lowest Head	Highest Head	Variations in Head	
0.05	94.1189	104.2004	10.0815	94.1151	104.2025	10.0874	−0.06
0.1	89.068	109.2383	20.1703	89.0714	109.2462	20.1748	−0.02
0.15	84.0113	114.2732	30.2619	84.0277	114.2898	30.2621	0
0.2	78.948	119.3051	40.357	78.9840	119.3335	40.3495	0.02
0.25	73.8765	124.3347	50.4581	73.9403	124.3772	50.4369	0.04
0.3	68.7955	129.3619	60.5665	68.8966	129.4209	60.5243	0.07
0.35	63.7017	134.3872	70.6854	63.8530	134.4646	70.6116	0.1
0.4	58.5928	139.4107	80.8179	58.8093	139.5083	80.699	0.15

Rate of Change	Observation Well No. 3						
	MC Calculation Results			IUM Calculation Results			Relative Error(%)
	Lowest Head	Highest Head	Variations in Head	Lowest Head	Highest Head	Variations in Head	
0.05	94.1441	104.2213	10.0772	94.1418	104.2233	10.0815	−0.04
0.1	89.0965	109.2566	20.1601	89.1010	109.2640	20.163	−0.01
0.15	84.0431	114.2889	30.2458	84.0603	114.3048	30.2445	0
0.2	78.9832	119.3184	40.3351	79.0195	119.3455	40.326	0.02
0.25	73.9152	124.3456	50.4304	73.9788	124.3863	50.4075	0.05
0.3	68.8376	129.3707	60.5331	68.9380	129.4270	60.489	0.07
0.35	63.7481	134.3939	70.6458	63.8973	134.4678	70.5705	0.11
0.4	58.6436	139.4154	80.7719	58.8565	139.5085	80.652	0.15

Rate of Change	Observation Well No. 4						
	MC Calculation Results			IUM Calculation Results			Relative Error (%)
	Lowest Head	Highest Head	Variations in Head	Lowest Head	Highest Head	Variations in Head	
0.05	95.4547	105.5433	10.0886	95.4466	105.5402	10.0936	−0.05
0.1	90.4155	110.6001	20.1846	90.3998	110.5871	20.1873	−0.01
0.15	85.3804	115.6652	30.2848	85.3530	115.6339	30.2809	0.01
0.2	80.3488	120.7401	40.3913	80.3062	120.6807	40.3745	0.04
0.25	75.3202	125.8274	50.5071	75.2593	125.7275	50.4682	0.08
0.3	70.2943	130.9296	60.6353	70.2125	130.7743	60.5618	0.12
0.35	65.2703	136.0506	70.7803	65.1657	135.8211	70.6554	0.18
0.4	60.2483	141.1954	80.947	60.1189	140.8680	80.7491	0.24

Rate of Change	Observation Well No. 5						
	MC Calculation Results			IUM calculation results			Relative Error(%)
	Lowest Head	Highest Head	Variations in Head	Lowest Head	Highest Head	Variations in Head	
0.05	91.1783	101.5286	10.3503	91.1738	101.5374	10.3636	−0.13
0.1	85.9656	106.6861	20.7205	85.9920	106.7192	20.7272	−0.03
0.15	80.7276	111.8301	31.1025	80.8102	111.9010	31.0908	0.04
0.2	75.4604	116.9617	41.5013	75.6284	117.0828	41.4544	0.11
0.25	70.1571	122.0831	51.926	70.4466	122.2646	51.818	0.21
0.3	64.8106	127.1951	62.3845	65.2648	127.4464	62.1816	0.33
0.35	59.4111	132.2985	72.8874	60.0830	132.6282	72.5452	0.47
0.4	53.9448	137.3947	83.4499	54.9012	137.8100	82.9088	0.65

Table 7. RMSE calculation results of variations in head for Scenario Three.

Rate of Change	0.05	0.1	0.15	0.2	0.25	0.3	0.35	0.4
RMSE	0.2236	0.3162	0.3873	0.4472	0.5000	0.5477	0.5916	0.6325

The smallest head and the largest head for each observation well decrease and increase, respectively, as the hydraulic conductivity rate of change increases, and the variations in head and the relative error increase correspondingly. The results of the proposed method show that when the rate of change reaches 0.4, the maximum of the relative error is only 0.65%. This suggests that the relative error is generally limited to 1%.

There is a nice linear relationship between the theoretical head variation calculated from the MC method and the rate of change of the hydraulic conductivity, as can be seen from Figure 4. Interestingly enough, the head variation calculated by the proposed method of this study shows almost the same linear relationship with the rate of change in the hydraulic conductivity. This explains why the two methods have relatively small relative errors. It is obvious that, although the number of hydrogeological elements that are allowed to change in Scenario Three is larger (Scenario Three includes seven changing hydrogeological elements), the linearity of the system interval response is improved, and the accuracy of the method is improved as well. Therefore, the number of intervals does not appear to be the primary factor determining the accuracy of the proposed method. Because the degree of linearity of Scenario Three is higher than that of Scenario Two, it is reasonable to infer that the degree of linearity should be the main factor affecting the accuracy of the proposed method.

As in Scenario Two, the head for the observation well is not always the same as the minimum (or maximum) theoretical head when all the hydrogeological elements take their minimum (or maximum) values. The result of the general perturbation method is also different from that obtained from the method of this study. Once again, using observation well 4 as an example, with a rate of change of 0.1, where the rate of change is for all the seven changing hydrogeological elements, the head variation of the observation well is 19.90 m using the general perturbation method. This result is obviously different from the head variation of 20.19 m calculated by the method of this study. It can be seen from the three scenarios of numerical examples when three conditions, including the aquifer are confined aquifer, $\varphi_i(\alpha)$ and $\phi_i(\alpha)$ both are linear expressions and the interval of the changing elements is little change, are satisfied in the same time, the IUM can perform better, while any of the above three conditions are not satisfied, the IUM validity has not yet been checked. The inverse of matrix is the main factor that influences the calculation efficiency of the method. For the above problems, we will carry out research in the following study. In addition, the method can only obtain the changing interval of the head, while giving the statistical characteristics of the head.

5. Conclusions

In this study, in contrast to the stochastic methods used in many numerical simulations of groundwater flow and solute transport, a new concept involving interval mathematics is proposed to quantify the uncertainty analysis of groundwater flow and solute transport. This concept originates from the concept of interval uncertainty. The form of the interval-type groundwater flow governing equation is presented, and the interval parameter perturbation method is provided. Since this method only required matrix inversion to obtain the derivative of the head for the changing hydrogeological elements, the method is computationally much more efficient than the conventional MC method. Using this method, the interval response of the groundwater head can be obtained, which is different from the value obtained by the general perturbation method.

Three hypothetical scenarios with an increasing degree of hydrogeological complexity and uncertainty are used as examples to illustrate the performance of the proposed method, including its accuracy and robustness. It appears that the accuracy of the method depends on the linearity between the simulated head variation and the changing hydrogeological elements. In addressing practical problems, the proposed method is recommended when the rate of change for the changing hydrogeological elements is less than a certain value such as 0.2, as great linearity between the simulated head variation and the changing hydrogeological elements has been identified under such condition. Further research is needed to determine the upper limit of the rate of change for the changing hydrogeological elements when this method is used under a wide range of hydrogeological conditions.

Acknowledgments: We would like to acknowledge funding provided by the National Natural Science Foundation of China (41202179; 51209109) and the Essential Science Indicators (ESI) Fund of Geosciences in China University of Mining and Technology. Analysis was done while Guiming Dong was hosted as a visiting scientist in the Geology and Geophysics Department at Texas A&M University.

Author Contributions: Guiming Dong completed the program developing and wrote the first draft. Juan Tian and Rengyang Liu completed the translation of this paper and improved the content of this paper. Hongbin Zhan put forward the whole idea of the paper and carried out the improvement of the paper.

Conflicts of Interest: The authors declare no conflict of interest.

References

1. Xu, T.; Valocchi, A.J. A Bayesian approach to improved calibration and prediction of groundwater models with structural error. *Water Resour. Res.* **2015**, *51*, 9290–9311. [CrossRef]
2. Tartakovsky, D.M. Assessment and management of risk in subsurface hydrology: A review and perspective. *Adv. Water Resour.* **2013**, *51*, 247–260. [CrossRef]
3. Joodavi, A.; Zare, M.; Ziaei, A.N.; Ferre, T.P.A. Groundwater management under uncertainty using a stochastic multi-cell model. *J. Hydrol.* **2017**, *551*, 265–277. [CrossRef]
4. Marchant, B.; Mackay, J.; Bloomfield, J. Quantifying uncertainty in predictions of groundwater levels using formal likelihood methods. *J. Hydrol.* **2016**, *540*, 699–711. [CrossRef]
5. Rojas, R.; Kahunde, S.; Peeters, L.; Batelaan, O.; Feyen, L.; Dassargues, A. Application of a multimodel approach to account for conceptual model and scenario uncertainties in groundwater modelling. *J. Hydrol.* **2010**, *394*, 416–435. [CrossRef]
6. Brink, C.V.D.; Zaadnoordijk, W.J.; Burgers, S.; Griffioen, J. Stochastic uncertainties and sensitivities of a regional-scale transport model of nitrate in groundwater. *J. Hydrol.* **2008**, *361*, 309–318. [CrossRef]
7. Refsgaard, J.C.; Christensen, S.; Sonnenborg, T.O. Review of strategies for handling geological uncertainty in groundwater flow and transport modeling. *Adv. Water Resour.* **2012**, *36*, 36–50. [CrossRef]
8. Nilsson, B.; Hojberg, A.L.; Refsgaard, J.C. Uncertainty in geological and hydrogeological data. *Hydrol. Earth Syst. Sci.* **2007**, *11*, 2675–2706. [CrossRef]
9. Matott, L.S.; Babendreier, J.E.; Purucker, S.T. Evaluating uncertainty in integrated environmental models: A review of concepts and tools. *Water Resour. Res.* **2009**, *45*, 735–742. [CrossRef]
10. Freeze, R.A. A stochastic-conceptual analysis of one-dimensional groundwater flow innonuniform homogeneous media. *Water Resour. Res.* **1975**, *12*, 567.
11. Smith, L.; Freeze, R.A. Stochastic analysis of steady state groundwater flow in a bounded domain: 2. Two-dimensional simulations. *Water Resour. Res.* **1979**, *15*, 1543–1559. [CrossRef]
12. James, A.L.; Oldenburg, C.M. Linear and Monte Carlo uncertainty analysis for subsurface contaminant transport simulation. *Water Resour. Res.* **1997**, *33*, 2495–2508. [CrossRef]
13. Tonkin, M.; Doherty, J. Calibration-constrained Monte Carlo analysis of highly parameterized models using subspace techniques. *Water Resour. Res.* **2009**, *45*, 206–216. [CrossRef]
14. Graham, W.; Mclaughlin, D. Stochastic analysis of nonstationary subsurface solute transport: 1. Unconditional moments. *Water Resour. Res.* **1989**, *25*, 2331–2355. [CrossRef]
15. Zhang, D.; Winter, C.L.; Laboratory, N. Moment-Equation Approach to Single Phase Fluid Flow in Heterogeneous Reservoirs. *SPE J.* **1999**, *4*, 118–127.
16. Lu, Z.; Zhang, D. On stochastic modeling of flow in multimodal heterogeneous formations. *Water Resour. Res.* **2002**, *38*, 8-1–8-15. [CrossRef]
17. Winter, C.L.; Tartakovsky, D.M.; Guadagnini, A. Numerical solutions of moment equations for flow in heterogeneous composite aquifers. *Water Resour. Res.* **2002**, *38*, 13-1–13-8. [CrossRef]
18. Winter, C.L.; Tartakovsky, D.M.; Guadagnini, A. Moment Differential Equations for Flow in Highly Heterogeneous Porous Media. *Surv. Geophys.* **2003**, *24*, 81–106. [CrossRef]
19. Sagar, B. Galerkin Finite Element Procedure for analysing flow through random media. *Water Resour. Res.* **1978**, *14*, 1035–1044. [CrossRef]
20. Dettinger, M.D.; Wilson, J.L. First order analysis of uncertainty in numerical models of groundwater flow part: 1. Mathematical development. *Water Resour. Res.* **1981**, *17*, 149–161. [CrossRef]
21. Sitar, N.; Cawlfield, J.D.; Kiureghian, A.D. First-order reliability approach to stochastic analysis of subsurface flow and contaminant transport. *Water Resour. Res.* **1987**, *23*, 794–804. [CrossRef]

22. Connell, L.D. An analysis of perturbation based methods for the treatment of parameter uncertainty in numerical groundwater models. *Transp. Porous Media* **1995**, *21*, 225–240. [CrossRef]

23. Venue, M.L.; Andrews, R.W.; Ramarao, B.S. Groundwater travel time uncertainty analysis using sensitivity derivatives. *Water Resour. Res.* **1989**, *25*, 1551–1566. [CrossRef]

24. Serrano, S.E. A new approach in modelling groundwater pollution under uncertainty. *Probab. Eng. Mech.* **1989**, *4*, 85–98. [CrossRef]

25. Li, S.G.; Liao, H.S.; Ni, C.F. Stochastic modeling of complex nonstationary groundwater systems. *Adv. Water Resour.* **2004**, *27*, 1087–1104. [CrossRef]

26. Ni, C.F.; Li, S.G. Modeling groundwater velocity uncertainty in nonstationary composite porous media. *Adv. Water Resour.* **2006**, *29*, 1866–1875. [CrossRef]

27. Ni, C.F.; Li, S.G. Approximate Analytical Solution to Groundwater Velocity Variance in Unconfined Trending Aquifers in the Presence of Complex Sources and Sinks. *J. Hydrol. Eng.* **2009**, *14*, 1119–1125. [CrossRef]

28. Beven, K.; Freer, J. Equifinality, data assimilation, and uncertainty estimation in mechanistic modelling of complex environmental systems using the GLUE methodology. *J. Hydrol.* **2001**, *249*, 11–29. [CrossRef]

29. Vázquez, R.F.; Beven, K.; Feyen, J. GLUE Based Assessment on the Overall Predictions of a MIKE SHE Application. *Water Resour. Manag.* **2009**, *23*, 1325–1349. [CrossRef]

30. Hastings, W.K. Monte Carlo Sampling Methods Using Markov Chains and Their Applications. *Biometrika* **1970**, *57*, 97–109. [CrossRef]

31. Geman, S.; Geman, D. Stochastic Relaxation, Gibbs Distributions, and the Bayesian Restoration of Images. *IEEE Trans. Pattern Anal. Mach. Intell.* **1984**, *6*, 721–741. [CrossRef] [PubMed]

32. Gelfand, A.E. Sampling-Based Approaches to Calculating Marginal Densities. *J. Am. Stat. Assoc.* **1990**, *85*, 398–409. [CrossRef]

33. Gilks, W.R.; Best, N.G.; Tan, K.K.C. Adaptive Rejection Metropolis Sampling within Gibbs Sampling. *J. R. Stat. Soc.* **1995**, *44*, 455–472. [CrossRef]

34. Haario, H.; Saksman, E.; Tamminen, J. An Adaptive Metropolis Algorithm. *Bernoulli* **2001**, *7*, 223–242. [CrossRef]

35. Elsheikh, A.H.; Wheeler, M.F.; Hoteit, I. Nested sampling algorithm for subsurface flow model selection, uncertainty quantification, and nonlinear calibration. *Water Resour. Res.* **2013**, *49*, 8383–8399. [CrossRef]

36. Elsheikh, A.H.; Demyanov, V.; Tavakoli, R.; Christie, M.A.; Wheeler, M.F. Calibration of channelized subsurface flow models using nested sampling and soft probabilities. *Adv. Water Resour.* **2015**, *75*, 14–30. [CrossRef]

37. Yakirevich, A.; Pachepsky, Y.A.; Gish, T.J.; Guber, A.K.; Kuznetsov, M.Y.; Cady, R.E.; Nicholsone, T.J. Augmentation of groundwater monitoring networks using information theory and ensemble modeling with pedotransfer functions. *J. Hydrol.* **2013**, *501*, 13–24. [CrossRef]

38. Zeng, X.K.; Ye, M.; Burkardt, J.; Wu, J.C.; Wang, D.; Zhu, X.B. Evaluating two sparse grid surrogates and two adaptation criteria for groundwater bayesian uncertainty quantification. *J. Hydrol.* **2016**, *535*, 120–134. [CrossRef]

39. Huisman, J.A.; Rings, J.; Vrugt, J.A.; Sorg, J.; Vereecken, H. Hydraulic properties of a model dike from coupled bayesian and multi-criteria hydrogeophysical inversion. *J. Hydrol.* **2010**, *380*, 62–73. [CrossRef]

40. Lu, D.; Ye, M.; Curtis, G.P. Maximum likelihood bayesian model averaging and its predictive analysis for groundwater reactive transport models. *J. Hydrol.* **2015**, *529*, 1859–1873. [CrossRef]

41. Qiu, Z.; Elishakoff, I.; Starnes, J.H., Jr. The bound set of possible eigenvalues of structures with uncertain but non-random parameters. *Chaos Solitons Fractals* **1996**, *7*, 1845–1857. [CrossRef]

42. Qiu, Z.; Ma, L.; Wang, X. Unified form for static displacement, dynamic response and natural frequency analysis based on convex models. *Appl. Math. Model.* **2009**, *33*, 3836–3847. [CrossRef]

43. Qiu, Z.; Ni, Z. An inequality model for solving interval dynamic response of structures with uncertain-but-bounded parameters. *Appl. Math. Model.* **2010**, *34*, 2166–2177. [CrossRef]

44. Muhanna, R.L.; Mullen, R.L. Bounds of Structural Response for All Possible Loading Combinations. *J. Struct. Eng.* **1999**, *125*, 98–106.

45. Chen, L.; Shen, Z.Y.; Yang, X.H.; Liao, Q.; Yu, S.L. An interval-deviation approach for hydrology and water quality model evaluation within an uncertainty framework. *J. Hydrol.* **2014**, *509*, 207–214. [CrossRef]

46. Yang, G.Q.; Guo, P.; Li, M.; Zhang, L.D. An improved solving approach forinterval-parameter programming and application to an optimal allocation ofirrigation water problem. *Water Resour. Manag.* **2016**, *30*, 701–729. [CrossRef]

47. Li, M.; Fu, Q.; Singh, V.P.; Liu, D. An interval multi-objective programming model for irrigation water allocation under uncertainty. *Agric. Water Manag.* **2018**, *196*, 24–36. [CrossRef]

48. Siddiqui, S.; Azarm, S.; Gabriel, S. A modified benders decomposition method for efficient robust optimization under interval uncertainty. *Struct. Multidiscip. Optim.* **2011**, *44*, 259–275. [CrossRef]

49. Jiang, C.; Hana, X.; Liu, G.P. A nonlinear interval number programming method for uncertain optimization problems. *Eur. J. Oper. Res.* **2008**, *188*, 1–13. [CrossRef]

50. Wu, H.C. The Karush–Kuhn–Tucker optimality conditions in an optimization problem with interval-valued objective function. *Eur. J. Oper. Res.* **2007**, *176*, 46–59. [CrossRef]

51. Qiu, Z.P. *Convex Method Based on Non-Probabilistic Set-Theory and Its Application*; National Defence Industry Press: Beijing, China, 2005; pp. 59–79. (In Chinese)

52. Li, P.T. Inversion of Hydrogeological Parameters Based on the Nested Bayesian Method. Master's Thesis, China University of Mining and Technology, Xuzhou, China, 2016; pp. 23–38. (In Chinese)

water

MDPI

Article

Mathematical Modeling of Non-Fickian Diffusional Mass Exchange of Radioactive Contaminants in Geological Disposal Formations

Anna Suzuki [1,*,†], **Sergei Fomin** [2], **Vladimir Chugunov** [3] **and Toshiyuki Hashida** [4]

[1]   Institute of Fluid Science, Tohoku University, Sendai, Miyagi 980-8577, Japan
[2]   Department of Mathematics and Statistics, California State University, Chico, CA 95929,
      USA; sfomin@csuchico.edu
[3]   Institute of Mathematics, Informatics and Natural Sciences, Moscow City University, 129226 Moscow,
      Russia; CHugunovVA@mgpu.ru
[4]   Fracture & Reliability Research Institute, School of Engineering, Tohoku University, Sendai, Miyagi 980-8579,
      Japan; hashida@rift.mech.tohoku.ac.jp
*   Correspondence: anna.suzuki@tohoku.ac.jp; Tel.: +81-22-217-5284
†   Current address: 2-1-1 Katahira, Aoba-ku, Sendai, Miyagi 980-8577, Japan.

Received: 30 July 2017; Accepted: 23 January 2018; Published: 29 January 2018

Abstract: Deep geological repositories for nuclear wastes consist of both engineered and natural geologic barriers to isolate the radioactive material from the human environment. Inappropriate repositories of nuclear waste would cause severe contamination to nearby aquifers. In this complex environment, mass transport of radioactive contaminants displays anomalous behaviors and often produces power-law tails in breakthrough curves due to spatial heterogeneities in fractured rocks, velocity dispersion, adsorption, and decay of contaminants, which requires more sophisticated models beyond the typical advection-dispersion equation. In this paper, accounting for the mass exchange between a fracture and a porous matrix of complex geometry, the universal equation of mass transport within a fracture is derived. This equation represents the generalization of the previously used models and accounts for anomalous mass exchange between a fracture and porous blocks through the introduction of the integral term of convolution type and fractional derivatives. This equation can be applied for the variety of processes taking place in the complex fractured porous medium, including the transport of radioactive elements. The Laplace transform method was used to obtain the solution of the fractional diffusion equation with a time-dependent source of radioactive contaminant.

Keywords: radioactive contaminant; fractional derivative; analytical solution

---

1. Introduction

High-level nuclear wastes are a by-product of nuclear power generation and other applications of nuclear fission or nuclear technology which must be shielded from humans and the environment for a long time. Subsurface nuclear waste repositories consist of engineered and geological barriers that isolate the radioactive materials from the human environment. If the radioactive contaminants leak to aquifers, the damage would be serious because it directly contaminates our drinking water. We need to answer how and when the contaminants leak from the power plants or the waste repositories unintentionally, and how much they affect human beings and the natural environment. Safe disposal of nuclear wastes requires an evaluation of the risks of contaminants for aquifers and prediction of the possible migration of contaminated groundwater.

Fluid flow and contaminant transport in aquifers are dominated by fractures and large pores. Numerous studies indicate that the real nature of solute transport in geological formations exhibits anomalous behavior [1–4]. Multiscale subsurface systems often produce power-law tails in breakthrough curves [5–8], as well as in a nuclear waste repository site [9]. The breakthrough curves are not adequately described by the typical advection-dispersion with an exponential residence time (e.g., [10,11]).

Problems of solute transport in a single fracture-matrix system have been addressed, and the analytical solutions have been developed based on the advection-dispersion equation [12,13]. Alternative transport models are proposed to capture the effects of spatial heterogeneities in fractured rocks and the effects of flow channeling or velocity dispersion [14,15]. The models have extended to the mass transfer models with time- or space-dependent dispersion coefficients (e.g., [16]), the multi-rate mass transfer models [17], the continuous time random walk approach [18], the time-domain random walk approach, the fractional advection-dispersion equation approach [19], and the stochastic approach [20].

The fractional derivative can be understood as a convolution of an integer-order derivative with a memory function [21], and the time convolution can capture memory effects, allowing particles to reside for long periods. The temporal fractional derivatives can produce power law residence times of solute transport. Liu et al. [22] considered the time fractional advection-dispersion equation, and the solution was obtained by using variable transformation. Fomin et al. [23] studied mass transport in a fractured-porous aquifer (i.e., aquifer filled with porous blocks) and modeled the effects of interaction with porous blocks in the aquifer by temporal fractional derivatives. Numerical study shows that varying the variations of order of fractional derivatives enables the description of different power law decays obtained from a homogeneous porous medium to a fractured medium [24].

This study proposes a mathematical model of radioactive contaminant transport in a single fracture within a confining porous matrix. Usually, sources of radioactive contamination vary with time. For example, an exponentially decaying source boundary condition is frequently used in radioactive waste disposal or non-aqueous phase liquid sites [25]. We derive the universal equation of mass transport for dissolved molecular size contaminants within a fracture, which accounts for the complexity of the confining porous matrix and temporal decay of the contaminant concentration. In this equation, the specific features of mass transport in the surrounding matrix and mass exchange between the fracture and matrix are modeled by the special function $Q_1(t)$, which represents the integral of convolution. This paper provides the analytic derivation of the function $Q_1(t)$ in its most general form, so that the majority of the well documented models can be obtained as particular cases of the presented model. For example, in the absence of radioactive decay, the solution presented in this paper reduces to the solution obtained by Fomin et al. [23] for $\Lambda = 0$ and $Q_1 = Q_0$. When mass flux is Fickian with $\Lambda = 0$ and $\alpha = 1$, the solution was obtained by Tang et al. [12], which also follows as a particular case from the solution obtained in this paper.

2. Model

2.1. Governing Equation

We consider radioactive solute transport in a single fracture surrounded by porous rocks. The fracture does not contain any porous inclusions, which is different from the concept in the previous study [23]. A schematic sketch of a fracture and rock matrix is presented in Figure 1. A parallel plate fracture is confined by porous rocks, which have same physical properties for the upper and lower sides. Cartesian coordinates (x, y) are chosen in such a manner that fluid in the fracture flows in the x-direction and that the coordinate y is perpendicular to the x-direction. Transport processes can be symmetrical with regard to the median line of the fracture at $y = -h$ (dashed line in Figure 1). This leads to the mass flux of 0 at $y = -h$, and the solutions in the sub-domains below and above this line are identical. Thus, we can only consider the upper half of the domain ($y = -h$).

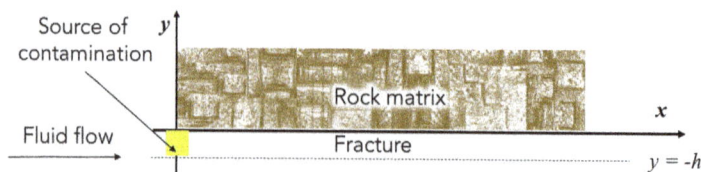

Figure 1. Schematic of a fracture surrounded with porous rocks.

Let c_1 and c_2 be the concentrations of the solute within the porous matrix and the fracture, respectively. We consider only dissolved molecular-size contaminants, not suspended particles in the aqueous phase radioactive particles. Because the thickness of the fracture is much smaller than its length, the mean concentration of the solute within the fracture can be given by $c = \frac{1}{h}\int_{-h}^{0} c_2 dy$. Mass transport in this system consists of (i) advection and (ii) diffusion in the fracture, (iii) absorption on the fracture walls, (iv) diffusion into the surrounding rocks, (v) adsorption on the walls of the pores in the surrounding rocks, and (vi) radioactive decay of radioactive contaminants both in fracture and porous matrix. Each governing equation within the fracture and within the rock matrix can be written as [12]:

$$\frac{\partial c}{\partial \tau} + \frac{1}{h}\frac{\partial s}{\partial \tau} = -v\frac{\partial c}{\partial x} + D\frac{\partial^2 c}{\partial x^2} - \lambda(c + \frac{1}{h}s) - \frac{q}{h}, \quad 0 < x < \infty, \tau > 0, \tag{1}$$

$$\frac{\partial c_1}{\partial \tau} + \frac{\rho_m}{\theta}\frac{\partial s_1}{\partial \tau} = -\frac{1}{\theta}\frac{\partial q}{\partial y} - \lambda(c_1 + \frac{\rho_m}{\theta}s_1), \quad 0 < y < \infty, \tau > 0, \tag{2}$$

where τ is time. s and s_1 are the mass of the solute adsorbed on the walls of the fracture and pores in the rock matrix, respectively. v is the average velocity of the solution in the fracture. D and D_1 are the effective diffusivities in the fracture and in the porous medium, respectively, which include dispersion and molecular diffusion in the fracture and in the porous medium. λ is a radioactive decay constant. q is the mass flux on the wall of the fracture. ρ_m is the density of the rock matrix, and θ is the matrix porosity.

It has been observed that pore spaces and micro-cracks in the rock matrix are distributed in various sizes and different orientations [26]. We assume that contaminants not only penetrate in the matrix due to molecular diffusion, but also migrate through micro-cracks due to advection. Thus, the effective diffusivity, D_1, accounts for dispersion and molecular diffusion. If micro-cracks within the surrounding matrix have an orientation perpendicular to the conducting fracture, contaminants may migrate long distances and lead to a faster process than diffusion (super diffusion). We should predict the worst scenario to evaluate the risks of contaminants' migration. Thus, in order to account for the advective process in the surrounding rocks, this study used the generalized Fickian mass flux in the matrix by introducing a fractional derivative, in the following form [27]:

$$q = -\theta D_1 \frac{\partial^\alpha c_1}{\partial y^\alpha}, \tag{3}$$

where α is the order of fractional derivative ($0 < \alpha < 1$). The value of $1/2 < \alpha < 1$ leads to faster (superdispersive) spreading, while the value of $0 < \alpha < 1/2$ causes slower (subdiffusive) spread [25]. Equation (3) describes Fickian diffusion when the index on space fractional derivative is $1/2$ (i.e., $\alpha = 1/2$). The fractional derivative can be defined by means of Laplace transformation $\mathcal{L}$, from the equation $\mathcal{L}[\frac{\partial^\alpha c_1}{\partial y^\alpha}] = p^{\alpha-1}(p\mathcal{L}[c_1] - c_1(\tau,0))$, which is equivalent to the Caputo definition [28], $\frac{d^\alpha c_1}{dy^\alpha} = \int_0^y \frac{(y-\xi)^{-\alpha}}{\Gamma(1-\alpha)}\frac{dc_1}{d\xi}d\xi$.

The relationship between c and s in Equation (1) and between c_1 and s_1 in Equation (2) can be assumed [12] as

$$s = K_f c, \tag{4}$$

$$s_1 = K_m c_1, \tag{5}$$

where K_f and K_m are given constants. Substituting correlations (4) and (5) into Equations (1) and (2) yields

$$R\frac{\partial c}{\partial \tau} = -v\frac{\partial c}{\partial x} + D\frac{\partial^2 c}{\partial x^2} - R\lambda c + \frac{\theta D_1}{h}\frac{\partial^\alpha c_1}{\partial y^\alpha}\Big|_{y=0}, \tag{6}$$

$$R_1\frac{\partial c_1}{\partial \tau} = D_1\frac{\partial}{\partial y}\left(\frac{\partial^\alpha c_1}{\partial y^\alpha}\right) - R_1\lambda c_1, \tag{7}$$

where $R_1 = 1 + \frac{\rho_m K_m}{\theta}$ and $R = 1 + \frac{K_f}{h}$ are retardation coefficients.

In general, concentration in the matrix c_1 is a function of both spatial coordinates, x and y: $c_1 = c_1(\tau, x, y)$. Let l and h be the characteristic scales for the length in the x-direction (along the aquifer) and y-direction, respectively. The scale l is defined by the distance of contaminant intrusion into the aquifer in the x-direction due to the advective transport, and the scale h is defined by the thickness of the aquifer. The characteristic values of the concentration gradient in x- and y-direction are C_0/l and C_0/h, respectively. Therefore, the ratio of the gradients can be estimated by the quotient of the length scales, h/l. Obviously, the scale l can be much greater than the scale h, and the ratio h/l can be very small. Hence, diffusion in the x-direction is negligibly small compared to the diffusion in y-direction. Thus, the derivative of c_1 with respect to x is ignored. This is the same assumption with Grisak and Pickens [29] and Tang et al. [12]. Dependence of c_1 on x is a consequence of the boundary conditions on the rock–fracture interface ($y = 0$), which couples c_1 with the mean concentration in the fracture, c.

In order to generalize the equations, the non-dimensional forms are derived with the proper characteristic scales. The scale for time represents the characteristic time for contaminant penetration in the rock matrix to the distance h, given by $\tau_m = \frac{h^{\alpha+1} R_1}{D_1}$. The scale for the variable x-coordinate along the fracture is the characteristic distance of contaminant migration by the characteristic time τ_m, described as $l = \frac{v\tau_m}{R} = \frac{h^{\alpha+1} v R_1}{R D_1}$. The scale for the y-coordinate is defined to be half of the aquifer, h. The initial concentration of solute at the inlet where the source of contamination is located, $c_0(0)$, can be used as the scale for solute concentration. Based on these scales, non-dimensional variables can be introduced as follows:

$$C = \frac{c}{c_0(0)}; C_1 = \frac{c_1}{c_0(0)}; Pe = \frac{vl}{D}; t = \frac{\tau}{\tau_m}; X = \frac{x}{l}; Y = \frac{y}{h}; \Lambda = \frac{\lambda R_1 h^{\alpha+1}}{D_1}. \tag{8}$$

Substituting the non-dimensional variables in Equation (8) into Equations (6) and (7) yields the following:

$$\frac{\partial C}{\partial t} - \frac{1}{Pe}\frac{\partial^2 C}{\partial X^2} + \frac{\partial C}{\partial X} + \Lambda C = \theta\frac{\partial^\alpha C_1}{\partial Y^\alpha}\Big|_{Y=0}, \quad 0 < X < \infty, t > 0, \tag{9}$$

$$\frac{\partial C_1}{\partial t} = \frac{\partial^{\alpha+1} C_1}{\partial Y^{\alpha+1}} - \Lambda C_1 \quad 0 < Y < \infty, t > 0. \tag{10}$$

The following boundary and initial conditions can be imposed:

$$t = 0, C = C_1 = 0; \tag{11}$$

$$X = 0, C = C_0(t); \tag{12}$$

$$X \to \infty, C \to 0; \tag{13}$$

$$Y \to \infty, C_1 \to 0; \tag{14}$$

$$Y = 0, C_1 = C; \tag{15}$$

where $C_0(t)$ is the non-dimensional concentration at the inlet of the fracture. Typically, the concentration distribution in the aquifer can be approximated by a parabola. Therefore, the maximum values of concentration can be reached in the middle of the aquifer, while the lowest values at the aquifer–matrix interface. In this case, assuming that the concentration C_1 on the interface $Y = 0$ is equal to the mean concentration C in the aquifer (the boundary condition (15)), which slightly overestimates the concentration on the border of the matrix. Therefore, the computed concentration C_1 in the region $(0 < Y < \infty)$ will be slightly overestimated. However, using the present model as a tool for predicting contamination in the real world situations, the slight overestimation of the possible hazardous contamination is a positive factor.

2.2. Analytical Solution

Equation (9) describes mass transport in a fracture, which contains the variables C and C_1 (i.e., concentration in the fracture and in the matrix, respectively). Let us consider the mass flux on the fracture–matrix interface on the right hand side in Equation (9), which can be written by:

$$Q = \frac{\partial^\alpha C_1}{\partial Y^\alpha}|_{Y=0}. \tag{16}$$

Based on an analogy of Duhamel's theorem [30], the solution for the concentration in the matrix, C_1, can be coupled with the concentration in the fracture, C, by the following equation:

$$C_1(t, X, Y) = e^{-\Lambda t} \frac{\partial}{\partial t} \int_0^t e^{\Lambda \tau} C(\tau, X) u_0(t - \tau, Y) d\tau, \tag{17}$$

where the function u_0 in Equation (17) is a solution of the following auxiliary problem:

$$\frac{\partial u_0}{\partial t} = \frac{\partial^{\alpha+1} u_0}{\partial Y^{\alpha+1}}; 0 < Y < \infty; t > 0, \tag{18}$$

$$t = 0, u_0 = 0; \tag{19}$$

$$Y = 0, u_0 = 1; \tag{20}$$

$$Y \to \infty, u_0 \to 0. \tag{21}$$

The mass flux in Equation (18) is given by:

$$Q_0(t) = -\frac{\partial^\alpha u_0}{\partial Y^\alpha}|_{Y=0}. \tag{22}$$

Mass flux differentiations, Q given by Equation (16) and Q_0 given by Equation (22), are performed with respect to the variable Y, whereas the differentiation and the integration in Equation (17) are performed with respect to the variables t and τ, respectively. We can change the order of fractional differentiation with respect to Y to the order of differentiation with respect to t and the order of integration with respect to τ. As a result, we obtain

$$Q = \frac{\partial^\alpha C_1}{\partial Y^\alpha}|_{Y=0} = e^{-\Lambda t} \frac{\partial}{\partial t} \int_0^t e^{\Lambda \tau} C(\tau, X) \frac{\partial^\alpha u_0}{\partial Y^\alpha}|_{Y=0} d\tau = -e^{-\Lambda t} \frac{\partial}{\partial t} \int_0^t e^{\Lambda t} C(\tau, X) Q_0(t - \tau) d\tau, \tag{23}$$

where $C(\tau, X)$ does not depend on Y. In addition, the mass flux Q given by Equation (23) can be rewritten in the form:

$$Q = -\frac{\partial}{\partial t} \int_0^t C(t - \tau, X) Q_1(\tau) d\tau, \tag{24}$$

where

$$Q_1 = e^{-\Lambda t} \frac{\partial}{\partial t} \int_0^t e^{\Lambda \tau} Q_0(t - \tau) d\tau. \tag{25}$$

The proof of the derivation of Equation (24) can be found in the Appendix.

Substituting Equations (23) and (24) into Equation (9) leads to the following boundary value problem for C (i.e., concentration in the fracture):

$$\frac{\partial C}{\partial t} - \frac{1}{Pe}\frac{\partial^2 C}{\partial X^2} + \frac{\partial C}{\partial X} + \Lambda C = -\theta\frac{\partial}{\partial t}\int_0^t C(\tau, X)Q_1(t - \tau)d\tau, \quad 0 < X < \infty, t > 0, \tag{26}$$

$$t = 0, C = 0; \tag{27}$$

$$X = 0, C = C_0(t); \tag{28}$$

$$X \to \infty, C \to 0. \tag{29}$$

Equation (26) describes the majority of the transfer processes in the fracture and the specific feature of mass exchange between the fracture and the matrix. The derivation can be done by constructing the appropriate function $Q_1(t)$ in Equation (25).

Applying the technique of the group analysis of differential equations [31], the solution of the formulated boundary-value problem (18)–(21) can be obtained in the following form [23]:

$$u_0(\eta) = 1 - \frac{\sum_{m=0}^{M-1}\frac{(-1)^m\Gamma(\frac{\alpha}{\alpha+1}+m)}{\Gamma[(\alpha+1)(m+1)]}\eta^{\alpha+m(\alpha+1)}}{\Gamma[1/(\alpha+1)]\Gamma[\alpha/(\alpha+1)]} + O(\eta^{\alpha+M((\alpha+1)}), \tag{30}$$

where $\eta = Yt^{-\frac{1}{1+\alpha}}$. $\Gamma(z)$ is Gamma function. This expression follows from (2.17)–(2.32) in [23] at $\mu = 0$.

From Equation (30), it follows that

$$\frac{\partial^\alpha u_0}{\partial Y^\alpha}\Big|_{Y=0} = -\frac{1}{\Gamma(1-\beta)}t^{-\beta}, \tag{31}$$

where $\beta = \alpha/(\alpha+1)$. Accounting for Equations (22) and (31), Equation (25) leads to the following expression:

$$Q_1 = \frac{e^{-\Lambda t}t^{-\beta}}{\Gamma(1-\beta)} + \frac{\Lambda^\beta}{\Gamma(1-\beta)}[\Gamma(1-\beta) - \gamma(1-\beta, \Lambda t)], \tag{32}$$

where $\gamma(a, z)$ is an incomplete Gamma function. Equation (32) represents the mass flux on the wall of the fracture when $C_1 = 1$ on the fracture wall.

Let us consider particular cases. If Q_0 and Q_1 were 0, mass exchange does not occur at the fracture–matrix interface. Then, the problem is reduced to the problem in [12]. If no radioactive decay occurs (i.e., $\Lambda = 0$) and diffusion is Fickian (i.e., $\alpha = 1, \beta = 1/2$), $Q_1 = \frac{1}{\sqrt{\pi t}}$. According to the definition of the fractional derivative [28], Equation (24) can be written as $Q = \frac{\partial^{1/2}C}{\partial t^{1/2}}$. If no radioactive decay occurs (i.e., $\Lambda = 0$) but diffusion is described by the generalized Fick's law (3) (i.e., $0 < \alpha < 1$ ($0 < \beta < 1/2$)), the mass fluxes can be given as $Q_1 = t^{-\beta}/\Gamma(1-\beta)$ and $Q = \frac{\partial^\beta C}{\partial t^\beta}$. When Λ is small and time is $t = O(1)$, formula (32) approaches the following asymptotic representation:

$$Q_1 = \frac{t^{-\beta}}{\Gamma(1-\beta)} - \frac{\Lambda t^{1-\beta}}{\Gamma(1-\beta)}[1 - \frac{\Gamma(1-\beta)}{\Gamma(2-\beta)}] + O(\Lambda^2 t^{2-\beta}). \tag{33}$$

The accuracy of this asymptotic formula can be easily verified by simple numerical computations. Our computations show that the difference between the values of Q_1 computed by Equations (32) and (33) is negligibly small within the relatively long time interval from 0 to $1/\Lambda$ and $\beta \geq 0.1$ [23]. Thus, Equation (33) can be used as a good approximation for Q_1, the exact value of which is given by

Equation (32). Using formula (33), the mass flux Q defined by Equation (24) can be presented through the fractional derivatives:

$$Q = \frac{\partial^\beta C}{\partial t^\beta} - \Lambda[1 - \frac{\Gamma(1-\beta)}{\Gamma(2-\beta)}](1-\beta)\frac{\partial^\beta}{\partial t^\beta}\int_0^t C(\tau, X)d\tau + O(\Lambda^2). \tag{34}$$

Let us turn to solution of the boundary value problem (26)–(29). In the case where $\Lambda = 0, \beta = 1/2$, and converting to originals, we obtain

$$C(T, X') = H(T - X')\text{erfc}[\frac{X'}{2\sqrt{T-X'}}]. \tag{35}$$

Or, in the case where $\Lambda \neq 0, \beta = 1/2$, then

$$C(T, X') = \frac{e^{-\Lambda X'}H(T-X')}{2}(e^{-X'\sqrt{\Lambda}}\text{erfc}[\frac{X'}{2\sqrt{T-X'}} - \sqrt{\Lambda(T-X')}] + e^{X'\sqrt{\Lambda}}\text{erfc}[\frac{X'}{2\sqrt{T-X'}} + \sqrt{\Lambda(T-X')}]). \tag{36}$$

Equations (35) and (36) are well-known solutions in [12,14]. Solution (36) does not follow from (35), while solution (35) follows from (36) at $\Lambda = 0$.

The function Q_1 at the right hand side in Equation (26) is given by Equation (33). If the time range is from 0 to $1/\Lambda$ for Equation (26), it is convenient to rescale Equation (26) with new time variable $T = \Lambda t$. In this case, the new spatial variable should be defined as $X' = X(\Lambda + \theta\Lambda^\beta)$. With these new non-dimensional variables, Equation (26) can be presented in the following form:

$$w_1\frac{\partial C}{\partial T} + \frac{\partial C}{\partial X'} - \epsilon\frac{\partial^2 C}{\partial X'^2} + C = -w_2\frac{\partial}{\partial T}\int_0^T C(T-\tau, X')\Psi_\beta(\tau)d\tau, \tag{37}$$

where $w_1 = \frac{\Lambda}{\Lambda+\theta\Lambda^\beta}$, $w_2 = \frac{\theta\Lambda^\beta}{\Lambda+\theta\Lambda^\beta}$, $\epsilon = \frac{\Lambda+\theta\Lambda^\beta}{Pe}$, and $\Psi_\beta = Q_1\Lambda^{-\beta} - 1$. Note that the expression for Ψ_β can be presented as $\Psi_\beta = [T^{-\beta}e^{-T} - \gamma(1-\beta, T)]/\Gamma(1-\beta)$, according to Equation (33). The third term at the left hand side in Equation (37) including ϵ describes diffusion in the fracture. Because $\Lambda << 1$ and $Pe = O(1)$, parameter ϵ in Equation (37) is small ($\epsilon << 1$). Hence, in major cases within time of the order of $1/\Lambda$, the effects of the diffusive transport in the fracture is negligible. Equation (37) can be rewritten as follows:

$$w_1\frac{\partial C}{\partial T} + \frac{\partial C}{\partial X'} + C = -w_2\frac{\partial}{\partial T}\int_0^T C(T-\tau, X')\Psi_\beta(\tau)d\tau. \tag{38}$$

Applying Laplace transformation $\mathcal{L}$ with respect to time to the Equation (38), we obtain

$$\frac{d\overline{C}}{dX'} + (sw_1 + 1 + sw_2\overline{\Psi}_\beta)\overline{C} = 0, \ 0 < X' < \infty, \tag{39}$$

$$X' = 0, \ \overline{C} = \overline{C}_0, \tag{40}$$

where $\overline{C} = \mathcal{L}[C], \overline{\Psi}_\beta = \mathcal{L}[\Psi_\beta]$. Substituting $\overline{\Psi}_\beta = [(s+1)^\beta - 1]/s$ into Equation (39) and integrating it accounting for the boundary condition (40) yields

$$\overline{C} = \overline{C}_0 e^{-[(s+1)w_1 + (s+1)^\beta w_2]X'}. \tag{41}$$

3. Results

First, we consider the case where the concentration at the inlet is constant (i.e., $C_0 = 1$ and $\overline{C}_0 = 1/s$). For this case, let us describe the concentration as C_c. The inverse Laplace transformation leads to the following expression:

$$\begin{aligned} C_c(T, X') &= L^{-1}[\frac{1}{s}\exp(-w_1X')\exp(-sw_1X')\exp(-w_2X'(s+1)^\beta)] \\ &= e^{-w_1X'}H(T - w_1X')G(T - w_1X', X'), \end{aligned} \tag{42}$$

where $H(z)$ is a Heaviside step function and

$$
\begin{aligned}
G(T, X') &= L^{-1}\left[\frac{\exp(-w_2 X'(s+1)^\beta)}{s}\right] \\
&= e^{-w_2 X'} - \frac{1}{\pi}\int_0^\infty e^{-T(\xi+1)}\exp[-\xi^\beta w_2 X'\cos(\pi\beta)]\sin[\xi^\beta w_2 X'\sin(\pi\beta)]\frac{d\xi}{\xi+1}.
\end{aligned}
\tag{43}
$$

when concentration in the inlet is an arbitrary function of T (i.e., $C_0 = C_0(T)$), then concentration in the fracture can be obtained by utilizing Duhamel's theorem [30]:

$$
C(T, X') = \frac{\partial}{\partial T}\int_0^T C_0(T-\tau)C_c(\tau, X')d\tau,
\tag{44}
$$

where C_c is the concentration when the concentration at the inlet is constant, defined by Equation (42). If the radioactivity decays exponentially at the inlet, the boundary concentration and its Laplace form are given as $C_0 = e^{-T}$ and $\overline{C}_0 = 1/(s+1)$. The solution C can be obtained by the inverse Laplace transform directly from Equation (41) as:

$$
\begin{aligned}
C(T, X') &= L^{-1}\left[\frac{1}{s+1}\exp[-w_1 X'(s+1)]\exp[-w_2 X'(s+1)^\beta]\right] \\
&= e^{-T}H(T - w_1 X')G'(T - w_1 X', w_2 X'),
\end{aligned}
\tag{45}
$$

where

$$
G'(T, X') = L^{-1}\left[\frac{1}{s}e^{-X's^\beta}\right] = 1 - \frac{1}{\pi}\int_0^\infty \frac{e^{-\xi T}}{\xi}\exp[-X'\xi^\beta\cos(\pi\beta)]\sin[X'\xi^\beta\sin(\pi\beta)]d\xi.
\tag{46}
$$

The effects of the order of fractional derivatives on radioactive contaminant transportation are shown in Figures 2 and 3. Analytical solutions for the constant concentration at the inlet given by Equation (42) are plotted for different β in Figure 2a,b. The initial concentration is $C(T = 0, X') = 0$, and the boundary source concentration at $x = 0$ is $C_0 = 1$. As shown in Figure 2a, concentrations at far points from the inlet were larger for $\beta = 0.5$ than that for $\beta = 0.1$. For the case of larger β, contaminants are most likely to migrate through the fracture. In contrast, small β describes transport with longer delays, which derive from diffusion into the surrounding rocks or adsorption and desorption to the fracture walls. Thus, smaller β describes a longer memory effect.

The mass flux on the fracture–matrix interface in Equation (3) is taken account for the generalized Fick's law with fractional spatial derivative, and does not describe the effect of temporal memory. However, since we deal with the mass flux on the fracture–matrix interface where the concentration of the transported contaminant significantly depends on time, it is physically obvious that the mass flux at the given moment of time depends not only on concentration at this moment of time, but also on how this concentration varied in the previous moment of time. This feature can be called the effect of temporal memory and mathematically described by the convolution integral in Equation (24). Equation (31) also explains that the generalized Fick's law with fractional spatial derivative accounts for the memory effect. Incidentally, Equation (24) allows the mixed problems of calculating concentration on the fracture–matrix interface to be split into separate problems of calculating concentration in the matrix and calculating concentration in the fracture, respectively. The latter significantly simplifies the analysis of mass transport in the matrix–fracture system.

As shown in Figure 3a, concentration for constant boundary source approached certain values of concentration, which did not reach the injected concentration ($C_0 = 1$). We can see that larger β shows larger concentration at the late time, showing the phenomenon of longer memory as discussed above.

Analytical solutions for the time-dependent boundary sources given by Equation (45) are plotted for different β in Figures 2b and 3b. With the same conditions as above, the initial concentration is $C(T = 0, X') = 0$. The boundary source was set to time-dependent $C_0 = \exp(-T)$. Concentration at $x = 0$ was 1 for constant boundary source (Figure 2a), while concentration for time-dependent boundary source started from 0.37 when $t = 1$ (Figure 2b). In the case of the time-dependent boundary source, concentrations decayed at the late time.

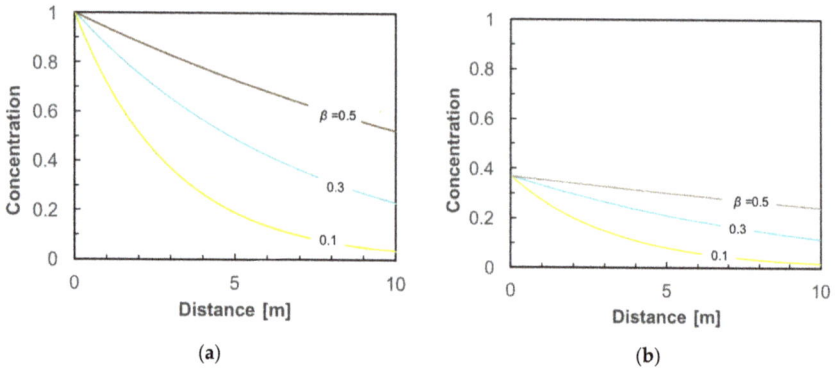

(a) (b)

Figure 2. Effects of beta on spatial distribution at $t = 1$ with $\Lambda = 0.5$ and $\theta = 0.01$. (**a**) Constant boundary source concentration and (**b**) time-dependent boundary source concentration.

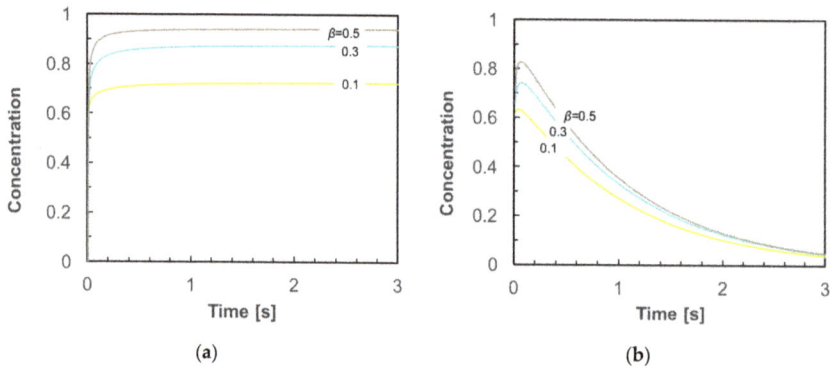

(a) (b)

Figure 3. Effects of beta on time history at $x = 1$ with $\Lambda = 0.5$ and $\theta = 0.01$. (**a**) Constant boundary source concentration and (**b**) time-dependent boundary source concentration.

4. Conclusions

We derived the analytical solution of the boundary value problem (26)–(29) for radioactive contaminant transport through a fracture. By deriving the formula (24), This formula allows the mixed problems of calculating concentration on the fracture–matrix interface to be split into separate problems of calculating concentration in the matrix and calculating concentration in the fracture, respectively. The latter significantly simplifies the analysis of mass transport in the matrix–fracture system. When the solute concentration at the inlet is constant, the concentration in the fracture can be obtained by Equation (42). When the source boundary concentration at the inlet decays exponentially, the concentration in the fracture can be obtained by Equation (45). We plotted the analytical solutions for different values of β, which indicate that the value of β allows the evaluation of the residence time of contaminants in the aquifer. This analysis, based on the analytical solutions of fractional diffusion equation, can provide simple and quick results to evaluate solute transport in fractured rocks. Most of the cases where radioactive contaminants cause troubles would be in unexpected situations. At that time, the simple and quick analysis proposed by this research will help instant management strategies.

Acknowledgments: This work was supported by the Japan Society for the Promotion of Science, under Grant-in-Aid for Young Scientists(A)(717H049760), whose support is gratefully acknowledged.

Author Contributions: Sergei Fomin, Vladimir Chugunov, and Anna Suzuki contributed to the development of the mathematical model and derivation of the analytical solutions. Anna Suzuki and Toshiyuki Hashida solved numerically the analytical solutions, and Anna Suzuki wrote the paper.

Conflicts of Interest: The authors declare no conflict of interest.

Appendix A

By substituting $t - \tau = \xi$ in the integral in Equation (23), the mass flux Q can be written in the form:

$$
\begin{aligned}
Q &= -e^{-\Lambda t} \frac{\partial}{\partial t} \int_0^t e^{\Lambda(t-\xi)} C(t-\xi, X) Q_0(\xi) d\xi \\
&= -\Lambda \int_0^t e^{-\Lambda \xi} C(t-\xi, X) Q_0(\xi) d\xi - \frac{\partial}{\partial t} \int_0^t e^{-\Lambda \xi} C(t-\xi, X) Q_0(\xi) d\xi
\end{aligned}
\tag{A1}
$$

This is equivalent to the mass flux Q on the left hand side in Equation (24). Now we consider the differentiation and the integration on the right hand side in Equation (24). Let's introduce the following function:

$$
Q_1 = e^{-\Lambda t} \frac{\partial}{\partial t} \int_0^t e^{\Lambda \tau} Q_0(t - \tau) d\tau
\tag{A2}
$$

In the same way as Q in Equation (A1), the function Q_1 is transformed as:

$$
Q_1 = \Lambda \int_0^t e^{-\Lambda \xi} Q_0(\xi) d\xi + e^{-\Lambda t} Q_0(t)
\tag{A3}
$$

By substituting function (A3), Equation (24) can be written in the form:

$$
\begin{aligned}
Q &= -\frac{\partial}{\partial t} \int_0^t C(t-\tau, X) Q_1(\tau) d\tau \\
&= -\frac{\partial}{\partial t} \int_0^t C(t-\tau, X)(\Lambda \int_0^t e^{-\Lambda \xi} Q_0(\xi) d\xi + e^{-\Lambda t} Q_0(t)) d\tau \\
&= -\frac{\partial}{\partial t} [\int_0^t C(t-\tau, X) e^{-\Lambda t} Q_0(t)) d\tau] - \Lambda \frac{\partial}{\partial t} [\int_0^t C(\eta, X)(\int_0^{t-\eta} e^{-\Lambda \xi} Q_0(\xi) d\xi) d\tau] \\
&= -\frac{\partial}{\partial t} [\int_0^t C(t-\tau, X) e^{-\Lambda t} Q_0(t)) d\tau] - \Lambda \int_0^t C(\eta, X) e^{-\Lambda(t-\eta)} Q_0(t-\eta) d\eta \\
&= -\Lambda \int_0^t e^{-\Lambda \xi} C(t-\xi, X) Q_0(\xi) d\xi - \frac{\partial}{\partial t} \int_0^t e^{-\Lambda \xi} C(t-\xi, X) Q_0(\xi) d\xi
\end{aligned}
\tag{A4}
$$

The latter expression in the right hand side in Equation (A4) is equivalent to the expression in Equation (A1). Thus, Equation (24) can be derived.

References

1. Coats, K.; Smith, B.D. Dead-end pore volume and dispersion in porous media. *Soc. Petrol. Eng. J.* **1964**, *4*, 73–84.
2. Long, J.C.S.; Remer, J.S.; Wilson, C.R.; Witherspoon, P.A. Porous media equivalents for network of discontinuous fractures. *Water Resour. Res.* **1982**, *18*, 645–658.
3. Sahimi, M. Flow phenomena in rocks: From continuum models to fractals, percolation, cellular automata, and simulated annealing. *Rev. Mod. Phys.* **1993**, *65*, 1395–1534.
4. Edery, Y.; Guadagnini, A.; Scher, H.; Berkowitz, B. Origins of anomalous transport in heterogeneous media: Structural and dynamic controls. *Water Resour. Res.* **2014**, *50*, 1490–1505.
5. Hatano, Y.; Hatano, N. Dispersive transport of ions in column experiments: An explanation of long-tailed profiles. *Water Resour. Res.* **1998**, *34*, 1027–1033.
6. Haggerty, R.; Wondzell, S.M.; Johnson, M.A. Power-law residence time distribution in the hyporheic zone of a 2nd-order mountain stream. *Geophys. Res. Lett.* **2002**, *29*, 1640.
7. Radilla, G.; Sausse, J.; Sanjuan, B.; Fourar, M. Interpreting tracer tests in the enhanced geothermal system (EGS) of Soultz-sous-Forêts using the equivalent stratified medium approach. *Geothermics* **2012**, *44*, 43–51.
8. Cardenas, M.B.; Slottke, D.T.; Ketcham, R.A.; Sharp, J.M., Jr. Navier-Stokes flow and transport simulations using real fractures shows heavy tailing due to eddies. *Geophys. Res. Lett.* **2007**, *34*, L14404.

9. Hodgkinson, D.; Benabderrahmane, H.; Elert, M.; Hautojärvi, A.; Selroos, J.O.; Tanaka, Y.; Uchida, M. An overview of Task 6 of the Aspö Task Force: Modelling groundwater and solute transport: Improved understanding of radionuclide transport in fractured rock. *Hydrogeol. J.* **2009**, *17*, 1035–1049.

10. Becker, M.W.; Shapiro, A.M. Interpreting tracer breakthrough tailing from different forced-gradient tracer experiment configurations in fractured bedrock. *Water Resour. Res.* **2003**, *39*, doi:10.1029/2001WR001190.

11. Tsang, Y.W. Study of alternative tracer tests in characterizing transport in fractured rocks. *Geophys. Res. Lett.* **1995**, *22*, 1421–1424.

12. Tang, D.H.; Frind, E.O.; Sudicky, E.A. Contaminant transport in fractured porous media: Analytical solution for a single fracture. *Water Resour. Res.* **1981**, *17*, 555–564.

13. Sudicky, E.; Frind, E. Contaminant transport in fractured porous media: Analytical solutions for a system of parallel fractures. *Water Resour. Res.* **1982**, *18*, 1634–1642.

14. Neretnieks, I. A note on fracture flow dispersion mechanisms in the ground. *Water Resour. Res.* **1983**, *19*, 364–370.

15. Becker, M.W.; Shapiro, A.M. Tracer transport in fractured crystalline rock: Evidence of nondiffusive breakthrough tailing. *Water Resour. Res.* **2000**, *36*, 1677–1686.

16. Yates, S.R. An analytical solution for one-dimensional transport in heterogeneous porous media. *Water Resour. Res.* **1990**, *26*, 2331–2338.

17. Haggerty, R.; Gorelick, S.M. Multiple-rate mass transfer for modeling diffusion and surface reactions in media with pore-scale heterogeneity. *Water Resour. Res.* **1995**, *31*, 2383–2400.

18. Berkowitz, B.; Cortis, A.; Dentz, M.; Scher, H. Modeling non-fickian transport in geological formations as a continuous time random walk. *Rev. Geophys.* **2006**, *44*, RG2003.

19. Benson, D.A.; Wheatcraft, S.S.W.; Meerschaert, M.M. Application of a fractional advection-dispersion equation. *Water Resour. Res.* **2000**, *36*, 1403–1412.

20. Morales-Casique, E.; Neuman, S.P.; Guadagnini, A. Non-local and localized analyses of non-reactive solute transport in bounded randomly heterogeneous porous media: Theoretical framework. *Adv. Water Resour.* **2006**, *29*, 1238–1255.

21. Schumer, R.; Meerschaert, M.M.; Baeumer, B. Fractional advection-dispersion equations for modeling transport at the Earth surface. *J. Geophys. Res. Earth Surf.* **2009**, *114*, doi:10.1029/2008JF001246.

22. Liu, F.; Anh, V.V.; Turner, I.; Zhuang, P. Time fractional advection-dispersion equation. *J. Appl. Math. Comput.* **2003**, *13*, 233–245.

23. Fomin, S.; Chugunov, V.; Hashida, T. The effect of non-Fickian diffusion into surrounding rocks on contaminant transport in a fractured porous aquifer. *Proc. R. Soc. A Math. Phys. Eng. Sci.* **2005**, *461*, 2923–2939.

24. Suzuki, A.; Fomin, S.A.; Chugunov, V.A.; Niibori, Y.; Hashida, T. Fractional diffusion modeling of heat transfer in porous and fractured media. *Int. J. Heat Mass Transf.* **2016**, *103*, 611–618.

25. Zumofen, G.; Klafter, J. Scale-invariant motion in intermittent chaotic systems. *Phys. Rev. E* **1993**, *47*, 851–863.

26. Penttinen, L.; Siitari-kauppi, M.; Ikonen, J. *Forsmark Site Investigation Determination of Porosity and Micro Fracturing Using the C-PMMA Technique in Samples Taken from Forsmark Area*; Technical Report; Svensk Kärnbränslehantering AB: Stockholm, Sweden, 2006.

27. Fomin, S.A.; Chugunov, V.A.; Hashida, T. Non-Fickian mass transport in fractured porous media. *Adv. Water Resour.* **2011**, *34*, 205–214.

28. Samko, S.G.; Kilbas, A.A.; Marichev, O.I. *Fractional Integrals and Derivatives: Theory and Applications*; Gordon and Breach Science Publishers: Philadelphia, PA, USA, 1993. doi:10.1515/fca-2016-0032.

29. Grisak, G.; Pickens, J. An analytical solution for solute transport through fractured media with matrix diffusion. *J. Hydrol.* **1981**, *52*, 47–57.

30. Carslaw, H.S.; Jaeger, J.C. *Conduction of Heat in Solids*; Clarendon Press: Oxford, UK, 1959.

31. Ibragimov, N.H. *CRC Handbook of Lie Group Analysis of Differential Equations: Volume 2: Applications in Engineering and Physical Sciences*; CRC Press: Boca Raton, FL, USA, 1994.

water

MDPI

Article

New Comparative Experiments of Different Soil Types for Farmland Water Conservation in Arid Regions

Yiben Cheng [1,2,*], Yanli Li [1], Hongbin Zhan [2,*], Hairong Liang [3], Wenbin Yang [1], Yinming Zhao [4] and Taojia Li [3]

[1]   Institute of Desertification Studies, Chinese Academy of Forestry, Beijing 100091, China; liyanli013@163.com (Y.L.); nmlkyywb@163.com (W.Y.)
[2]   Department of Geology and Geophysics, Texas A&M University, College Station, TX 77840, USA
[3]   Inner Mongolia Academy of Forestry, Hohhot, Inner Mongolia 010010, China; lkylhr@sina.com (H.L.); lijiatao_1020@163.com (T.L.)
[4]   Desert Forestry Experiment Center, Chinese Academy of Forestry, Dengkou, Inner Mongolia 015204, China; zhaoyingming2004@aliyun.com
*   Correspondence: chengyiben07@gmail.com (Y.C.); zhan@geos.tamu.edu (H.Z.)

Received: 27 November 2017; Accepted: 7 March 2018; Published: 9 March 2018

Abstract: Irrigated farmland is the main food source of desert areas, and moisture is the main limiting factor of desert farmland crop productivity. Study on the influence of irrigation on desert farmland soil moisture can guide the agricultural water resource utilization and agricultural production in those regions. At present, the efficiency of irrigation water usage in Northwest China is as low as approximately 40% of the irrigated water. To understand the response of farmland soil moisture in different soil types on irrigation in the Ulan Buh Desert of Inner Mongolia of China, this experimental study takes advantage of different infiltration characteristics and hydraulic conductivities of sand, clay, and loam to determine an optimized soil combination scheme with the purpose of establishing a hydraulic barrier that reduces infiltration. This study includes three comparative experiments with each consisting of a 100 cm thick of filled sand, or clay, or loam soil underneath a 50 cm plough soil, with a total thickness of 150 cm soil profile. A new type of lysimeter is installed below the above-mentioned 150 cm soil profile to continuously measure deep soil recharge (DSR), and the ECH2O-5 soil moisture sensors are installed at different depths over the 150 cm soil profile to simultaneously monitor the soil moisture above the lysimeter. The study analyzes the characteristics of soil moisture dynamics, the irrigation-related recharge on soil moisture, and the DSR characteristics before and after irrigation, during the early sowing period from 2 April to 2 May 2017. Research results show that: (1) Irrigation significantly influences the soil moisture of 0–150 cm depths. The soil moisture increase after the irrigation follows the order from high to low when it is in the order of loam, sand, and clay. (2) Irrigation-induced soil moisture recharge occurs on all three soil combinations at 0–150 cm layers, and the order of soil moisture recharge from high to low is: clay (54.3 mm, 43.39% of the total irrigation), loam (39.83 mm, 31.83% of the total irrigation), and sand (33.47 mm, 26.75% of the total irrigation). (3) After the irrigation event, DSR below 150 cm occurs for all three soil combinations. This study reveals the characteristics of irrigation-induced soil moisture recharge and DSR, and it shows that farmland consisting of an upper 50 cm plough soil and a lower 100 cm filled clay soil can save more water resource at the study site, which is useful in agricultural control measure and water resource management in arid regions.

Keywords: Ulan Buh Desert; DSR; infiltration; desert farmland; irrigation; sustainable development; water resource utilization efficiency

1. Introduction

The great temperature difference between day and night in arid regions is beneficial for the accumulation of photosynthetic products and the reduction of respiratory effects losses [1,2]. Therefore, agricultural production in these regions are of a larger amount and higher quality compared to semi-arid and humid regions [3,4]. In general, there are more agricultural development potential in arid region if the issue of water supply for irrigation can be coped with [5,6]. From an ecological standpoint, crop planting area in arid regions is where water and fertilizer are utilized intensively and, thus, may effectively reduce the invalid soil surface evaporation [7,8]. Previous studies show that reclaiming farmland in desert regions can help improve the soil microbial structure and soil quality [9–11] and, thus, is benign for desert ecological conservation [12,13]. Over the past six decades, there is a continuous evolution of oasis and desertified land in China [14,15]. Since the 1950s, the area of oasis in China expanded from 25,000 to 104,000 km^2 [16]. Meanwhile, lands that are undergoing desertification expanded from 53,000 to 114,000 km^2 [16]. The expansion of oasis benefits the regional ecological environment and provides more space for anthropogenic activities [17,18]. However, the water resource balance of the oasis is often disrupted and regional environment begins to worsen because of irresponsible development and poor understanding of the ecohydrological system of oasis, resulting in undesirable ecological problems, such as desertification and salinization [19–21]. As water resources are the main factor of ecological balance in arid regions, better understanding the water budget is indispensable for sustainable eco-agricultural development in those regions [22–24].

The eastern edge of Ulan Buh Desert, located at the northwestern inland area of China, is a transitional area between pastoral and agricultural areas [25,26]. It is an important part of the Hetao Plain, and an important agricultural area and food base of the Inner Mongolian Autonomous Region [27]. Under natural conditions, the moisture requirement of crops cannot be met because of sporadic and unevenly distributed precipitation. Therefore, irrigation is crucial for agricultural activities in this region [14]. As soil moisture is an important factor for crops to grow [28,29], and it usually changes after irrigation and precipitation, the study on the influence of irrigation on soil moisture can guide the agricultural water resource utilization and agricultural production.

Until presently, scientists have conducted numerous research on the correlation of vertical soil moisture distribution and corn production [30,31]. Some investigators have also conducted research on the relationship of amount of irrigation and quality of irrigated water on soil salt redistribution and spring corn water consumption [32,33]. However, such studies are rarely focused on the comparison of soil moisture responses to irrigation for different soil types, with even fewer field experiments on the real-time evolution of deep soil recharge (DSR) [34,35]. Soil moisture content is related to moisture pressure head [36] and soil unsaturated hydraulic conductivity [37]. When soil moisture content is relatively high, soil with larger particle sizes has a higher hydraulic conductivity; when soil moisture content is relatively low, soil with smaller particle sizes has a higher hydraulic conductivity [38,39]. Such a ubiquitous feature of unsaturated hydraulic conductivity and the soil moisture relationship can be taken into consideration for reducing infiltration loss in irrigated farmland in arid regions. Using the difference of hydraulic conductivities of two different soil particle sizes, one may put a different soil type (such as sand, clay, or loam) underneath the plough soil layer to effectively reduce the infiltration loss, achieving the goal of saving water resources in arid regions. This new measure can replace the current practice of using impermeable plastic films at certain depths of soil to prevent irrigation-induced infiltration in the arid regions, which is expensive and not environmental benign as a large quantity of plastic film is being used.

The objective of this study is to take advantage of different infiltration characteristics and hydraulic conductivities of sand, clay, and loam to determine an optimized soil combination scheme with the purpose of establishing a hydraulic barrier that reduces infiltration in arid farmlands. This study includes three comparative experiments with each consisting of a 100 cm thick filled sand, or clay, or loam soil underneath a 50 cm plough soil at the eastern edge of Ulan Buh Desert of China, with a total thickness of 150 cm soil profile. A new type of lysimeter is installed below the above-mentioned

150 cm soil profile to continuously measure DSR, and the ECH2O-5 soil moisture sensor is installed at different depths over the 150 cm soil profile to simultaneously monitor the soil moisture above the lysimeter. The responses of soil moisture and DSR to irrigation will be analyzed. The study provides the basis for sustainable water resource management in arid farmlands, such as the eastern edge of Ulan Buh Desert of China.

2. Study Area and Methods

2.1. Overview of the Study Area

The study site as shown in Figure 1 is at the Field I of the Desert Forestry Experimental Center administrated by the Chinese Academy of Forestry in Dengkou County at the northeastern part of Ulan Buh Desert of China. The geographical coordination is 40°19'7.81" N, 106°56'2475" E, with an altitude of 1043.0 m above mean sea level (m.s.l.). The study site has a typical temperate continental climate and a multi-year average temperature of 7.8 °C. The average annual sunshine is 3181 h, and the average annual frost-free period is 146 days. The multi-year average precipitation is 140.3 mm, and the site has a typical arid climate. The main soil type is irrigation silt, and this region has ample surface water resource supply from Yellow River [40,41].

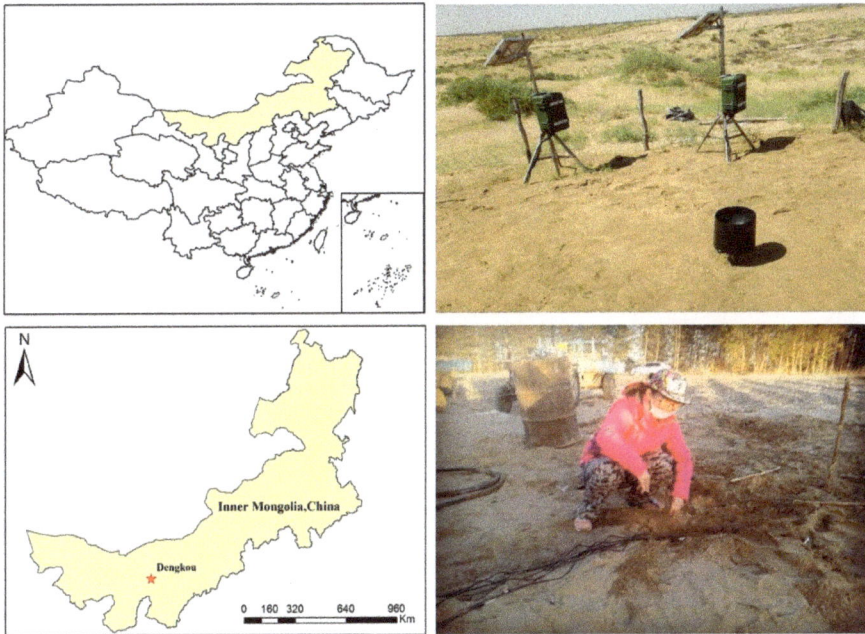

Figure 1. Overview of the experiment plot.

2.2. Experimental Design

2.2.1. Sample Settings

The shallow soil (0–50 cm) are under constant influence of cultivation, resulting in different soil physical and chemical properties (soil bulk density and mechanical composition, etc.) between uncultivated and cultivated soils, thus further affecting the dynamic changes of soil moisture. To simulate different soil types, sand, loam, and clay are used to replace the native soil below 50 cm depth, where the particle size distributions of the filled sand, loam and clay are listed in Table 1.

Table 1. Mechanical composition of three soil profile types.

Soil Type	Soil Mechanical Composition/μm									
	0.71–1.00	1.00–2.00	2.00–5.00	5.00–10.00	10.00–20.00	20.00–50.00	50.00–100.00	100.00–200.00	200.00–500.00	500.00–1000.00
Plough layer	22.84%	53.47%	23.07%	0.62%	0	0	0	0	0	0
Sand	1.05%	2.49%	1.03%	0.02%	0.04%	0.68%	2.59%	70.41%	21.33%	0.36%
Loam	50.41%	49.56%	0.03%	0	0	0	0	0	0	0
Clay	69.18%	30.82%	0	0	0	0	0	0	0	0

To compare and analyze the characteristics of soil moisture change and DSR for different soil types under the irrigation condition, comparative experiments have been run for three soil types of sand, loam, and clay. Using a typical cultivated land as the study site, after excavating an experimental pit of 200 cm long, 200 cm wide, and 300 cm deep, a DSR recorder is installed 150 cm below ground surface. The layers from the depth of 50 cm down to 150 cm are replaced by the sand, or loam, or clay listed in Table 1. To ensure the accuracy of experiments, plastic films are placed on four vertical surfaces of the experimental pit to separate the filled soil from the native soil outside. By doing so, one can prevent any possible lateral soil moisture migration because of soil heterogeneity. Caution has been taken to make sure no preferential flow occurring between the vertical plastic films and the soil. The used vertical plastic films will not affect any vertical migration of soil moisture in this experiment.

To ensure the normal growth of crops, excavated soil over the upper 50 cm depth (which is named the plough soil hereinafter) is put back on the top of filled soils, and the ground surface is leveled afterwards, as shown in Figure 2.

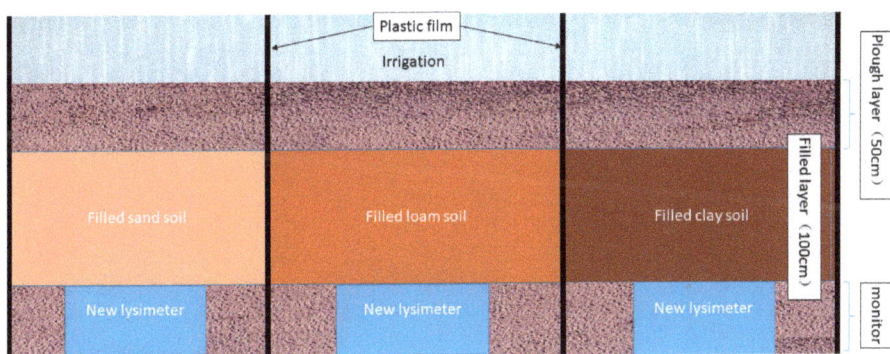

Figure 2. Three kinds of soil substitutions were carried out for the original soil. Each plot is separated by plastic film as an independent system.

2.2.2. Determination Indicators and Methods

1. DSR monitoring

The new DSR recorder (or lysimeter) is used to monitor the DSR of different soil types [42]. From the bottom up, the recorder consists of a drainage part (15 cm), a measuring part (35 cm), a flux collecting part (5 cm, filled with gravel and ceramic), and a capillary water holding part (65 cm filling with the tested soil). The measurement resolution is 0.2 mm, and the measurement accuracy is ±2%. After putting the DSR recorder in place, one needs to wait one to two months for the soil to settle naturally.

2. Soil moisture monitoring

The ECH$_2$O-5 soil moisture sensor (±3% accuracy, Decagon, Pullman, WA, USA) is used to monitor the soil moisture. For the monitoring of soil moisture and temperature dynamic changes at the upper 150 cm layer, ECH$_2$O-5 soil moisture sensors are placed at 5 cm, 50 cm, 100 cm, and 150 cm below ground surface.

3. Soil moisture storage calculation

Soil moisture storage, or the soil storage capacity (mm) of a certain soil thickness, is calculated using the following equation:

$$W = \sum_{i=1}^{n} \theta_i h_i \tag{1}$$

where W is the soil storage capacity (mm) of a given soil thickness, θ_i is the average moisture of the i-th soil layer (dimensionless), h_i is the thickness (cm) of the i-th soil layer ($i = 1, 2, 3, \ldots, n$), and n is the number of soil layers to the given soil thickness.

4. Soil particle size distribution

When installing the DSR recorder, a soil texture analysis is conducted for the upper 150 cm soil, and soil particle size and porosity samples are collected. The soil mechanical composition is determined using a Malvern mastersizer 200 laser particle size analyzer (England, accuracy: 0–1000 μm), as shown in Table 1. Soil moisture holding capacity and the field moisture holding capacity are measured by the cutting ring method as baseline data.

As shown in Table 1, the particle size distribution differences are obvious for the four types of soil including the plough soil, and the filled sand, loam and clay soils. The soil particle size in plough soil is mainly of 1–2 μm, which represents 53.47% of the total particles of this soil type. The soil particle size in sand is mainly of 100–200 μm, which represents 70.41% of the total particles of this soil type. The soil particle size of loam is mainly of 0.71–1 μm and 1–2 μm, which are 50.41% and 49.56% of the total particles of this soil type, respectively. The soil particle size of clay is mainly of 0.71–1 μm, which is 69.18% of the total particles of this soil type.

The study site mainly relies on Yellow River for supplying irrigation water. This experiment uses the typical flooding irrigation method that is commonly utilized by the local farmers. During the irrigation process, a portable LS300-A flow meter is used to measure the amount of irrigation. Pre-seeding irrigations were conducted on 17 April and 19 April 2017 in the study site, lasting for 109 min and 41 min, respectively. The amounts of irrigation were 118.64 mm and 6.5 mm, with a total of 125.14 mm.

3. Results and Analysis

3.1. The Dynamic Response of Different Type Soil Moisture on Irrigation

The soil moisture of the cultivated land is influenced by multiple factors, such as precipitation, irrigation, and crop growth. Since the study site is located in Northwestern China with scarce precipitation, and there are no precipitation events during the experimental period, the influence of precipitation on soil moisture can be ignored. As the experimental period proceeds the seeding season, the influence of crop growth on soil moisture can also be ignored.

To analyze the dynamic response of soil moisture to irrigation for different soil types, this experiment use 31-day soil moisture data from 2 April to 2 May 2017 to analyze the temporal variation of vertical soil moisture at different soil layers. During the experimental period, the maximum and minimum soil moistures at different layers as a function of time can be seen in Table 2. Statistical analysis shows that the 125.14 mm irrigation amount has a significant effect on soil moisture of the upper 150 cm soil layer.

Figure 3A–C shows the daily dynamic change of soil moisture at the upper 150 cm soil layer under different experimental treatments. It shows that the antecedent soil moisture levels of three different types of soil are relatively low before the irrigation event. This is mainly because the experimental field is located in an arid region with very limited winter and spring precipitations and no irrigation recharge.

The soil moistures in three types of soils at the upper 150 cm all fluctuated 15 days after irrigation, but with quite different variational patterns. The coefficient of variation (C.V.) can be compared by the degree of dispersion of the three sets of data. A larger C.V. means a larger irrigation influence on soil moisture. From the C.V. listed in Figure 3D, one can draw the following conclusions. Firstly, for soil moisture variation at the upper 50 cm layer, the degree of irrigation influence declines when the filled soil type changes from sand to loam, and then to clay. Secondly, for soil moisture variation at depths of 50–150 cm, the degree of irrigation influence declines from when the filled soil type changes from loam, to sand, and then to clay.

Table 2. Time change peak value of soil volume water content in the study site.

Soil Depth	Soil Type	Date (Month-Day)	Water Content Maximum/%	Date (Month-Day)	Water Content Minimum/%
	Sand	4-17	31.35	5-2	18.11
5 cm	Loam	4-17	31.31	4-12	19.11
	Clay	4-17	30.91	4-12	24.56
	Sand	4-17	25.29	4-2	7.158
50 cm	Loam	4-17	28.96	4-16	14.72
	Clay	4-20	30.83	4-16	25.37
	Sand	4-17	20.56	4-2	8.87
100 cm	Loam	4-18	31.29	4-2	14.2
	Clay	4-20	32.53	4-2	24.24
	Sand	4-20	22.12	4-2	12.11
150 cm	Loam	4-18	25.21	4-2	12.93
	Clay	4-23	27.41	4-2	20.5

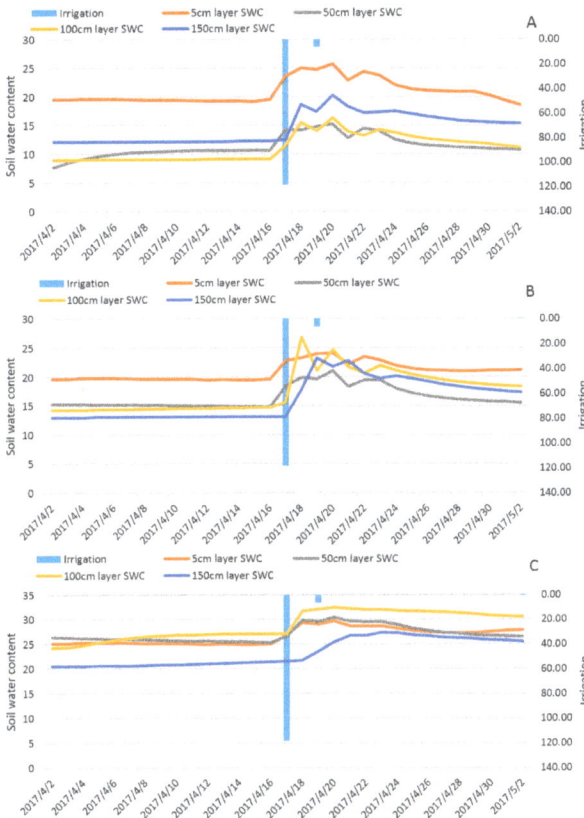

Figure 3. *Cont.*

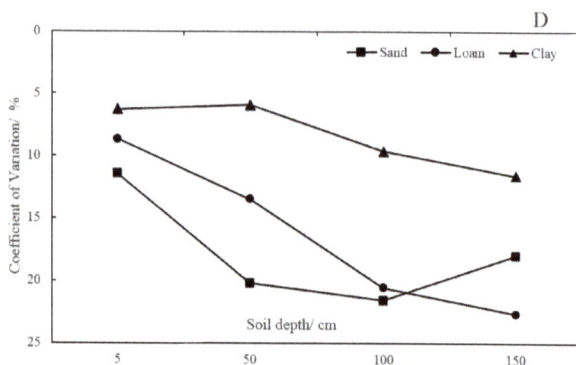

Figure 3. Different depth soil volume water content dynamic. (**A**) Sand soil; (**B**) loam soil; (**C**) clay soil; (**D**) C.V.

The study shows that after the fill-in of three different soil types below the upper 50 cm plough soil, under the same irrigation condition, different soil types show very different hydraulic conductivity and water storage capacity. Because the particle size of plough soil is close to that of loam, irrigation water can infiltrate faster into loam and increase the soil moisture content of loam. The particle size difference between clay and the plough soil is the most distinctive, so irrigation water infiltrates the least into the clay layer.

One can see that in the sand soil on the left plot in Figure 2, the particle sizes are mainly distributed in the range of 100–200 micrometers, which represents 70.41% of the total particles of sand soil. For such large grain sizes, a unique hydrological feature is notable. When the sand soil moisture content is relatively high (close to saturation), the hydraulic conductivity of such a type of soil is, of course, much greater than those of loam soil and clay soil. However, when the sand soil moisture content is relatively low, the hydraulic conductivity becomes even smaller than those of loam soil and clay soil at the same soil moisture content. Such a dramatic reduction of hydraulic conductivity of sand soil after the declines of soil moisture has been recognized for a long time in soil physics. This means that when the soil moisture content of the fill sand soil is relatively low, it can actually serve as a hydrologic barrier to prevent further loss of soil moisture from the above plough soil [43,44]. This explains why the sand soil in Figure 2 is included in the comparison study. Nevertheless, if the soil moisture content in the sand soil has increased above a certain level, such a hydrologic barrier effect will disappear and water can flow downward quite easily through such a fill sand soil layer.

In summary, above analysis implies that if one replaces the original cultivated soil layer right below the upper plough soil with clay, one can effectively reduce irrigation water infiltration below the plough soil layer, thus achieving the goal of saving water resource in arid regions.

From Figure 3A–C, one can see that during a certain time period after irrigation, the soil moistures at different layers for different soils universally increase. As time goes on, soil moisture at different layers gradually decreases. For the sand soil, the soil moisture at the upper 50 cm layer is obviously higher than that below 50 cm. For the loam soil, the soil moisture is distributed relatively evenly over the upper 50 cm. For the clay soil, the soil moisture over the 50–150 cm depth is obviously higher than other layers.

This study concerns the early sowing period between 2 April and 2 May. Thus, the result represents the features of spring; the results of other periods of the year need more research.

3.2. Recharging Effects of Recharge on Soil Moisture of 0–150 cm Layers

Compared to soil moisture, soil water storage can intuitively reflect the regional soil moisture supply capacity. Figure 4 shows the dynamic changes of soil water storage of the upper 150 cm layers for different soils from 2 April to 2 May 2017. From this table, one can see that after a certain time period since the cease of irrigation, the soil water storage of different soils have all altered greatly, demonstrating that the irrigation has significantly influenced the soil water storage of different soils at the 150 cm soil layer. The average post-irrigation soil water storage of the upper 150 cm layer for different soils are 426.10 mm for clay, 299.67 mm for loam, and 243.67 mm for sand.

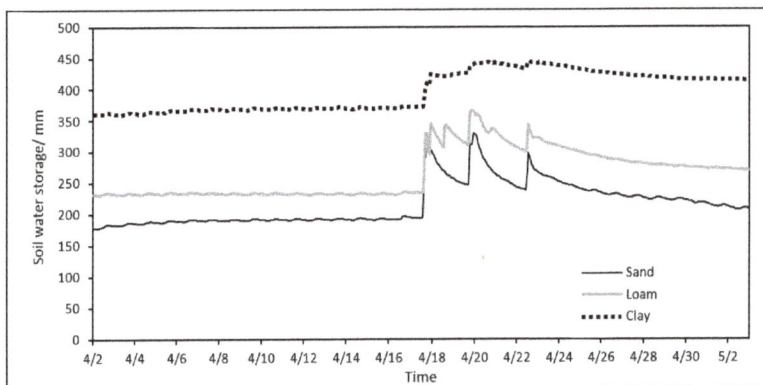

Figure 4. Soil water storage dynamic variation of different type dunes.

Table 3 shows soil water storage and their changes for different soils at the upper 150 cm layers for both pre-irrigation (2 April) and post-irrigation (2 May). This table shows that after the 2 April irrigation event, the soil water storage of sand, loam and clay soils all increase by 2 May. This means that irrigation has certain recharging effect on all the different soils at the upper 150 cm layers, and the amounts of recharge decline in the order of clay (54.3 mm), loam (39.83 mm), and sand (33.47 mm). For the upper 150 cm layer, sand, loam, and clay soils respectively store 26.75%, 31.83%, and 43.39% of the irrigated water during the period from 2 April to 2 May 2017. This leads to the conclusion that one replaces the original cultivated soil below the 50 cm plough soil with filled clay to effectively increase soil water storage for the entire 150 cm soil profile.

The soil moisture recharge variation under the same irrigation event at the upper 150 cm layer is mainly caused by different filled soil types, with soil water retention capacity closely correlated with soil clay content. Specifically, a higher soil clay content leads to a stronger soil water retention capacity. Table 1 shows that the clay content is the greatest in the filled clay soil, the least in the filled sand soil. Therefore, the field water storage capacity declines in the order of clay, loam, and sand. The post-irrigation water storage capacity also follows this pattern.

Table 3. Irrigation on soil water supply with different soil textures.

Soil Type	2 April 2017 Storage/mm	2 May 2017 Storage/mm	Difference of 2 April and 2 May 2017/mm
Sand	178.10	211.57	33.47
Loam	231.09	271.91	39.83
Clay	360.45	414.75	54.30

3.3. DSR Characteristics of Different Soil Types

Figure 5 shows the daily variational characteristics of DSR below different 150 cm soil profile. This figure shows that during the 15 days post-irrigation period, DSR appears for all three soil types. For the sand soil, DSR happens 13 h after irrigation, and lasts for 157 h, with a total amount of 110.87 mm. For the loam soil, DSR happens 72 h after irrigation with a total amount of 12.2 mm. For the clay soil, DSR happens 257 h after irrigation with a total amount of 0.2 mm.

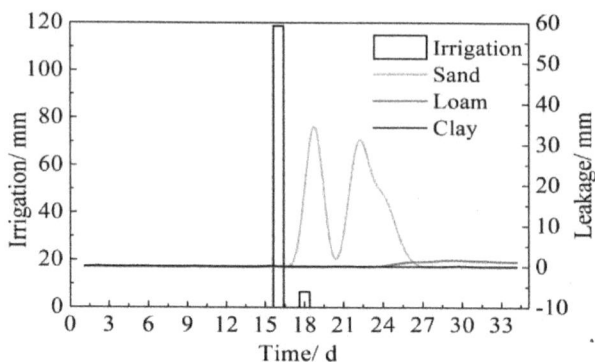

Figure 5. Daily variation of deep soil water leakage below 150 cm of different soil textures.

Soil clay content is an important factor influencing DSR, and a smaller soil clay content leads to a stronger infiltration capacity and greater DSR. Therefore, under the same irrigation condition, the amount of DSR should follow the declining order of sand, loam, and clay. Furthermore, DSR is negatively correlated to the post-irrigation soil water storage. Porosity is another important factor influencing DSR. Specifically, clay soil has a very poorly connected pore space and a very low effective porosity. In addition, colloids in clay soil swell after irrigation, leading to pore contraction and decreased soil permeability. All of this will result in a very small infiltration capacity and DSR for clay soil. Meanwhile, other factors such as the soil structure (layered, blocked, fragmented, etc.) also affect soil moisture infiltration capability and DSR.

The data used for this study come from farmland currently undergoing cultivation. Irrigation amount is based upon the routine cultivation use by local farmers. Therefore, one cannot guarantee that the irrigation amount currently in use is the optimized amount. The best irrigation amount should meet both the needs of crop growth and minimize the water resource waste such as DSR. In the future, a controlled experiment should be conducted to find out the best irrigation amount based on the soil profile.

4. Conclusions

The following conclusions can be obtained from this study:

1. Accumulative irrigation in the experimental field from 2 April to 2 May is 125.14 mm. For the upper 150 cm soil layer, irrigation has significant influence on soil moisture. For the upper 50 cm plough soil layer, the irrigation influence on different soils follows the declining order of sand, loam, and clay soils under the same irrigation strength and pattern. For the filled soil layer at depth of 50–150 cm, the irrigation influence on different soils follows the declining order of loam, sand, and clay soils under the same irrigation strength and pattern.

2. Irritation has recharging effect on soil moisture for all three types of soil at the upper 150 cm soil layer, and the recharge amounts follow the order of clay (54.3 mm, which is 43.39% of the total irrigation amount), loam (39.83 mm, which is 31.83% of the total irrigation amount), and sand (33.47 mm, which is 26.75% of the total irrigation amount).

3. Post-irrigation DSR appears in all three types of soil below 150 cm. The time when DSR occurs is 13 h after irrigation for sand, 72 h after irrigation for loam, and 257 h after irrigation for clay. The 15-day total DSR is 110.87 mm for sand, 12.2 mm for loam, and 0.2 mm for clay.

4. If one replaces the original cultivated soil layer right below the upper 50 cm plough soil whose particle sizes are mostly in the range of 1–2 μm with a 100 cm thick filled clay soil whose particles are primarily in the range of 0.71–1 μm, one can effectively reduce DSR below the 150 cm soil profile, thus achieving the goal of saving water resources for farmland in arid regions.

Acknowledgments: This study was supported with research grants from the National Natural Science Foundation of China (41661006, 41771206). The first author would like to thank Chinese Scholar Council for supporting his visit of Texas A&M University from 2016–2018. We thank four anonymous reviewers for their constructive comments which help improve the quality of the manuscript.

Author Contributions: Yiben Cheng designed the experiments, analyzed data and wrote the whole paper; Yanli Li collected field data and analyzed data; Hongbin Zhan guided the writing of the article and revised the article; Wenbin Yang designed the experiments and helped to select experimental area; Hairong Liang designed the experiments and acquired data; Yingming Zhao managed the experimental site; Taojia Li maintained the instrument.

Conflicts of Interest: The authors declare no conflict of interest.

References

1. Noy-Meir, I. Desert ecosystems: Environment and producers. *Annu. Rev. Ecol. Syst.* **1973**, *4*, 25–51. [CrossRef]

2. Flanagan, L.B.; Johnson, B.G. Interacting effects of temperature, soil moisture and plant biomass production on ecosystem respiration in a Northern Temperate Grassland. *Agric. For. Meteorol.* **2005**, *130*, 237–253. [CrossRef]

3. Wallace, J. Increasing agricultural water use efficiency to meet future food production. *Agric. Ecosyst. Environ.* **2000**, *82*, 105–119. [CrossRef]

4. Laurance, W.F.; Sayer, J.; Cassman, K.G. Agricultural expansion and its impacts on tropical nature. *Trends Ecol. Evol.* **2014**, *29*, 107–116. [CrossRef] [PubMed]

5. Reij, C.; Waters-Bayer, A. *Farmer Innovation in Africa: A Source of Inspiration for Agricultural Development*; Routledge: London, UK, 2014.

6. Rockström, J.; Falkenmark, M. Increase water harvesting in Africa. *Nature* **2015**, *519*, 283. [CrossRef] [PubMed]

7. Dai, J.; Dong, H. Intensive cotton farming technologies in China: Achievements, challenges and countermeasures. *Field Crops Res.* **2014**, *155*, 99–110. [CrossRef]

8. Canakci, M.; Topakci, M.; Akinci, I.; Ozmerzi, A. Energy use pattern of some field crops and vegetable production: Case study for Antalya Region, Turkey. *Energy Convers. Manag.* **2005**, *46*, 655–666. [CrossRef]

9. Köberl, M.; Müller, H.; Ramadan, E.M.; Berg, G. Desert farming benefits from microbial potential in arid soils and promotes diversity and plant health. *PLoS ONE* **2011**, *6*, e24452. [CrossRef] [PubMed]

10. Yin, X.; Song, B.; Dong, W.; Xin, W.; Wang, Y. A review on the eco-geography of soil fauna in China. *J. Geogr. Sci.* **2010**, *20*, 333–346. [CrossRef]

11. Yoder, R.E. A direct method of aggregate analysis of soils and a study of the physical nature of erosion losses. *Agron. J.* **1936**, *28*, 337–351. [CrossRef]

12. Wikelski, M.; Cooke, S.J. Conservation physiology. *Trends Ecol. Evol.* **2006**, *21*, 38–46. [CrossRef] [PubMed]

13. Dolan, B.F. Water developments and desert bighorn sheep: Implications for conservation. *Wildl. Soc. Bull.* **2006**, *34*, 642–646. [CrossRef]

14. Wang, X.; Chen, F.; Hasi, E.; Li, J. Desertification in China: An assessment. *Earth-Sci. Rev.* **2008**, *88*, 188–206. [CrossRef]

15. Hao, X.; Chen, Y.; Xu, C.; Li, W. Impacts of climate change and human activities on the surface runoff in the Tarim River Basin over the last fifty years. *Water Resour. Manag.* **2008**, *22*, 1159–1171. [CrossRef]

16. Tao, W. Some issues on oasification study in China. *J. Desert Res.* **2010**, *5*, 995–998.

17. Zha, Y.; Gao, J. Characteristics of desertification and its rehabilitation in China. *J. Arid Environ.* **1997**, *37*, 419–432. [CrossRef]

18. Liu, J.; Zhang, Z.; Xu, X.; Kuang, W.; Zhou, W.; Zhang, S.; Li, R.; Yan, C.; Yu, D.; Wu, S. Spatial patterns and driving forces of land use change in China during the early 21st century. *J. Geogr. Sci.* **2010**, *20*, 483–494. [CrossRef]

19. Ezcurra, E. *Global Deserts Outlook*; UNEP/Earthprint: Hertfordshire, UK, 2006.

20. Abtew, W.; Melesse, A. Landscape Dynamics and Evapotranspiration. In Proceedings of the World Environmental and Water Resources Congress, West Palm Beach, FL, USA, 22–26 May 2016.

21. Oestigaard, T. *Water Scarcity and Food Security along the Nile: Politics, Population Increase and Climate Change*; Nordiska Afrikainstitutet: Uppsala, Sweden, 2012.

22. Schwinning, S.; Sala, O.E. Hierarchy of responses to resource pulses in arid and semi-arid ecosystems. *Oecologia* **2004**, *141*, 211–220. [CrossRef] [PubMed]

23. West, N.E. Structure and function of microphytic soil crusts in wildland ecosystems of arid to semi-arid regions. *Adv. Ecol. Res.* **1990**, *20*, 179–223.

24. Jolly, I.D.; McEwan, K.L.; Holland, K.L. A review of groundwater-surface water interactions in arid/semi-arid wetlands and the consequences of salinity for wetland ecology. *Ecohydrology* **2008**, *1*, 43–58. [CrossRef]

25. Laity, J.J. *Deserts and Desert Environments*; John Wiley & Sons: Hoboken, NJ, USA, 2009.

26. Zhao, H.; Li, G.; Sheng, Y.; Jin, M.; Chen, F. Early–middle Holocene lake-desert evolution in northern Ulan Buh Desert, China. *Palaeogeogr. Palaeoclimatol. Palaeoecol.* **2012**, *331*, 31–38. [CrossRef]

27. Chen, J.; He, D.; Cui, S. The response of river water quality and quantity to the development of irrigated agriculture in the last 4 decades in the Yellow River Basin, China. *Water Resour. Res.* **2003**, *39*. [CrossRef]

28. McKendry, P. Energy production from biomass (part 1): Overview of Biomass. *Bioresour. Technol.* **2002**, *83*, 37–46. [CrossRef]

29. Grubb, P.J. The maintenance of species-richness in plant communities: The importance of the regeneration niche. *Biol. Rev.* **1977**, *52*, 107–145. [CrossRef]

30. Kravchenko, A.N.; Bullock, D.G. Correlation of corn and soybean grain yield with topography and soil properties. *Agron. J.* **2000**, *92*, 75–83. [CrossRef]

31. Hassan, N.A.; Drew, J.V.; Knudsen, D.; Olson, R.A. Influence of soil salinity on production of dry matter and uptake and distribution of nutrients in barley and corn: I. Barley (*Hordeum vulgare* L.). *Agron. J.* **1970**, *62*, 43–45. [CrossRef]

32. Fereres, E.; Soriano, M.A. Deficit irrigation for reducing agricultural water use. *J. Exp. Bot.* **2006**, *58*, 147–159. [CrossRef] [PubMed]

33. Ayars, J.; Phene, C.; Hutmacher, R.; Davis, K.; Schoneman, R.; Vail, S.; Mead, R. Subsurface drip irrigation of row crops: A review of 15 years of research at the Water Management Research Laboratory. *Agric. Water Manag.* **1999**, *421*, 1–27. [CrossRef]

34. Herkelrath, W.; Hamburg, S.; Murphy, F. Automatic, real time monitoring of soil moisture in a remote field area with time domain reflectometry. *Water Resour. Res.* **1991**, *27*, 857–864. [CrossRef]

35. Scott, R.L.; Shuttleworth, W.J.; Keefer, T.O.; Warrick, A.W. Modeling multiyear observations of soil moisture recharge in the semiarid American Southwest. *Water Resour. Res.* **2000**, *36*, 2233–2247. [CrossRef]

36. Klute, A. *Water Retention: Laboratory Methods*; Soil Science Society of America, American Society of Agronomy: Madison, WI, USA, 1986.

37. Eching, S.; Hopmans, J.; Wendroth, O. Unsaturated hydraulic conductivity from transient multistep outflow and soil water pressure data. *Soil Sci. Soc. Am. J.* **1994**, *58*, 687–695. [CrossRef]

38. Campbell, G.S. A simple method for determining unsaturated conductivity from moisture retention data. *Soil Sci.* **1974**, *117*, 311–314. [CrossRef]

39. Van Genuchten, M.T. A closed-form equation for predicting the hydraulic conductivity of unsaturated soils. *Soil Sci. Soc. Am. J.* **1980**, *44*, 892–898. [CrossRef]

40. Zhang, X.Y.; Arimoto, R.; An, Z.S. Dust emission from Chinese desert sources linked to variations in atmospheric circulation. *J. Geophys. Res.* **1997**, *102*, 28041–28047. [CrossRef]

41. Chun, X.; Hen, F.; Fan, Y.; Xia, D.; Zhao, H. Formation of Ulan Buh Desert and its environmental evolution. *J. Desert Res.* **2007**, *6*, 005.

42. Cheng, Y.; Zhan, H.; Yang, W.; Dang, H.; Li, W. Is annual recharge coefficient a valid concept in arid and semi-arid regions? *Hydrol. Earth Syst. Sci.* **2017**, *21*, 5031. [CrossRef]

43. Modaihsh, A.S.; Horton, R.; Kirkham, D. Soil water evaporation suppression by sand mulches. *Soil Sci.* **1985**, *139*, 357–361. [CrossRef]

44. Liu, X.P.; Zhang, T.H.; Zhao, H.L. Influence of dry sand bed thickness on soil moisture evaporation in mobile dune. *Arid Land Geogr.* **2006**, *29*, 523–526.

MDPI

St. Alban-Anlage 66

4052 Basel

Switzerland

Tel. +41 61 683 77 34

Fax +41 61 302 89 18

www.mdpi.com

Water Editorial Office

E-mail: water@mdpi.com

www.mdpi.com/journal/water